Re-envisioning A
in Remote Se

Re-envisioning Advances in Remote Sensing

Urbanization, Disasters and Planning

Edited by
Ripudaman Singh

CRC Press
Taylor & Francis Group
Boca Raton London New York

CRC Press is an imprint of the
Taylor & Francis Group, an **informa** business

First edition published 2022
by CRC Press
6000 Broken Sound Parkway NW, Suite 300, Boca Raton, FL 33487-2742

and by CRC Press
2 Park Square, Milton Park, Abingdon, Oxon, OX14 4RN

CRC Press is an imprint of Taylor & Francis Group, LLC

ISBN: 978-1-032-12457-5 (hbk)
ISBN: 978-1-032-12460-5 (pbk)
ISBN: 978-1-003-22462-4 (ebk)

DOI: 10.1201/9781003224624

Typeset in Times LT Std
by SPi Technologies India Pvt Ltd (Straive)

Contents

Preface

Re-envisioning Advances in Remote Sensing: Urbanization, Disasters and Planning has sprouted from the publication of a previous book published by Taylor & Francis Group Publishers, with the title *Re-envisioning Remote Sensing Applications: Perspectives from Developing Countries* (CRC Press, 2021). A greater response from the contributors brought double the number of chapters, which could not be published altogether in the same volume. It led to the idea of bringing out twin books on remote sensing applications. Continuing from its predecessor volume, applications in three broader areas (in the fields of urbanization, disaster management and planning perspectives) are presented through sixteen chapters, covering remote sensing applications in all such fields from various cities of India, such as Aligarh, Amritsar, Lucknow, Ludhiana; its various regions, such as Central Haryana, Central Rajasthan, Kandi region of Punjab, Kashmir Valley, Kotropi (Himachal Himalayas), Western Ghats and India as region; and various international contributions in remote sensing applications from the capital city of Dhaka and Rangamati District (of Bangladesh), the capital city of Harare (Zimbabwe), Free State Province (of South Africa) and the African continent as a region.

Advances in technological innovations, since the beginning of the Industrial Revolution, have really exceeded previous discoveries and inventions. Progress being made in the field of remote sensing is accumulating and widening to reach new horizons. Current innovation in geospatial technologies and their combination with ICT developments are transforming the nature of geospatial data, and its analysis as well. Google Earth has popularized the usage of maps and images in those subjects also which were reluctant to have spatial connotations earlier. Usage of GIS, GPS and remote sensing is increasing day by day, and everyone is using them irrespective of their subject affiliations. Current innovations in artificial intelligence (AI), big data, the internet of things (IoT) and machine learning, coupled with advances in ICT, are expanding the domains and prevalence of remote sensing applications in broader perspectives with wider reach.

Re-envisioning Advances in Remote Sensing foresees such innovative applications in diverse fields. The interface of remote sensing with smart digital devices such as mobile apps and tracking sensors and the availability of very big spatial data are changing the learning environment and smart learning. Open-source geospatial data availability has furthered the remote sensing applicability with programming languages viz. AI, R, Python. 'Spatio-Environmental Distribution of Drought Disaster Events: A Space-based Approach using Terra-MODIS Vegetation Index' used R programming to analyse EVI (MOD13Q1.006) and climate data obtained from AppEEARS and POWER databases (Chapter 5). Machine learning and varied algorithms are being used for analysing geospatial data and widening the remote sensing applications. Similarly, 'Geospatial Techniques to Quantify Urban Change: The Case of Harare, Zimbabwe' analysed land use classification using the random forest machine learning classifier (Chapter 10). Enhanced technological advances in smartphones, on the other hand, are taking remote sensing to these smart devices. One such study is presented through use of a smartphone to map noise pollution in Ludhiana city (Chapter 2). Smart remote sensing applications are being enhanced by diverse civilian GPS uses in automobiles and smartphones. Advances among drones, UAVs and other such machines provide platforms for nano processors and smart gadgets. Smart and fusion remote sensing applications are upcoming. 'Feasibility Analysis of a Railway Route using Remote Sensing/Geoinformatics and Analytical Hierarchy Process (AHP) Techniques' analysed the usefulness of integrating remote sensing/geoinformatics and AHP in feasibility studies (Chapter 13).

Certainly, with the rapidly changing technologies with increased geospatial data creation, and every minute improving programming languages, sophisticated techniques, advancing geospatial tools – with all such integrations, remote sensing is bound to be recreating itself and moving beyond its traditional boundaries through exploring new frontiers. Remote sensing of micro spaces as well as of extra-terrestrial spheres is anticipated to widen its applications at micro levels and piercing into extra-terrestrial explorations. Undoubtedly, a post remote sensing phase is anticipated where it will be progressing into new realms of advanced applications far beyond the traditional usages, moving toward new horizons and reaching new heights. The present volume is expected to bring out innovative ideas regarding remote sensing applications with better methodologies, newer techniques, smarter programming languages, and beyond. Re-envisioned advances in remote sensing are forwarded to researchers to discover the new frontiers and explore the applications which are still unknown.

Ripudaman Singh

Acknowledgements

First and foremost, I acknowledge the Indian Institute of Remote Sensing and Indian Society of Remote Sensing, Dehradun for providing this opportunity to initiate and accomplish this publication endeavour. This volume sprouted from the first book, already published by the Taylor & Francis Group (*Re-envisioning Remote Sensing Applications: Perspectives from Developing Countries*, CRC Press, 2021); as the invited chapters exceeded the available space in that book, it evolved into twin books. I am indebted to IIRS and ISRS, Dehradun for extending my work and allowing this publication to be accomplished. I am grateful to the Taylor & Francis Group in general and CRC Press in particular for accepting my twin book proposal and presenting it to the world.

I extend my gratitude to all my teachers, colleagues and friends, who from time to time have encouraged and motivated me to accomplish this task. As acknowledged earlier, I pay my sincere gratitude to my teachers, Principal (retd.) Dr Darshan Singh Manku Ji, Professor Emeritus Gopal Krishan Ji, Professor Surya Kant Ji, who stimulated the geographer in me and always inspired me to look beyond the horizon and explore the unknown. I am grateful to Shri P.L.N. Raju Ji, Director NESAC, for continuously supporting me for *any* queries regarding IIRS and the broader domain of remote sensing. From time to time there have been interactions and suggestions from senior colleagues from the Punjab Remote Sensing Center, Ludhiana, Dr Brijendra Pateriya, Director PRSC, Dr V. K. Verma, Dr Anil Sood, Dr D. C. Loshali, Dr Ajay Mathur, Sh. P. K. Litoria, and Dr R. K. Setia. I am thankful to all the contributors and reviewers for their timely submissions and support. I am grateful to all my friends and well-wishers for their encouragement and motivation.

Finally, I pay my reverence and my indebtedness to my parents, who brought me into this world. Whatever and wherever I am today is because of their care and upbringing. I am indebted to my mother (Mrs Baljit Kaur), a pious homemaker and my inspiration, my father (the late) Professor Baldev Singh Vig, who brought me into the field of teaching and scholarship. Endless thanks are to my better half (Dr Manpreet Kaur) for her greatest support and providing me with ample time to accomplish this task during the pandemic situation. It is anticipated that the researchers in the arena of remote sensing, budding geographers and all explorers of the unknown, will treasure these resources and further their cause of research through re-envisioned advances in remote sensing.

Ripudaman Singh

The Editor

Dr Ripudaman Singh works as Professor of Geography at Amity University, Noida. After completing his doctoral research from Panjab University, Chandigarh, through UGC-JRF/SRF fellowships, he has taught at various institutions, including Post Graduate Government College, Sector-11, Chandigarh (2005–2010); Central University of Punjab, Bathinda (2010–2011); Lyallpur Khalsa College, Jalandhar (2011–2012), and Lovely Professional University, Phagwara (2012–2021). He has also served at Government College, Hoshiarpur (2004) on an honorary basis and at Panjab University, Chandigarh, as Guest Faculty (2005–2006). He has been a recipient of the 'Young Geographers Award' from the National Association of Geographers, India (NAGI) in 2005.

Dr Singh has a special interest in remote sensing and GIS, development geography, regional development and planning, geography of India, political geography, applied geography and disaster management. Since 2012, Dr Singh has been serving as the Coordinator of IIRS-Edusat Programs at Lovely Professional University. He has published more than sixty research papers and articles in reputed international/national journals and has four books to his credit. His other two books are under publication. In addition to various seminars, Dr Singh has convened an International Conference at LPU and edited two special issues of Scopus-listed journals through the papers called through that conference.

Dr Singh specializes in the field of regional development. He has been working on regional disparities in India and analysing how these disparities have evolved and how their nature, extent and intensity have varied over time and space. His PhD thesis on 'Trends in Regional Disparities in India since Independence: A Geographical Analysis' was highly acclaimed and widely published as *Regional Disparities in Pre and Post Reform India* (Saptrishi, Chandigarh, 2016). He has recently completed an ICSSR IMPRESS major research grant project on 'Revisiting Regional Disparities in India 2020' and is under the process of publication by international publishers. Remote sensing applications in various fields of geography including urban and regional planning are among his core research areas.

Contributors

Abdullah-Al- Faisal
Department of Urban & Regional Planning
Rajshahi University of Engineering and Technology (RUET), Rajshahi and
GIS Centre, Operational Centre Amsterdam (OCA), Médecins Sans Frontières (MSF)
Cox's Bazar, 4750, Bangladesh

Abdulla-Al Kafy
Department of Urban & Regional Planning
Rajshahi University of Engineering & Technology (RUET), Rajshahi, and
ICLEI South Asia, Rajshahi City Corporation, Rajshahi-6203, Bangladesh

Wani Suhail Ahmad
Research Scholar, Department of Geography
Aligarh Muslim University (AMU)
Aligarh, UP, India

Mohd. Shamsul Alam
Professor, Department of Geography & Environment
Jahangirnagar University
Savar, Dhaka, Bangladesh

V. S. Arya
Director, Haryana Space Applications Centre (HARASAC)
Hisar, Haryana, India

Bhawna Bali
Assistant Professor, TERI School of Advanced Studies
New Delhi, India

Kapil Batra
Department of Quality Assurance,
Lovely Professional University
Punjab, India

Johannes A. Belle
Disaster Management Training and Education Centre for Africa
University of the Free State
Bloemfontein, South Africa

Ashok Beniwal
Associate Professor, Department of Geography
Government College
Adampur, Hisar, Haryana, India

M Sultan Bhat
Department of Geography and Disaster Management
University of Kashmir,
Srinagar, India

Chandan Kumar Boraiaha
Ore Research & Exploration Group (ORE-G), Department of Geology
Central University of Kerala
Kasaragod, Kerala, India

Rashmi Chandan
Ore Research & Exploration Group (ORE-G), Department of Geology
Central University of Kerala
Kasaragod, Kerala, India

Kanwar Deepak
Associate Professor and Head,
Department of Geography
DAV College
Jalandhar, Punjab, India

Dheeraj Gambhir
Punjab Remote Sensing Centre, PAU Campus
Ludhiana, Punjab, India

Webster Gumindoga
Department of Civil Engineering
University of Zimbabwe
Mt Pleasant, Harare, Zimbabwe

Nur Hussain
Department of Geography and Earth Sciences
McMaster University, Ontario, Canada

Saleha Jamal
Assistant Professor, Department of Geography
Aligarh Muslim University (AMU), Aligarh, UP, India

Rajesh Jolly
Assistant Professor in Geography, Lovely Professional University (LPU)
Phagwara, Punjab, India

Dheera Kalota
Former Faculty, Department of Geography
Lovely Professional University (LPU), Phagwara, Punjab, India
Independent Researcher
Faridabad, Haryana, India

Neeru Kaushal
Lovely Professional University
Phagwara, Punjab, India

Mohammad Firoz Khan
Professor (Retired), Department of Geography
Jamia Millia Islamia University, New Delhi, India

Sandeep Kumar
Assistant Professor of Geography, Government College, Adampur, Haryana, India

Tafadzwanashe Mabhaudhi
School of Agricultural, Earth and Environmental Sciences University of KwaZulu-Natal, Pietermaritzburg, South Africa and
International Water Management Institute, (IWMI-GH),
West Africa, Kumasi, Ghana

James Magidi
Geomatics Department
Tshwane University of Technology
Pretoria, South Africa

Ishfaq Hussain Malik
Department of Geography
Aligarh Muslim University
Aligarh, Uttar Pradesh, India

Paidamwoyo Mhangara
School of Geography, Archaeology and Environmental Studies
University of the Witwatersrand
Johannesburg, South Africa

Manoj
Assistant Professor of Geography, CRM Jat College
Hisar, Haryana, India

Durdanah Mattoo
Department of Geography and Disaster Management
University of Kashmir
Srinagar, India

Sajad Ahmad Mir
Department of Geography and Disaster Management
University of Kashmir,
Srinagar, India

Sylvester Mpandeli
Water Research Commission of South Africa (WRC)
Lynnwood Manor, Pretoria, South Africa
School of Environmental Sciences
University of Venda
Thohoyandou, South Africa

Tahir Hussain Muntazari
Department of Civil Engineering
National Institute of Technology
Srinagar, India

Walter Musakwa
Department of Urban and Regional Planning
University of Johannesburg
Johannesburg, South Africa

Luxon Nhamo
Water Research Commission of South Africa (WRC)
Lynnwood Manor, Pretoria
Centre for Transformative Agricultural and Food Systems (CTAFS), School of Agricultural, Earth and Environmental Sciences
University of KwaZulu-Natal
Scottsville, Pietermaritzburg, South Africa

Olusola O. Ololade
Centre for Environmental Management
University of the Free State
Bloemfontein, South Africa

Israel R. Orimoloye
Centre for Environmental Management
University of the Free State
Bloemfontein, South Africa
Disaster Management Training and Education Centre for Africa
University of the Free State
Bloemfontein, South Africa

Simran Purswani
MTech UDM, TERI School of Advanced Studies
New Delhi, India

Abdur Rouf Khan
Department of Urban & Regional Planning
Rajshahi University of Engineering and Technology (RUET
Rajshahi, Bangladesh
BRAC Institute of Educational Development
BRAC University, Dhaka, Bangladesh

Nusrat Rafique
Department of Geography and Disaster Management
University of Kashmir, Srinagar, India

Md Mostafizur Rahman
Department of Urban & Regional Planning
Rajshahi University of Engineering and Technology (RUET)
Rajshahi, Bangladesh
Department of Photogrammetry and Geoinformatics
Budapest University of Technology and Economics, Budapest, Hungary

Mohd. Shahinoor Rahman
Department of Earth and Environmental Sciences
New Jersey City University
Jersey City, NJ, USA

G M Rather
Department of Geography and Disaster Management
University of Kashmir
Srinagar, India

Muhammad Rizwan
MSc Researcher, Department of Geography and Earth Sciences
McMaster University
Ontario, Canada

Sumita Roy
Department of Urban & Regional Planning
Rajshahi University of Engineering and Technology (RUET)
Rajshahi, Bangladesh

Harpinder Singh
Punjab Remote Sensing Centre, PAU Campus
Ludhiana, Punjab, India

Ripudaman Singh
Professor of Geography, Amity Institute of Social Sciences (AISS)
Amity University
Noida, Uttar Pradesh, India

Li Suju
Chief of the Satellite Remote Sensing Department, National Disaster Reduction Center of China
Ministry of Emergency Management of P.R.C.
Beijing, China

Introduction

Re-envisioning Advances in Remote Sensing

Ripudaman Singh
Amity University, Noida, Uttar Pradesh, India

CONTENT

Almost two and a half millennia ago, Socrates (469–399 BCE) described how human beings must rise above the earth – to the top of the atmosphere and beyond – for only thus will they fully understand the world in which they live. Today, Google Earth has become an incredible resource because, from hundreds of miles in space, we can zoom in, and find things. Through Google Earth, everyone looks for their house first, and that is the tip of the iceberg with remote sensing (Parcak, 2009). Moving beyond one's house and looking into the physical environment and at cultural patterns, and tracking disasters, through the thousands of satellites which are revolving around the earth, day and night, acting as human eyes in the sky, remote sensing has really moved beyond human imagination and thought (Singh, 2021).

Technological advances made since the Industrial Revolution have surpassed previous discoveries and inventions in their cumulative progression. Although advances in remote sensing can be traced to the depiction of photogrammetry principles, electromagnetic spectrum and related discoveries, remote sensing became established as a subject some six decades ago (Table 0.1). Cold War rivalries between the United States and the Soviet Union (1950–1990) led to the early developments of the space age and making advances in remote sensing. During the Cold War, both these superpowers were trying to dominate each another through technological advances, wherein the Soviet Union and the USA launched the Sputnik and Explorer satellites into space in 1957 and 1958 respectively. Simultaneously, first use of the term 'remote sensing' goes to *Ms Evelyn Pruitt*, a geographer in the US Office of Naval Research, who coined that term for the newly emerging field, in the late 1950s, which was far advanced from photointerpretation of aerial photographs (Reeves, 1975). Table 0.1 depicts some of the key milestones in the advancement of remote sensing since 1759 to the present day. Through these years, developments in photogrammetry, cameras, aerial photography and satellite imageries, all have advanced remote

DOI: 10.1201/9781003224624-1

TABLE 0.1
Key Milestones in Remote Sensing Advances

1759 Perspectiva Liber by *Lambert* (Photogrammetry)	**1957** First Satellite *Sputnik* launched by USSR
1800 Infrared discovered by *Sir W. Herchel*	**1958** First American Satellite *Explorer 1* launched
1827 Photograph taken by *Nicéphore Niépce*	**1959** *First photograph of Earth* from Space
1839 *Photography* begins as a profession	**1960** *TIROS* Satellite launched (Remote Sensing use)
1847 Infrared Spectrum by *JBL Foucault*	**1964** Multiband photography for earth resource (NASA)
1858 Air Photograph by *Gaspard-Felix Tournachon*	**1972** *Earth Resources Technology Satellite* by US
1859 *Photography through Balloons* starts	**1973** *Skylab* US Space station launched
1873 Electromagnetic Spectrum by *JC Maxwell*	**1982** *Thematic Mapper* with 30 m resolution & 7 bands
1882 Aerial Photography through Kites (*ED Archibald*)	**1986** *Hyper spectral Sensors* developed
1902 Airplane designed by the *Wright brothers*	**1986** *SPOT 1* satellite launched by France
1903 Pigeon carrier camera designed *(J. Neubronner)*	**1988** Indian Remote Sensing Satellite *IRS-1A* launched
1909 *Photography* starts through *Airplanes*	**1990** *RADARSAT* projects in Canada
1916 Aerial Reconnaissance in World War I	**1991** European Remote Sensing Satellite (*ERS-1*)
1934 American Society of Photogrammetry established	**1999** *Landsat 7* with Enhanced Thematic Mapper Plus
1935 Radar developed in Germany	**1999** *IKONOS* with 1 m resolution launched
1940 Applications of Non-Visible Part of EMS	**2001** High resolution commercial satellite *QuickBird*
1944 Manual of Photogrammetry by American Society	**2006** *Cloud Computing EC2* by Amazon
1947 Spectral Reflectance of natural material by *Krinov*	**2008** *GeoEye 1* (0.41 to 1.65 m resolution) launched
1950 Military Research and Development in Cold War	**2009** *Bhuvan*: Gateway to geospatial world launched
1950s *Evelyn Pruitt* coined the term Remote Sensing	**2019** *Google Earth Engine* for satellite images data
1956 *Colwell* used infrared in plant disease detection	**2020** *EagleView* processed its 100 millionth image

Source: Singh, 2021

sensing to the higher and higher zenith. Since the beginning of the space age, thousands of satellites have been flown to gather information regarding terrestrial and extra-terrestrial surfaces. From the days of Sputnik, lots of these satellite imageries have been captured. Besides, 2020 marking the 100 millionth image processing by EagleView, a private company and a leading technology provider of aerial imagery, data analytics and GIS solutions (EagleView, 2020).

Present age of information and use of data have been transforming our lives, work and leisure. The data volume in a single file could be depicted by unit of byte. Any information data in remote sensing or others, could be understood from its size (Table 0.2). Increased data production and its dispensation lies at the core of digital

TABLE 0.2
Data Volumes and its other Attributes

Unit	Value	Abbreviation	Size (in bytes)
Bit	0/1	b	1/8th of a byte
Bytes	8 bits	B	1 byte
Kilobyte	1,000 bytes	KB	1,000 bytes
Megabyte	$1,000^2$ bytes	MB	1,000,000 bytes
Gigabyte	$1,000^3$ bytes	GB	1,000,000,000 bytes
Terabyte	$1,000^4$ bytes	TB	1,000,000,000,000 bytes
Petabyte	$1,000^5$ bytes	PB	1,000,000,000,000,000 bytes
Exabyte	$1,000^6$ bytes	EB	1,000,000,000,000,000,000 bytes
Zettabyte	$1,000^7$ bytes	ZB	1,000,000,000,000,000,000,000 bytes
Yottabyte	$1,000^8$ bytes	YB	1,000,000,000,000,000,000,000,000 bytes

Source: Adapted from https://mynasadata.larc.nasa.gov/basic-page/data-volume-units (Credits Roy Williams)

transformations and digitization processes. Thus, the data created, captured or replicated all over the world becomes the global datasphere, which is expected to grow from 33 to 175 zettabytes during years 2018 to 2025, respectively (Reinsel et al./IDC, 2018). It has further been found that during the year 2020, 64 zettabytes of data have been generated and the digital data going to be created in the next five years will be greater than twice the size of all the data created since the beginning of digital storage in the computer age (IDC, 2021). In comparison, the remotely sensed data and information being produced is relatively small; however, remote sensing is also passing through a makeshift and being intermingled with deep learning through big data, the Internet of Things (IoT) and artificial intelligence (AI).

Since the launch of Sputnik in 1957, thousands of satellites have been placed in various orbits encircling the earth. An estimate by United Nations Office for Outer Space Affairs, accounts around 7,900 satellites launched in the space including probes, rockets and other such devices. During the year 2019, 28 states and 1 international intergovernmental organization registered more than 400 space objects, wherein new countries like Bhutan, Kenya have also registered one satellite each and India registered 21 satellites (13 functional, 7 non-functional and 1 re-entry satellite), making it the sixth largest fleet of satellites in the space (Table 0.3) after United States, China, Russia, United Kingdom and Germany (UNOOSA, 2020).

Among the group of countries which are having fleets of remote sensing satellites, India is among the few countries which produces economical remotely sensed data, having a vide coverage. Soon after the flight of its first satellite Aryabhatta in 1975, India launched its remote sensing earth observation satellite Bhaskara-I in 1979. Since then, more than hundred satellites have been successfully launched and placed in the earth observation orbits. Table 0.4 depicts the list of India's all major space programs and particularly its remote sensing satellites. Within ten years of Bhaskara-I (1979), India successfully launched its indigenously built IRS-1A (Indian Remote Sensing-1A) satellite in 1988 and at present the total number of such RS and Earth observation (EO) satellites is about five dozen. This list culminates with the launch of Amazônia-1,

TABLE 0.3
Space Object Registration by Countries, 2019

State of Registry	Functional	Non-functional	Re-entry	Change	Total
Argentina	1	0	0	0	
Australia	7	0	0	1	
Azerbaijan	1	0	0	0	
Belarus	1	0	0	0	
Belgium	0	0	19	0	
Bhutan	1	0	0	0	
China	88	0	0	0	
Columbia	1	0	0	0	
Denmark	1	0	0	0	
Egypt	1	0	0	0	
European Space Agency	4	0	0	1	
Finland	5	0	0	0	
France	2	11	4	7	
Germany	16	0	1	1	
India	**13**	**7**	**1**	**0**	
Indonesia	1	0	0	0	
Israel	1	0	0	0	
Japan	8	1	3	3	
Kenya	1	0	0	0	
Lithuania	2	0	0	0	
Mexico	1	0	0	1	
New Zealand	0	13	6	0	
Norway	0	0	0	5	
Poland	1	0	0	0	
Russian federation	34	0	17	0	
South Africa	1	0	0	0	
United Arab Emirates	2	0	0	0	
United Kingdom	18	0	0	6	
United States	145	20	93	0	
Functional Registered	**357**				
Non-functional registered		**52**			
Satellite re-entry notifications			144		
Change in status				25	
Total Space objects registered					**409**

Source: UNOOSA, 2020

in early 2021, being India's first dedicated commercial mission of NewSpace India Ltd. (NSIL), under the Department of Space (DoS). Amazônia-1 is the optical EO satellite of Brazil's National Institute for Space Research (INPE), which has been successfully launched by ISRO and NSIL of India, for monitoring deforestation in the Amazon basin and agricultural diversification in Brazil (ISRO, 2021).

Among India's heavy satellites, the GSAT-11 weighing about 5854 kg, launched in 2018, became India's heaviest, most advanced multiband communication satellite (ISRO, 2018). Comparably, Kalamsat-V2 satellite (weighing just 1.26 kg) launched along with Microsat-R, through PSLV C44, in 2019 got the distinction of being the lightest satellite ever put into the earth's orbit (BBC, 2019). Earlier, ISRO had flown 104 satellites in one go, through PSLV-C37 in February 2017, making another world

TABLE 0.4
India's Satellites (Communication, Navigational and Remote Sensing)

Communication/Navigational Satellites	Remote Sensing and EO Satellites
Aryabhatta (1975) first Indian Satellite	**Bhaskara-I** (1979) first Experimental RS EO Satellite
Rohini Technology Payload (1979) for measuring performance of launch vehicle, SLV	**Rohini RS-I** (1980) first Indigenously launched by SLV
Ariane Passenger Payload Experiment (APPLE) (1981) Relay of TV programmes	**Rohini RS-D1** (1981) Solid state camera for RS applications
INSAT-1A (Indian National Satellite) (1982) first multipurpose communication and meteorology satellite	**Bhaskara-II** (1981) first Indian satellite for EO from orbit
INSAT-1B (Indian National Satellite) (1983) 11 C-band and 2 S-band transponders for TV communications, weather and disaster warning service	**Rohini RS-D2** (1983) smart sensor camera for capturing 2500+ pictures
INSAT-1C (Indian National Satellite) (1988) for more All-India Radio and Doordarshan, DoS, IMD services	**Stretched Rohini Satellite Series (SROSS-1)** (1987) Astrophysics, Earth Remote Sensing and upper atmospheric monitoring experiments
INSAT-1D (Indian National Satellite) (1990) for Communication & Meteorological Observations	**IRS-1A** (Indian Remote Sensing-1A) (1988) first Indigenous remote sensing capability
INSAT-2DT (Indian National Satellite) (1992) communication satellite known for Arabsat-1C and INSAT-2R	**Stretched Rohini Satellite Series (SROSS-2)** (1988) carried RS payload of German space agency, Gamma Ray Astronomy payload
INSAT-2A (Indian National Satellite) (1992) first Indian multipurpose communication and meteorology satellite-based search and rescue	**IRS-1B** (Indian Remote Sensing-1A) (1991) first Remote Sensing Satellite by ISRO using imagery generated by RS technology
INSAT-2B (Indian National Satellite) (1993) second satellite in NSAT-2 series	**Stretched Rohini Satellite Series (SROSS-C)** (1992) Astrophysics, Earth Remote Sensing and upper atmospheric experiments
INSAT-2C (Indian National Satellite) (1993) updated improved communication in remote areas such as North East, Andaman & Nicobar Island	**IRS-1E** (Indian Remote Sensing) (1993) Indian Experimental EO Satellite to develop Earth imagery
INSAT-2D (Indian National Satellite) (1997) Identical to INSAT-2C. improved communication	**Stretched Rohini Satellite Series (SROSS-C2)** (1994) Astrophysics, Earth Remote Sensing and upper atmospheric experiments. Identical to SROSS-C
INSAT-2E (Indian National Satellite) (1999) Geostationary communications and weather satellite, communication provisions to Asian countries, Australia	**IRS-P2** (Indian Remote Sensing) (1994) Indian RS Satellite for managing natural resources
INSAT-3B (Indian National Satellite) (2000) Indian satellite for developmental and mobile communication	**IRS-1C** (Indian Remote Sensing) (1995) Indian second-generation operational RS Satellite
GSAT-1 (GramSat-1) (2001) Experimental satellite for first GSLV developmental flight	**IRS-P3** (Indian Remote Sensing) (1996) Experimental EO Satellite, X-ray astronomy and C-band transponder
TES (Technology Experiment Satellite) (2001) for validating altitude and orbit control system	**IRS-1D** (Indian Remote Sensing) (1997) Similar to IRS-1C in resolution, spectral bands, imaging etc.
INSAT-3C (Indian National Satellite) (2002) for voice, video and digital data services for India and neighbouring countries	**Oceansat-1 (IRS-P4)** (1999) first Indian satellite for Ocean applications with Ocean Colour Monitor (OCM) and MSMR

(Continued)

TABLE 0.4 (Continued)

Communication/Navigational Satellites	Remote Sensing and EO Satellites
INSAT-3A (Indian National Satellite) (2003) Multipurpose Geostationary satellite for telecommunication, broadcasting, search & rescue opt.	**MetSat (Kalpana-1)** (2002) Indian Meteorological Satellite, renamed in memory of Kalpana Chawla
GSAT-2 (GramSat-2) (2003) Experimental satellite for second GSLV developmental flight	**Resource Sat-1 (IRS-P6)** (2003) enhancing data quality provided by IRS-1C and IRS-1D satellites
INSAT-3E (Indian National Satellite) (2003) Communication satellite to augment the existing INSAT System.	**Cartosat-1** (2005) Stereoscopic EO satellite covers entire globe in 1,867 orbits on 126-day cycle
GSAT-3 (GramSat-3/EduSat) (2004) first Indian satellite for educational services through distant mode.	**Cartosat-2** (2007) Advanced RS satellite carrying panchromatic camera capable of scene specific spot images
HamSat (2005) Microsatellite for amateur radio communication for India and international radio operator	**Space Capsule Recovery Experiment (SRE-1)** (2007) Experiments in micro gravity conditions
INSAT-4A (Indian National Satellite) (2005) Advanced satellite for telecommunication and broadcasting services	**Cartosat-2A** (2008) Indian Military satellite carrying panchromatic camera capable of capturing B&W pictures of visible region of EMS
INSAT-4C (Indian National Satellite) (2006) Indian satellite for telecommunication, for I-2K satellite bus.	**IMS-1 (Third World Satellite-TWsat)** (2008) first Indian satellite to use ISRO's Mini Satellite
INSAT-4B (Indian National Satellite) (2007) Indian satellite for telecommunication, for I-3K satellite bus.	**Chandrayaan-1** (2008) India's first Moon Mission satellite to explore the moon
INSAT-4CR (Indian National Satellite) (2007) Replacement satellite of INSAT-4C. DTH services	**RISAT-2 (Radar Imaging Satellite)** (2009) Indian radar reconnaissance satellite to monitor borders
ANUSAT (Anna University Satellite) (2009) Student's research microsatellite for amateur radio experiments	**Oceansat-2 (IRS-P4)** (2009) service continuity for Oceansat-1 for OCM
GSAT-4 (GramSat-4/HealthSat) (2010) first Indian satellite to employ ion propulsion, experimental communication and navigation satellite	**Cartosat-2B** (2010) EO satellite carrying panchromatic camera capable of capturing B&W pictures of visible region of EMS
StudSat (Student Satellite/CubeSat) (2010) first Indian picosatellite, a miniaturized satellite designed by undergraduate students across India	**Resource Sat-2** (2011) enhanced multispectral resolutions and spatial coverage
GSAT-5P/INSAT-4D (2010) Geosynchronous satellite as a replacement of INSAT-3E	**Megha-Tropiques** (2011) developed by ISRO and France's CNES to study water cycle in tropical atmosphere
YouthSat (2011) Indo-Russian educational artificial satellite based on ISRO's Mini Satellite-1 bus.	**Jugnu** (2011) RS CubeSat Satellite operated in IIT Kanpur for agriculture and disaster monitoring
GSAT-8/INSAT-4G (2011) first satellite to carry GAGAN payload. Indian communication satellite	**SRMSat (Sri Ramaswamy Memorial Satellite)** (2011) a nanosatellite developed by SRM University to monitor atmospheric greenhouse gases
GSAT-12 (GramSat-12) (2011) Replacement for INSAT-3B, providing services like tele-education, tele-medicine, disaster management support and satellite internet access	**RISAT-1** (Radar Imaging Satellite) (2012) Indian RS satellite, heaviest EO satellite by India

Communication/Navigational Satellites	Remote Sensing and EO Satellites
GSAT-10 (GramSat-10) (2012) second satellite to carry GAGAN payload. Indian communication satellite	**SARAL** (Satellite with ARGOS and ALTIKA) (2013) Joint Indo-French satellite for altimetric measurements to study Ocean circulation and sea surface elevation
IRNSS-1A (Indian Regional Navigation Satellite System) (2013) first navigation satellite in IRNSS series for GPS purposes.	**Mangalyaan-1** (Mars Orbiter Mission-MOM) (2013) for developing technologies for designing, planning and management operations for interplanetary missions
INSAT-3D (Indian National Satellite) (2013) Meteorological Satellite with advanced weather monitoring payloads	**Astrosat** (2015) first Indian satellite with multiwavelength space observatory
GSAT-7/INSAT-4F (2011) Military communication satellite for Indian Navy	**Cartosat-2C** (2016) EO satellite carrying panchromatic camera capable of capturing minute long video of a fixed spot
GSAT-14 (GramSat-14) (2014) Indian communication satellite to replace GSAT-3	**SathyabamaSat** (2016) micro experimental satellite from Sathyabama University to collect greenhouse gases data
IRNSS-1B (Indian Regional Navigation Satellite System) (2014) Navigation satellite for tracking and mapping services	**Pratham** (2016) India's ionospheric research satellite
IRNSS-1C (Indian Regional Navigation Satellite System) (2014) Navigation satellite for tracking and mapping services having CDMA ranging payload, LRR	**PISat** (PESIT Imaging Satellite) (2016) RS nanosatellite by PES Institute of Technology, Bangaluru, with 80 m resolution
GSAT-16 (GramSat-16) (2014) Indian communication satellite with increased transponders, telecommunication and VSAT services	**ScatSat-1** (Scatterometer Satellite) (2016) Miniature satellite by ISRO for weather forecasting, cyclone tracking
IRNSS-1D (Indian Regional Navigation Satellite System) (2015) Only satellite in constellation for navigation services through L5 and S-band	**Resource Sat-2A** (2016) enhanced multispectral resolutions and spatial coverage
GSAT-6/GramSat-6 (2015) Multimedia communication satellite (S-DMB)	**Cartosat-2D** (2017) EO satellite carrying various nanosatellites and making a record for taking 101 satellites along in one go
GSAT-15/GramSat-15 (2015) Multimedia communication satellite for more bandwidth for DTH and VSAT services	**INS-1A** (ISRO Nano Satellite) (2017) carries Surface BRDF Radiometer (SBR) and Single Event Upset Monitor (SEUM)
IRNSS-1E (Indian Regional Navigation Satellite System) (2016) for navigational services	**INS-1B** (ISRO Nano Satellite) (2017) carries Earth Exosphere Lyman Alpha Analyzer (EELA) and Origami Camera payload
IRNSS-1F (Indian Regional Navigation Satellite System) (2016) Sixth satellite in constellation for navigational satellite system series	**GSAT-19/GSAT-19E** (2017) Multimedia communication satellite with Geostationary Radiation Spectrometer (GRASP) payload to monitor and study the nature of charged particles
IRNSS-1G (Indian Regional Navigation Satellite System) (2016) Seventh and final satellite in constellation for navigational satellite system series	**NIUSat** (Noorul Islam University Satellite) (2017) by NICHE (Noorul Islam Centre for Higher Education, for agricultural applications
Swayam-1 (2016) A 1-Upicosatellite developed by College of Engineering, Pune to demonstrate passive attitude control	**Cartosat-2E** (2017) EO satellite for urban planning, infrastructural development, utilities planning and traffic management
INSAT-3DR (Indian National Satellite) (2016) Meteorological Satellite with advanced weather monitoring payloads	**MicroSat-TD** (Microsatellite) (2018) EO satellite and India's 100th satellite

(Continued)

TABLE 0.4 (Continued)

Communication/Navigational Satellites	Remote Sensing and EO Satellites
GSAT-18/GramSat-18 (2016) Multimedia communication satellite with 24 C-band, 12 extended C-band and 12 Ku-band transponders	**INS-1C** (ISRO Nano Satellite) (2018) nanosatellite to accompany bigger satellites on PSLV, carrying MMX-TD (Miniature Multispectral Imager-Technology Demonstrator) Payload
GSAT-9/South Asia Satellite (2017) Communication and meteorology satellite by ISRO for SAARC region	**HysIS** ISRO's Mini satellite-2 (IMS-2) (2018) EO satellite to study the Earth's surface in Visible, Near Infrared and Shortwave Infrared regions of EMS
GSAT-17/GramSat-17 (2017) Communication satellite and heaviest launched by ISRO	**ExseedSat-1** (2018) India's first privately built satellite along with 63 other satellites
IRNSS-1H (Indian Regional Navigation Satellite System) (2017) for replacing IRNSS-1A satellite in completing constellation for navigational satellite series	**Microsat-R** (2019) EO satellite by DRDO and ISRO for military use
GSAT-6A (2018) Communication satellite for developing various technologies	**KalamSAT-V** (2019) Named after President Dr. APJ Abdul Kalam, built by High School students as a Femto Satellite. World's lightest satellite
IRNSS-1I (Indian Regional Navigation Satellite System) (2017) Eighth satellite in navigational satellite system series	**Chandrayaan-2** (2019) India's second Lunar exploration mission
GSAT 29 (2018) Communication satellite to provide high speed bandwidth to VRC in rural areas	**EMISAT** (2019) Reconnaissance satellite intended for EMS measurement
GSAT-7A (2018) Advanced military communication satellite for Indian Air Force	**Cartosat-3** (2019) EO third generation agile advanced satellite with high resolution imaging capability
GSAT-11 (2018) Communication satellite and India's heaviest satellite built by ISRO	**RISAT-2BR1** (2019) Radar Imaging EO satellite having resolution of 0.35 meters
GSAT-31 (2019) High throughout telecommunication satellite	**EOS-01** (2020) EO satellite for application sin agriculture, forestry and disaster management
GSAT-30 (2020) communication satellite with enhanced I-3K Bus with C and Ku-bands	**Amazônia-1** (2021) India's first dedicated commercial mission of New Space India Ltd. (NSIL), Govt. of India's company under DoS. Optical EO satellite of INPE, Brazil, a joint venture with ISRO for monitoring deforestation in Amazon basin and agricultural diversification
CMS-01 (2020) 42nd communication satellite of India with extended C-band	

Source: ISRO, Govt. of India (https://www.isro.gov.in/list-of-spacecrafts)

record at that time. Through this mission ISRO put into orbit Cartosat-2 series satellite, and INS-1 and INS-2 (nanosatellites) along with international customer satellites from USA (96), and the Netherlands, Switzerland, Israel, Kazakhstan and UAE one each (ISRO, 2017).

Advances in satellite technologies have further led to diversification in their sizes, data capturing and applications. At present, the place of large and heavy satellites is being taken by small and miniature satellites, which are more economical and highly sophisticated. Varying sizes of satellites ranging from a small school bus size to a small lunchbox, monitoring the Earth from space, their respective sizes could be compared from Figure 0.2 (NOAA, 2015). Usually, SmallSats (Small Satellites/ Spacecrafts) are under 180 kilograms of weight, and according to their varied volumes, these are further categorized as Minisatellites, Microsatellites, Nanosatellites, Picosatellites and Femtosatellites (Table 0.5).

CubeSats, On the other hand, are classes of nanosatellites, which use a standardized size and form factor, where 1 unit (1U) measures 10×10×10 cms, which may be extendable to varied sizes of 1.5, 2, 3, 6, 12 etc. (Figure 0.1). Combinations of small/ nano and CubeSats have revolutionized the EO industry and the number of space objects has increased many times over with their advent. These have really advanced from being perceived as toys and only suitable for educational purposes, to vigorous platforms for conducting space missions (Straub et al., 2019). With the thousandth launch of a CubeSat, exponential growth in remote sensing is expected further (Villela et al., 2019). Simultaneously, market of small satellites is continuously growing with variety of applications in weather monitoring, surveillance, EO, navigation, communication, meteorology and others. Their future growth further lies in their preferences for Global Navigation and Satellite System (GNSS) and telecommunication services through IoT and others. With a tremendous increase in the information technology and (spatial) data creation, global space economy is expected to grow from 350 billion dollars currently to over 1 trillion dollars by 2040 (Morgan Stanley, 2021).

India's operational remote sensing commenced with launching of IRS-1A in 1988, and soon after the National Natural Resources Management System (NNRMS)

TABLE 0.5
Classes of Small Satellites

Small Satellites (SmallSats)	Below 180 Kilograms
Minisatellite	100 180 kilograms
Microsatellite	10–100 kilograms
Nanosatellite	1–10 kilograms
Picosatellite	0.01–1 kilogram
Femtosatellite	0.001–0.01 kilogram

Source: NASA (https://www.nasa.gov/content/what-are-smallsats-and-cubesats)

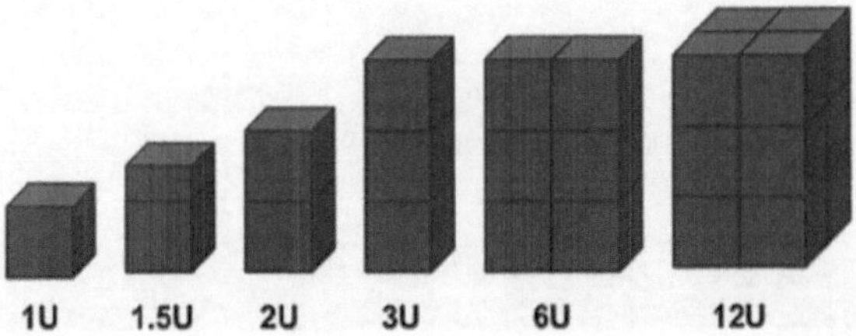

FIGURE 0.1 CubeSats of varied sizes

Source: NASA (https://www.nasa.gov/sites/default/files/thumbnails/image/what_are_cubesats.png)

was established as an institutional mechanism to utilize the spatial data from RS satellites (Figure 0.2). To tap the market of geospatial data and commercial exploitation of space products, technical consultancy services and technology transfer, Antrix Corporation Limited (ACL) was incorporated as private limited company by Government of India in 1992. Currently, ISRO is operating four series of satellites explicitly, Cartosat, Oceansat, Resourcesat and RISAT (Figure 0.3), whereas INSAT-3D/3DR satellites are providing meteorological services (ACL, 2021). It provides varied spatial data for a long period of time and doing efforts to improve its spatial, spectral and temporal resolutions. Some of the archived data from earlier IRS missions that have been made available through Antrix Corporation Limited (2021) are listed in Table 0.6.

Among the operational remote sensing satellites of India, Cartosat-1 was launched in 2005, which covers the entire Earth in 1,867 orbits in a 126-day cycle, having a spatial resolution of 2.5 m in panchromatic band. It has been followed by Cartosat-2 (2007), Cartosat-2A (2008), Cartosat-2B (2010), Cartosat-2C (2016), Cartosat-2D (2017), Cartosat-2E (2017), and Cartosat-3 (2019) respectively. Cartosat-2 series (Figure 0.4) got its improved resolution of 0.65 m and Cartosat-3 further improving to 0.28 m (Table 0.7).

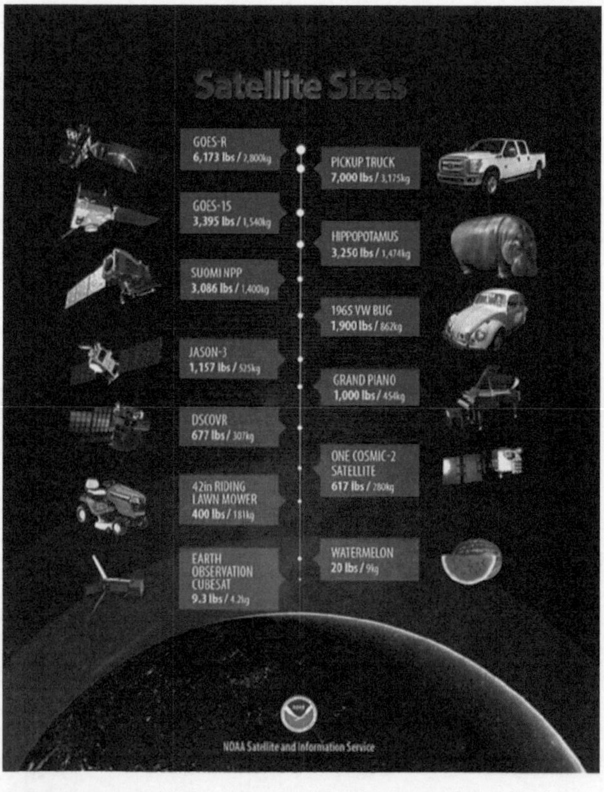

FIGURE 0.2 Satellite sizes

Source: NOAA. https://www.nesdis.noaa.gov/sites/default/files/satellite-comparison.jpg

FIGURE 0.3 Satellite series of India

Source: ACL, 2021

TABLE 0.6
Archived Data from Earlier IRS Satellites/Missions through ANTRIX

RS Satellite	Archived Data Available through ANTRIX
IRS-1A/IRS-1B	April 1988 to May 1991 & Oct 1991 to Sep 2001 respectively
IRS-1C/IRS-1D	Nov 1996 to Sep 2007 & Jan 1998 to Dec 2009 respectively
Resourcesat-1	Dec 2003 to Sep 2013
Oceansat-1	July 1999 to Aug 2010
RISAT-1	July 2012 to Sep 2016

Source: ACL, 2021

Resourcesat satellites (Resource Sat-1, 2003 and Resource Sat-2, 2011, Resource Sat-2A, 2016) were launched to enhance further the data quality, multispectral resolutions (Figure 0.5) and spatial coverage, provided by IRS-1C and IRS-1D earlier (Table 0.8).

Oceansat-1 (1999) and Oceansat-2 (2009) satellites have been successfully providing oceanic data, earlier through bands 6 and 7 and later 8 band OCM with multispectral operating camera in Visible Near IR spectral range, with geometric field view of 360 m covering a swath of 1420 kms, having repetivity of two days (Table 0.9 and Figure 0.6). Furthermore, Oceansat-3 (2019) has improved the number of bands

FIGURE 0.4 Pan-sharpened image from Cartosat-2S

Source: ACL, 2021

from 8 to 13, and a reduced bandwidth of 10/20 nm instead of 20/40 nm earlier. Oceansat-3A is expected to be flown later this year (ACL, 2021).

RISAT-1 (Radar Satellite), was the first indigenously designed microwave remote sensing satellite, launched in 2012. It has the capability of cloud penetration and day-night imaging capability with synthetic aperture radar (SAR) payloads operating in C-band. It has major applications in paddy crop monitoring and disaster management with respect to flooding and cyclonic disturbances. RISAT-1A and 1B are the repeat satellites of RISAT-1 and recently RISAT-2BR1 (2019) has been launched with improved resolution of 0.35 m (Table 0.10). On the other hand, HRSAT (high resolution satellite) is conceived as a constellation of 3 satellites in a single plane, configured for cartographic imaging, with ~ 1 m ground sampling distance (GSD), with 1 m Pan resolution and less than 4 m in multispectral mode having daily revisit (ACL, 2021).

Successful journey of Indian Space Research Organization could be assessed from its various achievements in successfully launching 111 spacecraft missions, including 3 nano satellites and 1 micro satellite; 80 launch missions including Scramjet-TD, RLV-TD, Crew Escape System; 12 student satellites; 2 re-entry missions; and launching 342 foreign satellites of 34 countries (ISRO, 2021).

The book, *Re-envisioning Remote Sensing Applications: Perspectives from Developing Countries* (2021), was intended to re-envision and present all the major applications of remote sensing in one place (Figure 0.7). However, owing to the limited size of that volume, the present book came out to accommodate the remaining themes and chapters under the title: *Re-envisioning Advances in Remote Sensing: Urbanization, Disasters and Planning*.

As the main idea behind compiling these volumes was prompted by the rapid changes in technological advances, which have impelled the remote sensing applications beyond the traditional boundaries of the subject, all these applications have been designed in these two volumes. The first, *Re-envisioning Remote Sensing*

TABLE 0.7
Cartosat: Specifications and Data Products

Cartosat-1

Payload Specifications		Cartosat-1 Data Products		
Parameters	**Specification**	**Sensor**	**Product Type**	**Coverage**
Resolution (m)	2.5	Pan Fore	Path-Row scene-based (Standard product: Mono Georeferenced/Georeferenced & radiometrically corrected)	27.5 × 27.5
Swath (km)	27.5			
Spectral Bandwidth (µm)	0.5 – 0.85			
Quantization (Bits)	10	Pan Aft	Path-Row scene-based (Precision product: Ortho rectified product)	27.5 × 27.5
Revisit (days)	5			
Equatorial Crossing Time	10:30 a.m.			

Cartosat 2S

Payload Specifications

	Sensors	
Parameters	**Pan**	**Multispectral**
Resolution (m)	0.65	1.6
Swath (km)	9.6	9.6
Spectral Bandwidth (µm)	0.45 – 0.9	Band 1: 0.45 – 0.52 Band 2: 0.52 – 0.59 Band 3: 0.62 – 0.68 Band 4: 0.77 – 0.86
Quantization (Bits)	11	11
Equatorial Crossing Time	9:30 a.m.	

Data Products

Sensor	Product Type	Coverage (sq km)	Products
Pan	Path-Row scene-based	9.6 × 9.6	Standard product – Georeferenced/Orthokit – Pan + Mx Bundle/Merged Precision product – Orthorectified
Mx	Path-Row scene-based	9.6 × 9.6	

Cartosat-3

Payload Specifications

	Sensors	
Parameters	**Pan**	**Multispectral (Mx)**
Resolution (m)	0.28	~1.00
Swath (km)	17	17
Spectral Bandwidth (µm)	0.4 – 0.95	Band 1: 0.45 – 0.52 Band 2: 0.52 – 0.59 Band 3: 0.62 – 0.68 Band 4: 0.77 – 0.86
Revisit (days)	4	4

Source: ACL, 2021

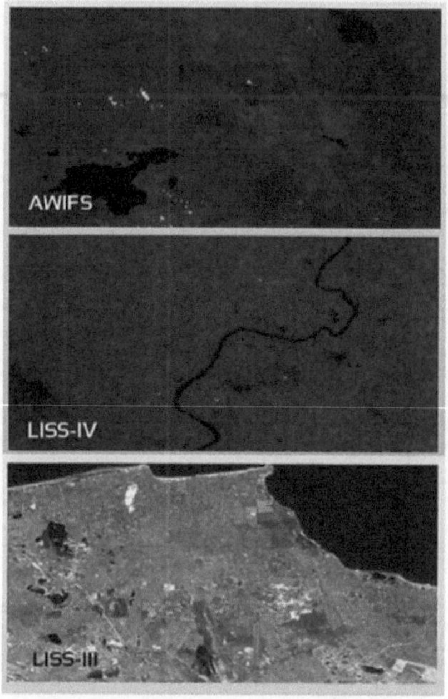

FIGURE 0.5 Resourcesat images

Source: ACL, 2021

Applications: Perspectives from Developing Countries (Singh, 2021), title for which has been kept as proposed earlier, but its contents have been restricted to the sections of Agricultural, Soil and Land Degradation Studies; Hydrology, Microclimates and Climate Change Impacts; Land Use/Land Cover Analysis Applications; Resource Analysis and Bibliometric Studies. Continuing from that forward, second volume embraced the title: *Re-envisioning Advances in Remote Sensing: Urbanization, Disasters and Planning*. Considering these broader themes, present book is proposed with following contents.

Introduction: Re-envisioning Advances in Remote Sensing
I Urbanization and its Impacts
II Geospatial Technology for Disaster Management
III Remote Sensing Applications in Models and Planning
Conclusion: Outlook to Future Research Agenda

The Introduction: Re-envisioning Advances in Remote Sensing depicts the overall structure and gives an overview of its contents in brief. Apart from depicting technological advances in remote sensing, a glimpse into India's remote sensing accomplishments is also given. The present volume contains 16 individual chapters in addition to the introductory and concluding parts.

Section I of the book – *Urbanization and its Impacts* – comprises four chapters with case studies of Indian cities of Aligarh, Ludhiana and Lucknow and one chapter on capital city of Dhaka. Chapter 1 on the dynamics of urban land use and its impact

TABLE 0.8
Resourcesat – Specifications and Data Products

Resourcesat Sampler – 3S & 3SA

Parameters	Sensor (Pan)	Sensors (Multispectral Mx)
Resolution (m)	1.25	2.5
Swath (km)	60	9.6
Spectral Bandwidth (μm)	0.4 – 0.95	Band 1: 0.45 – 0.52 Band 2: 0.52 – 0.59 Band 3: 0.62 – 0.68 Band 4: 0.77 – 0.86
Revisit (days)	4	4

Resourcesat – 2 & 2A

Parameters	LISS-4	LISS-3	AWiFS
Resolution (m)	5.8	23.5	56
Swath (km)	70/23	140	740
No. of Bands	3	4	4
Spectral Bandwidth (μm)	Band 2: 0.52 – 0.59 Band 3: 0.62 – 0.68 Band 4: 0.77 – 0.86 Band 3: default band in Mono	Band 2: 0.52 – 0.59 Band 3: 0.62 – 0.68 Band 4: 0.77 – 0.86 Band 5: 1.55 – 01.70	Band 2: 0.52 – 0.59 Band 3: 0.62 – 0.68 Band 4: 0.77 – 0.86 Band 5: 1.55 – 1.70
Quantization (Bits)	10	10	12
Revisit (days)	5 (Mono band)	24	5
Equatorial Crossing Time	10:30 a.m. +/– 10 min (at descending node)		

Resourcesat – 2 & 2A Data products

Sensor	Product Type	Coverage: sq km	Correction Level	Accuracy (m)
AWiFS	Path-Row Scene-based	370 × 370	Georeferenced & Geo-Ref + RPC	200 m
			Orthorectified	Better than 50
LISS-3	Path-Row Scene-based	140 × 140	Georeferenced & Geo-Ref + RPC	150 m
			Orthorectified	Better than 24
	Geocoded 15′ × 15′	28 × 28	Orthorectified	Better than 24
LISS-IV Mx	Path-Row Scene-based	70 × 70/23 × 23	Georeferenced & Geo-Ref + RPC	100 m
			Orthorectified	Better than 5
	Geocoded 7.5′ × 7.5′	14 × 14	Orthorectified	Better than 5

Resourcesat – 3 & 3A

	Sensors		
Parameters	ALISS-3 (A, B & C)	ALISS-3 (C)	ATCOR
Resolution (m)	20	10	240
Swath (km)	925	280	

(*Continued*)

TABLE 0.8 (CONTINUED)
Resourcesat – Specifications and Data Products

	Resourcesat Sampler – 3S & 3SA		
Spectral Bandwidth (μm)	Band 1: 0.45 – 0.52 Band 2: 0.52 – 0.59 Band 3: 0.62 – 0.68 Band 4: 0.77 – 0.86 Band 5: 1.55 – 1.70	Band 1: 0.45 – 0.52 Band 2: 0.52 – 0.59 Band 3: 0.62 – 0.68 Band 4: 0.77 – 0.86 Band 5: 1.55 – 1.70	0.4 – 1
Revisit (days)	4	11	

Source: ACL, 2021

on land surface temperature (LST) in Aligarh city, Uttar Pradesh, analyses land-use changes in the historical city of Aligarh, which have resulted in the creation of an urban heat island (UHI) and resultant changes in LST. A remarkable change in land use and land cover has been observed in thirteen years between 2005 and 2018, due to population growth and increasing rate of urbanization. The city has become a serious UHI which has increased the LST of the city, which in turn has affected the health of thousands of people. A significant portion of agricultural area has been transformed into built-up structures owing to the increased demands for greater industrial, residential and related settlements. The area under water bodies as well as plantation has shown steep decline due to rapid land use changes. It asks for protecting the water bodies from pollution as well as maintaining the ecological balance of the city. Chapter 2 on the use of a smartphone to map noise pollution – a case study of Ludhiana city – attempts to re-envision a remote sensing application, in the sense that it has tried to analyse noise and location data through the micro sensors of a smartphone. It also asks for utilizing cloud computing in such analysis, and planners/policy makers can analyse the maps/data to find remedies to reduce problems of urban traffic and noise. This study suggests that GIS, crowd sourcing, data analytics, cloud and IoT, when used together can provide effective solutions for noise and traffic management. Chapter 3, an assessment of urban green spaces under the smart cities mission: the case of Qaiserbagh ABD area of Lucknow city, attempts to highlight important aspects of urban green spaces in preserving the environment of an urban area. This study focuses on assessing urban green spaces in the city of Lucknow with detailed analysis of its Smart City Mission's Qaiserbagh ABD Area. The assessment of green spaces has been done using multiple data sets including satellite imageries, norms related to green spaces, observational mapping and survey of park users. The results obtained show considerable vegetation cover in the old city which has been getting reduced due to increased built-up area and vanishing open spaces over a period of nearly two decades. There is a complete absence of adherence to universal design guidelines for enabling access to differently abled persons. It asks for the need to adopt a holistic approach toward improving and maintaining urban green spaces that promotes sustainability, inclusiveness and livability of the city. Chapter 4 on the impact on land surface temperature and urban heat island due to land-use and land cover changes in the Dhaka metropolitan area, using remote sensing and GIS techniques, evaluates the impact of UHI by estimating LST in the Dhaka Metropolitan Area using quantitative multi-temporal thermal

TABLE 0.9
Oceansat – Specifications and Data Products

Oceansat-2

Parameters	OCM
Spectral Range	402–885 nm
Resolution across track	360 m
Resolution along track	236 m
Resolution (m)	360
Swath (km)	1420
Spectral Bandwidth (nm)	Band-1: 402 – 422
	Band-2: 433 – 453
	Band-3: 480 – 500
	Band-4: 500 – 520
	Band-5: 545 – 565
	Band-6: 610 – 630
	Band-7: 725 – 755
	Band-8: 845 – 885
Quantization	12 bit

Data Products

Product type	Aligned	Format Supported
Standard product	Path	HDF, LGSOWG
Georeferenced	North	HDF, LGSOWG

Oceansat 3 & 3A

Parameters	OCM	SSTM-1
Spectral Range	407–1020	
Resolution across track	Bands 1–10: 360 m	1080 m
	Bands 11–13: 1,080 m	
Resolution along track	236 m	
Swath (km)	~1500 km	
No. of bands	13	2 Bands
Spectral Bandwidth (nm)	Band-1: 407 – 417 nm	Band 1: 11 μm
	Band-2: 438 – 448 nm	Band 2: 12 μm
	Band-3: 485 – 495 nm	
	Band-4: 505 – 515 nm	
	Band-5: 550 – 560 nm	
	Band-6: 561 – 571 nm	
	Band-7: 615 – 625 nm	
	Band-8: 665 – 675 nm	
	Band-9: 677 – 685 nm	
	Band-10: 705 – 715 nm	
	Band-11: 775 – 785 nm	
	Band-12: 860 – 880 nm	
	Band-13: 1000 – 1020 nm	
Quantization	12 bit	

Source: ACL, 2021

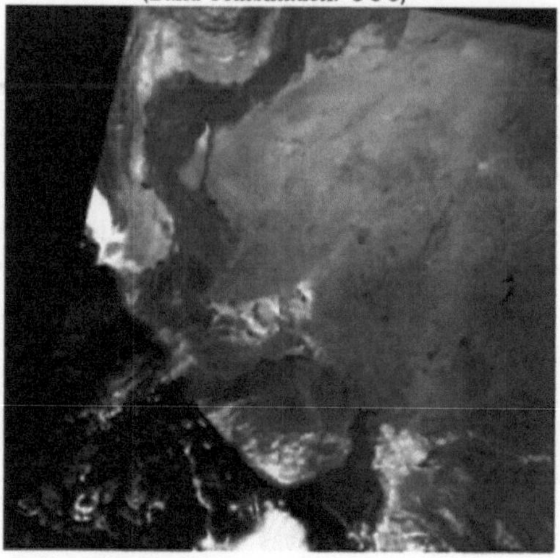

FIGURE 0.6 Oceansat 2 image

Source: ACL, 2021

remote sensing images and GIS techniques. This study analyses the trend of land use/land cover and LST change for the years 1991, 2001, 2011 and 2019, and to predict the future LST (2029) using the artificial neural network-based cellular automata (ANN-CA) algorithm. Its findings suggest an average increase of 3–5 degrees Celsius in LSTs over 28 years and creation of UHIs mainly linked to urban expansion. The built-up area has expanded from 30 per cent to more than 90 per cent between 1991 and 2019, indicating unbalanced urbanization between the built-up and unbuilt city regions.

Section II of the book: Geospatial Technology for Disaster Management, consist of six chapters dealing with disasters such as drought, waterlogging, floods, landslides and urban congestion. Chapter 5 on the spatio-environmental distribution of drought disaster events – a space-based approach using the Terra-MODIS Vegetation Index – analyses drought disaster events for three years 2016, 2017 and 2018 in the Free State Province of South Africa, have been evaluated using remote sensing data, 250m 16 days EVI (MOD13Q1.006) and climate data obtained from AppEEARS and POWER databases using R programming. It revealed drought disasters in 2016, 2017 and 2018 in the study area, which were more evident in the months of January, February, October, November and December. The southern regions were found to witness more drought disaster conditions with a drought index of below 20 per cent, especially in the affected months. These areas were more vulnerable to drought disasters, especially with a prolonged occurrence and a severe water dearth, decline in agricultural products, and loss of habitats and other natural ecosystems in the affected area. Information hotspots have been identified for environmental and ecosystem conservation for better environmental and disaster management. Chapter 6 on the sustainable management of a waterlogged and

TABLE 0.10
HRSAT and RISAT Specifications

HRSAT-1

Payload Specifications

Parameters	Sensors: Pan	Multispectral (Mx)	LWIR
Resolution (m)	1.0	<4	20
Swath (km)	15	15	6
Spectral Bandwidth (μm)	0.45 – 0.8	Band 2: 0.52 – 0.59 Band 3: 0.62 – 0.68 Band 4: 0.77 – 0.86	7.1–11
Revisit (days)	Daily (on AOI basis)		

RISAT-1

Payload Specification

Imaging Mode	Swath (km)	Slant Range Resolution (m)	Polarization
Coarse Resolution ScanSAR mode (CRS)	220	50	Linear/Dual/Circular
Medium Resolution ScanSAR mode (MRS)	115	25	Linear/Dual/Circular
Fine Resolution Stripmap mode (FRS-1)	25	3	Linear/Dual/Circular
Fine Resolution Stripmap mode-2 (FRS-2)	25	6/9	Quad/Circular

RISAT-1A & 1B

Payload Specification

Parameters	Specifications
Frequency	C-band (5.35 GHz)
Polarization	Single, Dual & Circular (Hybrid)
Modes	Strip map, CRS, MRS, Spotlight
Resolution	3 to 6 m, 25 m, 50 m
Swath	10 km to 240 km
Incidence Angles	20° – 49°
Repetivity	25 days for 240 km swath systematic

Source: ACL, 2021

salt-affected area through geospatial technology – a case study of central Haryana – analyses the waterlogged and salinity areas of central Haryana using remote sensing data. It has identified salt affected area in the region (63.22 sq. km.) which could be sub classified into saline soil (50.14 sq. km.) saline-sodic (10.40 sq. km) and sodic soils (2.68 sq. km). Detailed analysis of geospatial data also found area

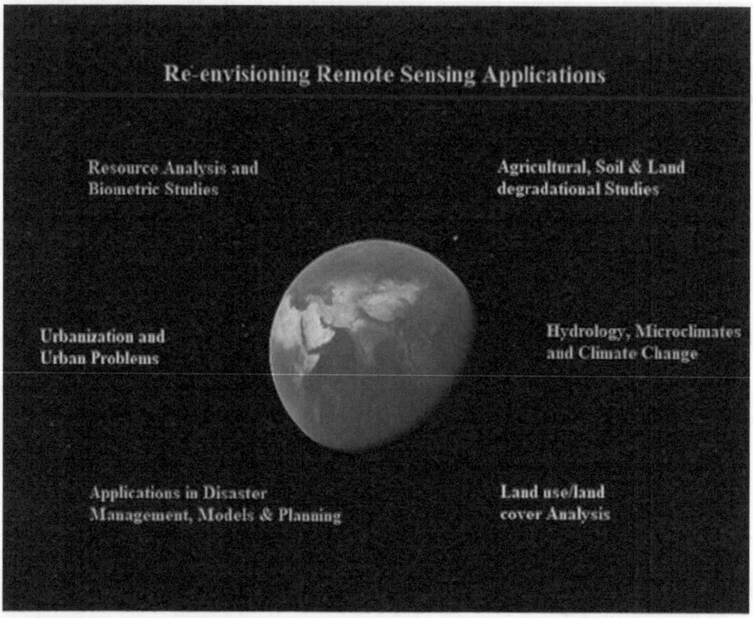

FIGURE 0.7 Re-envisioning Remote Sensing Applications

Source: Singh, 2021

under seasonal waterlogging of about 105.17 sq. km. and permanent waterlogged category of 4.38 sq. km. These areas need early remedy for their inclusion in agricultural land. Chapter 7 on flood risk assessment and analysis of the Kashmir Valley floor through remote sensing is an in-depth analysis of the Kashmir Valley floor in relation to recurring flood estimates, as the risk of floods in the valley has made the population vulnerable to different types of problems. It analyses the flood risk in the Kashmir Valley floor and demarcates the zones that are prone to high risk of flooding. Chapter 8 on remote sensing applications in landslide susceptibility index mapping of Rangamati district, Bangladesh, focused on the landslide susceptibility index (LSI) mapping of Rangamati district by using remote sensing and GIS-based weighted overlay model (WOM). Four different types of satellite data have been utilized to simulate WOM. Shuttle Radar Topography Mission (SRTM) data have been analysed to retrieve elevation, slope, and stream ordering through GIS-based hydrological analysis and digital elevation model (DEM). Landsat-5 data have been used to identify vegetation cover, calculated according to the normalized difference vegetation index (NDVI) model. Google Earth information has been utilized for land use recognition. Tropical Rainfall Measuring Mission (TRMM) data have been simulated in geospatial Kriging interpolation techniques to measure the rainfall intensity distribution. WOM-based landslide susceptibility area identification is found as an essential component of disaster management, risk reduction and sustainable development. Chapter 9 on remote sensing application in landslide assessments – the case of Kotropi in Himachal Himalayas – depicts the case of the Kotropi landslide. Satellite images were found useful in identifying the changing rock structure in the landslide zone during various points of time. It was found that less attached dirt and shear strength and dry thickness were exceptionally low in the landslide area. Chapter 10 of this section on geospatial techniques to quantify urban change – the case of Harare, Zimbabwe – quantified changes in the built-up area in the Zimbabwean capital city of Harare, between 1986 and 2018, using multispectral remote sensing. Land-use classification was carried out using Landsat images (urban and non-urban areas) using the random forest machine learning classifier. The built-up environment was found to have increased by more than 185 per cent between 1986 and 2018, at a rate of more than 5 per cent per annum growth.

Section III of the book – Remote Sensing Applications in Models and Planning – is comprised of six chapters focusing on remote sensing applications in various model buildings and planning exercises. Chapter 11 on remote sensing & GIS-based site identification for solid waste management of Amritsar municipal corporation, Punjab, India, attempts to identify the site for solid waste management in Amritsar, which is a historical and religious city of the state with an international airport. A number of variables like land use, groundwater, soils, drainage, slope, forest/plantation, roads etc. have been analysed through remote sensing data for considering the identification of suitable site for waste management. Chapter 12, on the use of remote sensing techniques in land degradation mapping, presents a review of remote sensing applications in the identification of various land degradations and mapping its extent and related aspects. Chapter 13, on the feasibility analysis of a railway route using remote sensing/geoinformatics and analytical hierarchy process (AHP) techniques, has come up with the feasibility analysis of a railway route between Shivamogga and Harihar towns of Karnataka state in India to highlight the usefulness of integrating remote sensing/geoinformatics and AHP in feasibility studies. This study finds the AHP consistency index of < 0.1 was found and the proposed route has been found moderate to most suitable for laying the railway track. Chapter 14, on comparing the performance of inverse distance weighting and modified Shepard's interpolation models with different sampling arrangements, has comparatively assessed the performance of IDW and MSI models in relation to different sampling densities and patterns, for which elevation data has been collected from Google Earth in three different sample patterns and densities from central Rajasthan. It found that modified Shepard performs better than IDW in five out of nine instances of sample patterns and density but with minor differences. Nevertheless, IDW was found to be a consistent model, displaying almost similar results in all cases of sample strategies. Chapter 15, on the application of remote sensing in the groundwater potential analysis of developing countries, presents remote sensing applications in groundwater resource mapping and exploration, particularly in developing countries, with special reference to the African continent. Chapter 16 on remote sensing applications in identification of villages on the basis of intensity of problems in physical resources in the Kandi region, Punjab, applies geospatial technologies for the identification of the problem intensity with severity level. It used drainage maps, soil erosion maps based on satellite data of IRS Resourcesat-1, LISS III. The villages have been identified based on the close proximity of the severity of these problems with respect to the physical resource base of the region. Study is found to be helpful in micro level planning.

Last, *Conclusion: Outlook to Future Research Agenda* envisioned the future research agenda for remote sensing applications. It is projected that there would be tremendous increase in the remote sensing applications with its interface with big data, cloud computing, and Internet of Things. It is further expected that open-sourced remote sensing with Artificial Intelligence and usage of various programming languages viz., Python, R and others would also be growing to the next levels.

REFERENCES

Antrix Corporation Limited. (2021). *Indian Remote Sensing (IRS) Satellite Data and Services for International Customers.* Remote Sensing Data & Services, Antrix Corporation Limited (ACL), Bengaluru. Retrieved from: https://www.antrix.co.in/sites/default/files/IRS%20%20Brochure.pdf

BBC. (2019). *India puts 'lightest satellite' Kalamsat V2 into orbit.* 25th January, 2019. Retrieved from: https://www.bbc.com/news/world-asia-india-46956595

EagleView. (2020). *EagleView Hits Key Milestone, Processing 100M Images So Far This Year.* Retrieved from https://www.eagleview.com/newsroom/2020/06/eagleview-hits-key-milestone-processing-100m-images-so-far-this-year/

International Data Corporation. (2021). *Data Creation and Replication Will Grow at a Faster Rate than Installed Storage Capacity, According to the IDC Global Data Sphere and Storage Sphere Forecasts.* International Data Corporation, United States Retrieved from: https://www.idc.com/getdoc.jsp?containerId=prUS47560321&utm_medium=rss_feed&utm_source=alert&utm_campaign=rss_syndication

ISRO. (2017). *PSLV-C37 Successfully launches 104 Satellites in a Single Flight.* Retrieved from: https://www.isro.gov.in/pslv-c37-successfully-launches-104-satellites-single-flight

ISRO. (2018). *GSAT-11 Mission.* Bangaluru: Indian Space Research Organization. Retrieved from: https://www.isro.gov.in/Spacecraft/gsat-11-mission

ISRO. (2021). *PSLV-C51/Amazonia-1: PSLV-C51, the first dedicated launch for NSIL, successfully launches Amazonia-1 and 18 Co-passenger satellites from Sriharikota.* Bangaluru: Indian Space Research Organization. Retrieved from: https://www.isro.gov.in/launcher/pslv-c51-amazonia-1

Stanley, M. (2021). *A New Space Economy on the Edge of Liftoff.* Retrieved from: https://www.morganstanley.com/Themes/global-space-economy

Parcak, S. H. (2009). *Satellite Remote Sensing for Archaeology.* New York: Routledge.

Reeves, R. G. (Ed.) (1975). *Manual of Remote Sensing.* Vol. 1, Virginia Falls Church: American Society of Photogrammetry, p. 27.

Reinsel, D., Grantz, J., and Rydning, J. (2018). *Data Age 2025: The Digitization of the World From Edge to Core.* An IDC White Paper No. US44413318, Sponsored by Seagate. International Data Corporation, USA. Retrieved from: https://www.seagate.com/files/www-content/our-story/trends/files/idc-seagate-dataage-whitepaper.pdf

Singh, R. (Ed.). (2021). *Re-envisioning Remote Sensing Applications: Perspectives from Developing Countries* (1st ed.). CRC Press. doi:10.1201/9781003049210

Straub, J., Swartwout, M., Nunes, M., and Lappas, V. (2019). Cubsats and Small Satellites. *International Journal of Aerospace Engineering*, Vol. 2019. doi:10.1155/2019/9451673

UNOOSA (2020). *Annual Report-2019.* United Nations, Vienna: United Nations Office for Outer Space Affairs. Retrieved from: https://www.unoosa.org/res/oosadoc/data/documents/2020/stspace/stspace77_0_html/UNOOSA_Annual_Report_2019.pdf

Villela, T., Costa, C.A., Brandao, A.M., Bueno, F. T., and Leonardi, R. (2019). Towards the Thousandth CubeSat: A Statistical Overview. *International Journal of Aerospace Engineering*, 2019. doi:10.1155/2019/5063145

Part I

Urbanization and Its Impact

1 Dynamics of Urban Land Use and Its Impact on Land Surface Temperature (LST) in Aligarh City, Uttar Pradesh

Saleha Jamal, Ishfaq Hussain Malik, and Wani Suhail Ahmad
Aligarh Muslim University (AMU), Aligarh, India

CONTENTS

1.1 INTRODUCTION

When the atmospheric and ground temperature conditions in urban areas are higher than in rural areas in its vicinity, it is known as an 'urban heat island (UHI)'. In more precise terms, the reduction in vegetation cover in the urban areas, characteristics of urban materials, geometry of urban areas, heat generated by human activities,

DOI: 10.1201/9781003224624-3

weather and location of urban areas are the pivotal factors responsible for the phenomenon of the urban heat island (UHI) (Voogt, 2004). According to EPA (2008), 'high emissions of air pollutants and greenhouse gases, consumption of energy at higher pace, degraded human health, comfort and contaminated quality of water are some of the major repercussions of UHI'. In the beginning, observations taken from fixed ground-based thermometers were used for the study of UHI. With the passage of time, satellite data obtained from thermal remote sensing have made feasible the remote observation of UHI. It had opened the gates for new opportunities for the monitoring of UHI through the amalgamation of GIS and thermal remote sensing. According to Weng, Lu and Schubring (2004), 'there is a sharp decrease in LST with the increase of green space in urban areas because it is a green space which can produce cool island effects by the processes of evapotranspiration and emissivity, in comparison to impenetrable surfaces'. Green spaces act as an umbrella that generates the beam of shade and conceals the land surfaces. In this way green spaces halt the direct radiation of sun rays on the ground surface which in turn protects the direct heating of the ground (Li et al., 2012). Singh et al. (2017) observed,

> Surface temperature in the main central part of the Lucknow city is much higher than open area which is located in its vicinity, those areas having a dense tall vertical building shows the higher temperature while the areas having a good vegetation cover and water bodies displayed the lower temperatures.

LST measurements which are taken by thermal remote sensing are generally used to study the phenomenon of the heat island. Landsat 8-OLI/TIRS data and number of geospatial approaches like multiresolution grid-based, urban rural gradient and spatial metrics-based techniques are mainly used for depth analysis and observation.

There is about 4°C difference in the LST of impervious surface (exposed surface) and of green spaces. It depicts the significant part played by green space in reducing the UHI effects thus, provides an essential service to urban ecosystems. Unplanned urbanization, rapid population growth in Aligarh city of Uttar Pradesh, which is one of the oldest cities of the world, has resulted the unsustainable development which became driving force in increasing the impervious area e.g. roads, buildings, asphalt, concretes etc., which in turn reduced the length and breadth of green spaces e.g. farming lands, tress, parks, gardens etc. in the city. In order to study the phenomena of UHI, temperature is one of the pivotal environment variables in order to comprehend the surface physical characteristic and climate which is observed and analysed by Earth-observing remote sensing systems (Worner, 2013). Tall vertical buildings, concerts, asphalt and anthropogenic activities like transport and industrial works in urban areas are some of the main reasons which cause the UHI. Construction of buildings and other infrastructures results in the heavy degradation of vegetation cover and thus causes the negative natural cooling effects. The tall vertical expansion of buildings and narrow streets and lanes reduce the pace of airflow and thus give impetus to high air temperature. Besides this, vehicles, industries and air conditioners also produce a large amount of heat and thus rise the temperature in the surrounding environment. Singh and Kalota (2019) while analysing the case of Ludhiana city have found that the rapid industrialization and influx of population is leading to urban sprawl and UHI generation there. This increasing temperature in the city becomes the cause of 'heat syncope, heat exhaustion, heat stroke and heat cramps'. This phenomenon of UHI affects entire biotic life. Increasing the rate of energy consumption for different cooling purposes reduces the quality of air which in turn affects the health of human beings (Taqi et al., 2017) and causes many chronic respiratory diseases like asthma. Thus, critical analysis of UHIs is obligatory which provides us information and a set of guidelines for the future city planning and sustainable development of the environment.

1.1.1 Objective

The objective of the present study is to analyse the spatial distribution of urban land use change and its impact on land surface temperature (LST) in Aligarh city.

1.1.2 Study Area

Aligarh is a historical city which has a cherished and significant history including Mughal, Maratha's and British rule. Aligarh lies in western part of North India in Uttar Pradesh state. It possesses several headquarters of the district and is famously known as an industrial and educational city. Aligarh Muslim University, lock industry and its nearness to the capital of India (New Delhi) make the city an important functional city. Aligarh city and its environs is spread over 6,000 hectares of land but the area under municipal limit is only 4,985 hectares. Out of the total developed area (67.48%), residential area accounts for 77.89 per cent, educational institutions 8.55 per cent, transport 3.60 per cent, industrial and commercial 3.18 per cent, open spaces and parks 0.06 per cent and for recreation purposes only 0.08 per cent. It has been divided into two parts by the Delhi-Kolkata railway line; the southern part is relatively older part of the city while northern part is newer, where Aligarh Muslim University is located. The city has a total population of 874,000 (Census of India, 2011). It has 70 municipal wards (Figure 1.1) with 102,004 households spread over 427 Mohallas with a total population of 874,000 (Census of India, 2011) (Figure 1.2).

1.1.3 Database and Methodology

Landsat 8-OLI and Landsat 4–5 TM data with spatial resolution of 30m for the month of March 2018 and 2005 of the study area were downloaded from Earth explorer, United States Geological Survey (http://earthexplorer.usgs.gov/). Improcessing techniques for both the data sets have been carried out. In order to get the false colour composite, layer stack of bands was carried out followed by subset of Aligarh city from the downloaded tile. For Landsat 8-OLI resolution merge was done to increase the spatial resolution from 30m to 15m. Supervised 'maximum likelihood classification' (Ganaie et al. 2020, Jamal et al. 2019, Khan et al., 2020) was carried out in which 100 samples (training sets) for each category were selected with the help of stratified random sampling. In addition to this, recode of the categories at specific places was also done wherever required after field validation. Thus, accuracy assessment was also performed for the validation of results in Aligarh city. Thus, final

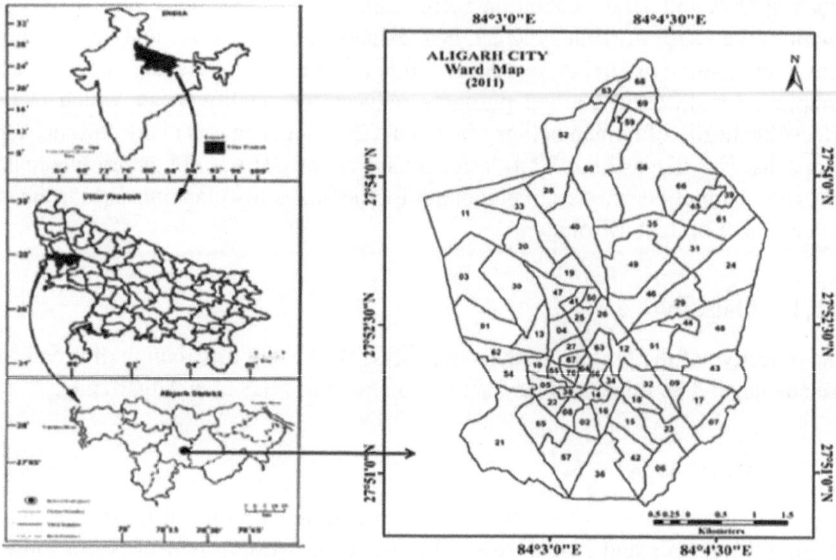

FIGURE 1.1 Aligarh City Ward Map

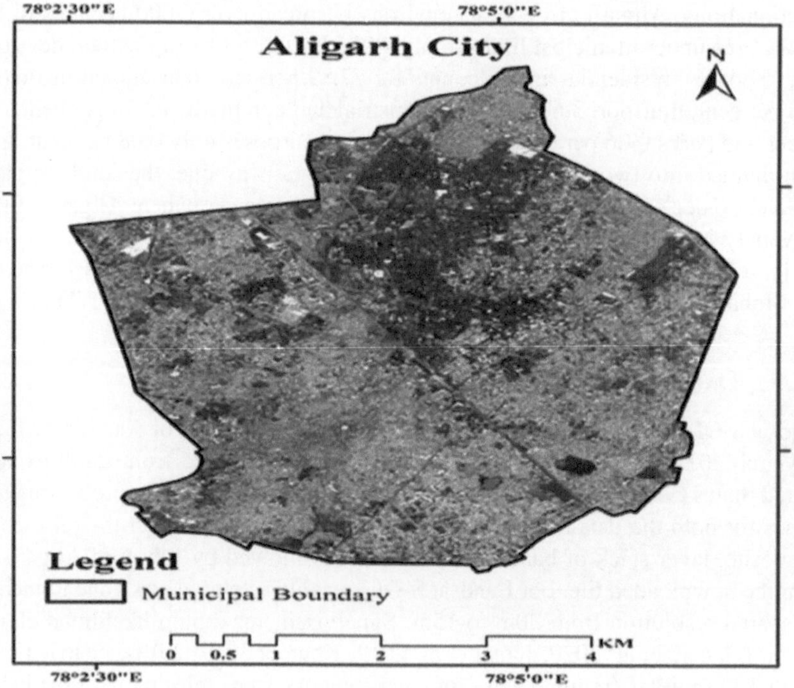

FIGURE 1.2 Aligarh City

multidate land use land cover maps were prepared. For 'normalized difference vegetation index', a measure of vegetation health was also carried out using NIR band 4 and RED band 5. For the estimation of land surface temperature, band 10 of Landsat 8-OLI was used. Further, emissivity, a measure of LST was also carried out. All such techniques were carried out by the use of GIS software, Arc GIS 10 and Erdas Imagine 14 (Figure 1.2a).

- **TOA (Top of Atmosphere) Spectral Radiance**
  For the calculation of TOA, it requires band 10 of Landsat 8-OLI with values of formula taken from satellite metadata. The formula for retrieving spectral radiance at the TOA is:

$$L\lambda = ML * Q\text{cal} + AL - Oi \tag{1.1}$$

Where, 'ML = band-specific multiplicative rescaling factor
Qcal = Band 10 image

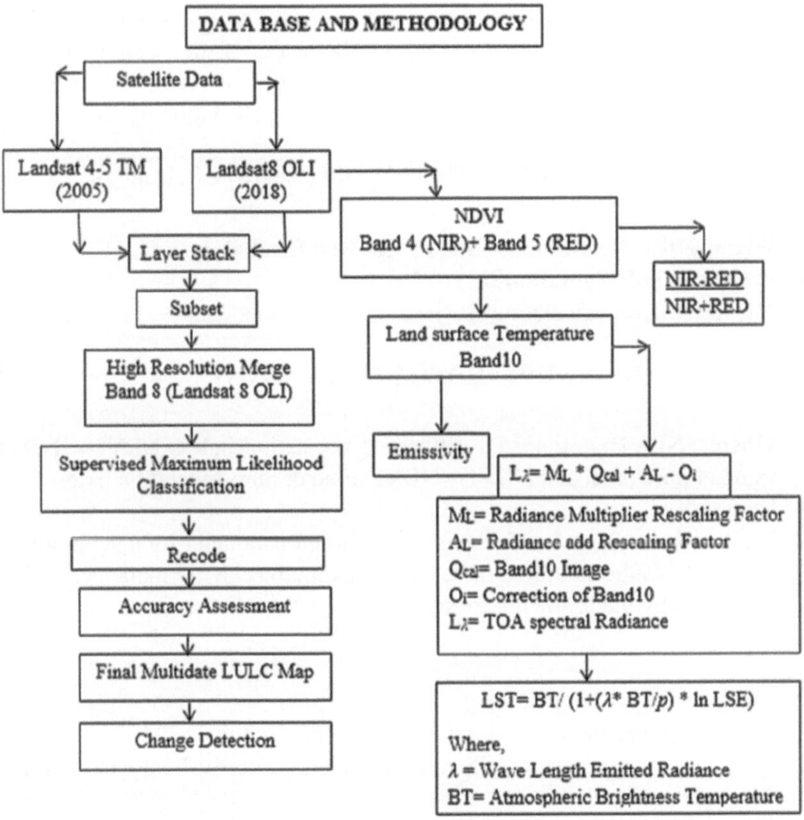

FIGURE 1.2A Database and Methodology

AL = band-specific additive rescaling factor
Oi = correction for Band 10'

Metadata of the Landsat 8-OLI satellite image
'Thermal constant, Band 10: $K1$= 1321.08 and $K2$= 777.89
Rescaling factor, Band 10: ML= 0.000342 and AL = 0.1
Correction, Band 10: Oi= 0.2'

- **Brightness temperature (BT)**
After the conversion of digital numbers to reflection, thermal constants were used from metadata file to convert 'TIRS' band data to BT. The formula used to convert reflection to BT is as follows:

$$\text{'BT} = K2/\ln[(K1/L\lambda)+1]-273.15 \tag{1.2}$$

Where, $K1$ and $K2$ = band-specific thermal conversion constants (satellite metadata file)'

- **Normalized Difference Vegetation Index (NDVI)**
NIR band 5 and Red band 5 of Landsat 8-OLI were used for calculating NDVI values. It is used to find out the health of the vegetation cover. Calculation of NDVI helps to calculate 'proportion of vegetation and emissivity'. All these measures are important for the calculation of LST.

$$\text{'NDVI} = \text{NIR}(\text{band }5) - R(\text{band }4)/\text{NIR}(\text{band }5) + R(\text{band }4) \tag{1.3}$$

Where, NIR= Near-infrared Band 5, R= Red Band 4'

- **Proportion of Vegetation (P_v)**
The formula for calculating 'P_v' is:

$$\text{'}P_v = (\text{NDVI} - \text{NDVI}s/\text{NDVI}_v - \text{NDVI}s)^2\text{'} \tag{1.4}$$

Where, 'NDVI= normalized difference vegetation index, NDVIs= NDVI value of pure pixel and NDVI$_v$= NDVI value of pure vegetation pixel'

- **Emissivity**
Land surface emissivity (LSE) is an important parameter for calculation of LST. LSE predicts emitted radiance by scaling blackbody radiance. LSE is calculated by the following formula:

$$\varepsilon\lambda = \varepsilon_{v\lambda P_v} + \varepsilon_s\lambda\ (1-P_v) + C\lambda \tag{1.5}$$

Where, '$\varepsilon_{v=}$ vegetation emissivity, $\varepsilon_{s=}$ soil emissivity
C = surface roughness, 0.005 constant value (C=0, for homogeneous and flat surfaces).'

- **Land surface temperature (LST)**
  Last step i.e., calculating LST is computed as

$$\text{'}T_s = \mathrm{BT}/\{1+[(\lambda \mathrm{BT}/\rho)\ln \varepsilon\lambda]\}\text{'} \tag{1.6}$$

Where, 'T_s = LST in Celsius (°C), BT= at-Sensor BT (°C)
λ = wavelength of emitted radiance (λ = 10.895), $\varepsilon\lambda$ = emissivity calculated in Equation (1.5), and

$$\rho = h * c/\sigma = 1.438 \times 10^{-2}\,\mathrm{m\,K} \tag{1.7}$$

Where,
σ = Boltzmann constant (1.38 × 10^{-23} J/K)
h = Planck's constant (6.626 × 10^{-34} J s)
c = velocity of light ((2.998 × 10^{8} m/s)'

1.2 RESULTS AND DISCUSSION

1.2.1 Land Use Land Cover and Land Use Land Cover Change Detection in Aligarh City

Figure 1.3 depicts the LST of Aligarh city, Figure 1.4 depicts Emissivity of Aligarh city, Figure 1.5 depicts density map of Aligarh city in 2001 while as Figure 1.6 depicts density map of Aligarh city in 2011. From these figures, it is clear that LST of Aligarh city varies among the different parts of the city and there has also been a remarkable change in the density of the city as the density of settlements has increased in some parts of the city.

A total of five land use land cover classes in Aligarh city have been classified, which are plantation, agriculture, settlement, wasteland and water bodies (Table 1.1). Maximum likelihood supervised classification was adopted for the classification of land use land classes (LULC) categories. From the analysis of Table 1.1, it is evident that settlement has registered a 100 per cent increase in Aligarh city in 13 years. According to the LULC statistics of the Aligarh city, settlement class in occupies an area of 6,695 hectares in the year 2005, which marked a steep increase in its growth rate and registered an area of 13,390 hectares in 2018. Settlement is the only classified category in the study area which registered an increasing growth rate. This phenomenal increase is attributed to population growth in the city, migration from its adjoining areas, industrial development and educational hub. This growth rate has occurred at the cost of fertile agricultural lands, water bodies, vegetation and vacant lands in Aligarh city. The other classes like water body, wasteland, plantation and agriculture registered a negative growth rate in their respective areas from 2005 to 2018 as calculated from satellite measurements. Water bodies registered the

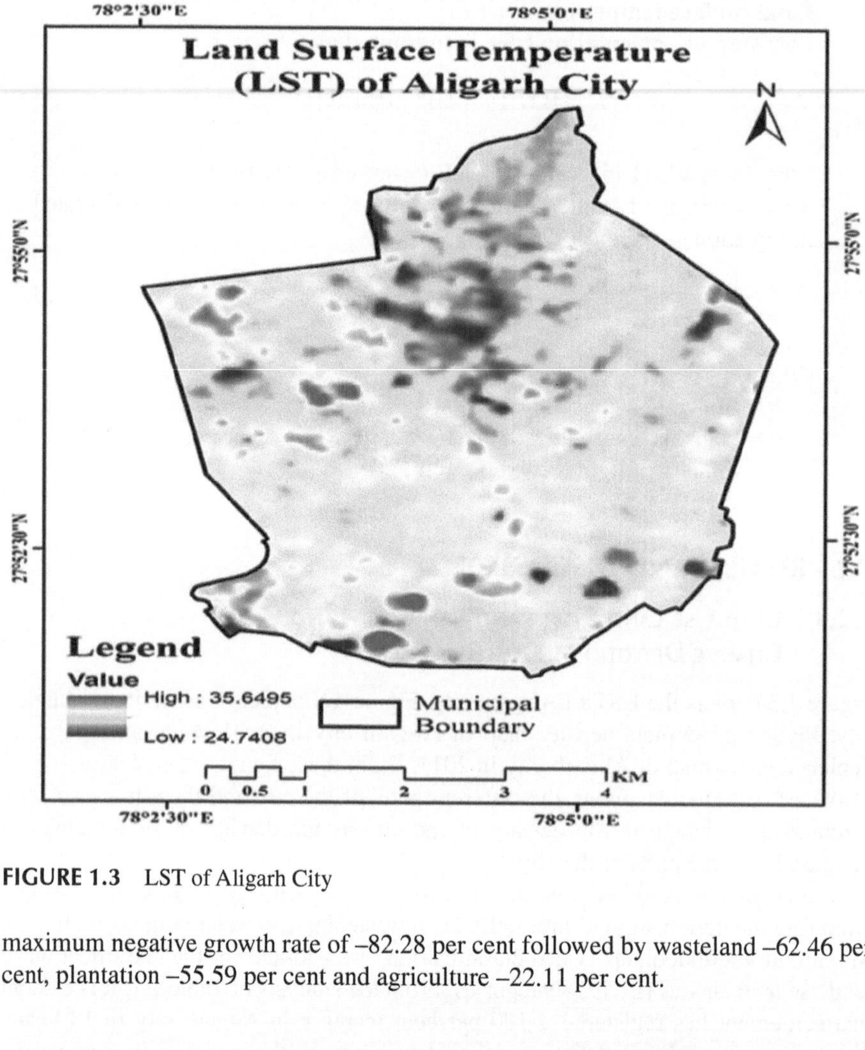

FIGURE 1.3 LST of Aligarh City

maximum negative growth rate of –82.28 per cent followed by wasteland –62.46 per cent, plantation –55.59 per cent and agriculture –22.11 per cent.

1.2.2 Normalized Difference Vegetation Index (NDVI) of Aligarh City

NDVI values range from –1 to +1, and is an index of quantitative measurement of vegetation health which ranges from 0–1, where 0 means no greenness or minimal greenness, whereas the value 1 means maximum greenness. NDVI quantifies vegetation phenology, cover, structure or health of leaves of any study area under investigation. NDVI is calculated using visible reflected light and near-infrared by the vegetation. If the vegetation under investigation reflects near-infrared rays in large portion and absorbs majority of the visible light which falls on it, it is considered as healthy vegetation. However, if it reflects in large amount the visible light and absorbs more near-infrared, then it is unhealthy. It uses the ratio between red light and near-infrared light. In the present study, NDVI value ranges were analysed from +1 to –1 (Figure 1.7). It is evident that +ve value of NDVI is found in area of low

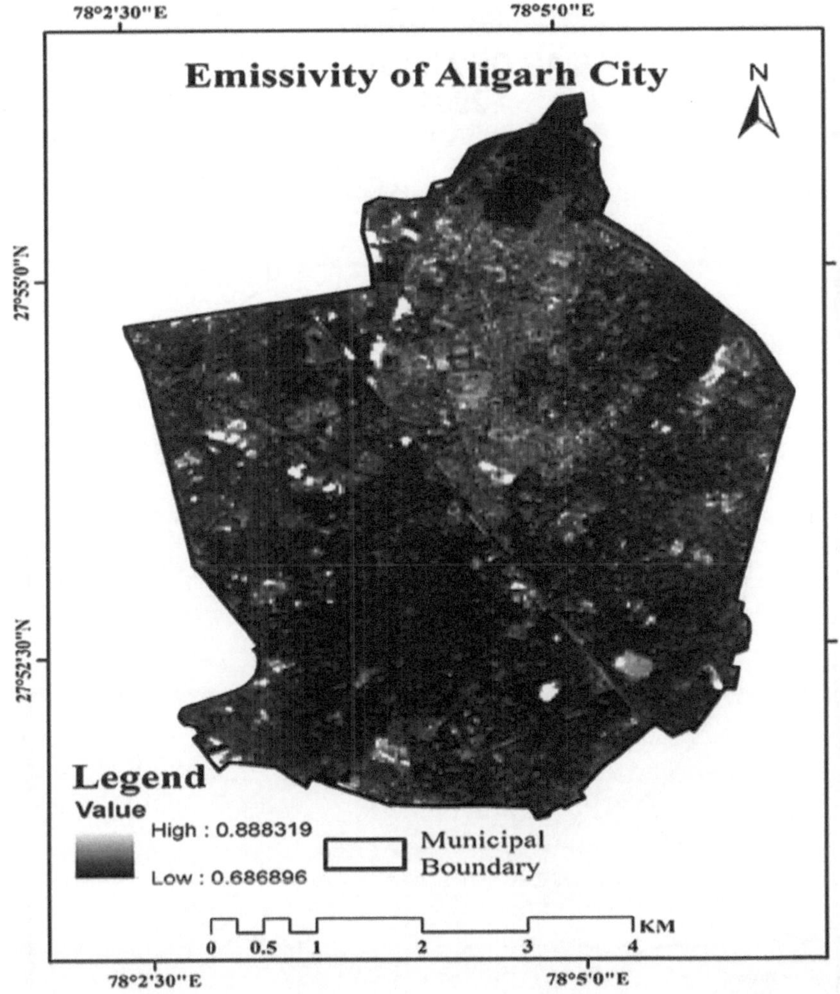

FIGURE 1.4 Emissivity of Aligarh City

density of settlements and those areas which are densely populated, and low vegetation cover depicts a low value (0 to –1). The area under university (Aligarh Muslim University) depicts a good value of healthy vegetation and the core city area where the settlements are very dense depicts a minimal greenness.

1.2.3 Relation Between Land Surface Temperature (LST), Normalized Difference Vegetation Index (NDVI) and Land Use Land Classes (LULC)

Relation between land use land cover dynamics and land surface temperature was assessed using the satellite data Landsat 4–5 TM and Landsat 8-OLI. For the purpose

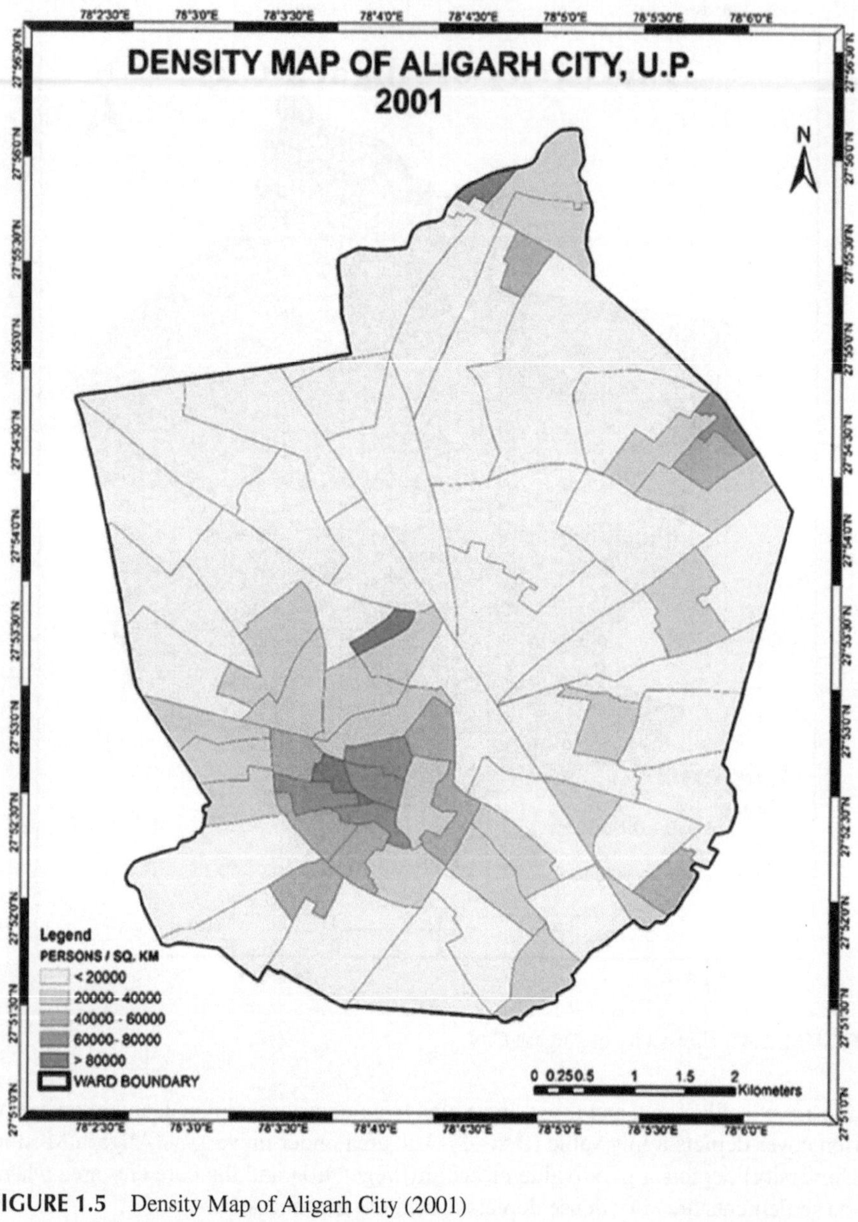

FIGURE 1.5 Density Map of Aligarh City (2001)

of studying LULC in Aligarh city, the LULC classes in March 2005 and 2018 were extracted by Landsat 4–5 TM and Landsat 8-OLI. However, for the estimation of LST, band 10 of Landsat 8 was taking into consideration. After analysis, it was found that there is significant relation of LST with LULC dynamics in Aligarh city. It is quite evident that vegetation cover in the study area has shown low radiant

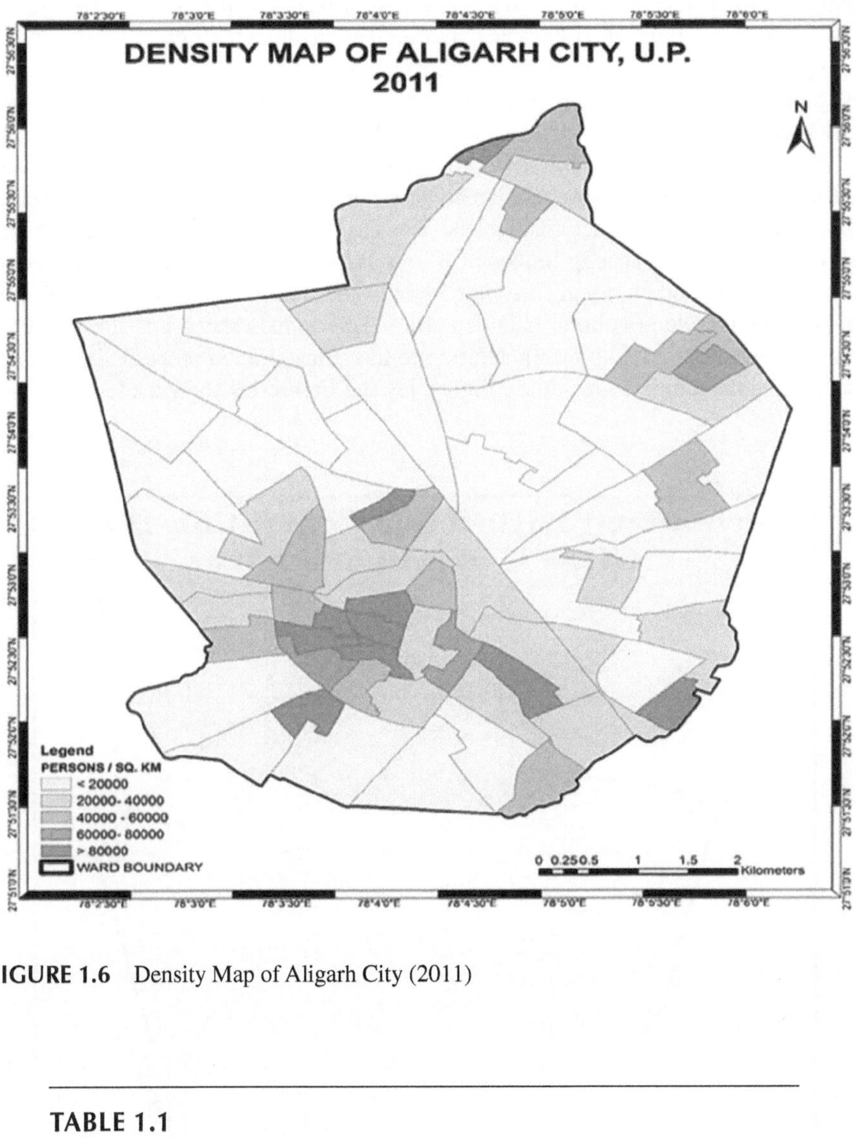

FIGURE 1.6 Density Map of Aligarh City (2011)

TABLE 1.1
LULC Statistics in Aligarh City

Classes	Area in Hectares (2005)	Area in Hectares (2018)	Change (%)
Plantation	4,635	2,058	–55.59
Agricultural	10,134	7,893	–22.11
Settlement	6,695	13,390	100
Wasteland	2,523	947	–62.46
Water bodies	367	65	–82.28
Total	**24,353**	**24,353**	

temperature, as the vegetation reduces amount of heat stored in surface and soil by the mechanism of transpiration. The maximum temperature through LST estimations in Aligarh city was found to be about 35.6°C, while the lowest temperature recorded is found to be 24.7°C (Figure 1.7). Thus, the average temperature found in Aligarh city is about 30.15°C.

The hottest area in Aligarh city is found to be concentrated in maximum density settlement areas where the vegetation cover is very minimal. In this area low mixing of air happens and the concretization also leads to UHI phenomena, due to much heat absorption and emittance in the form of long terrestrial radiations, thereby heating the atmosphere and increasing the temperature. The areas having good concentration of vegetation and agricultural land depict low LST values because it allows the air movements and mixing of air. In addition to this, transpiration process also leads to decreasing the temperature values. Figure 1.8 and Figure 1.9 show the land use land

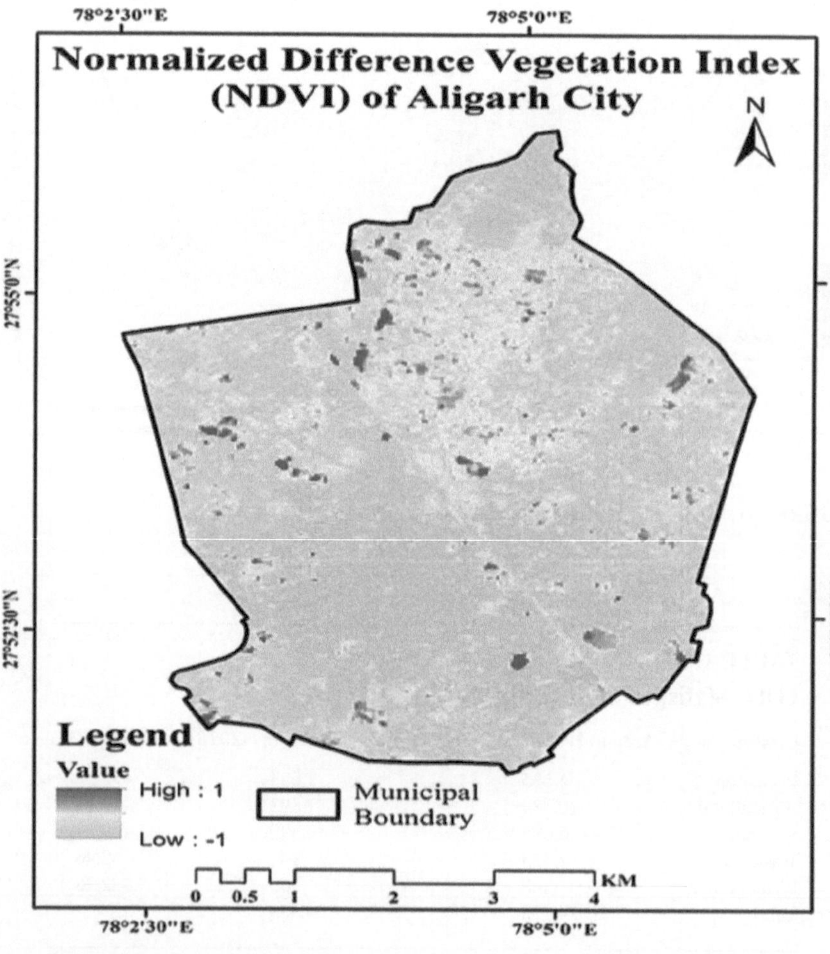

FIGURE 1.7 NDVI of Aligarh City

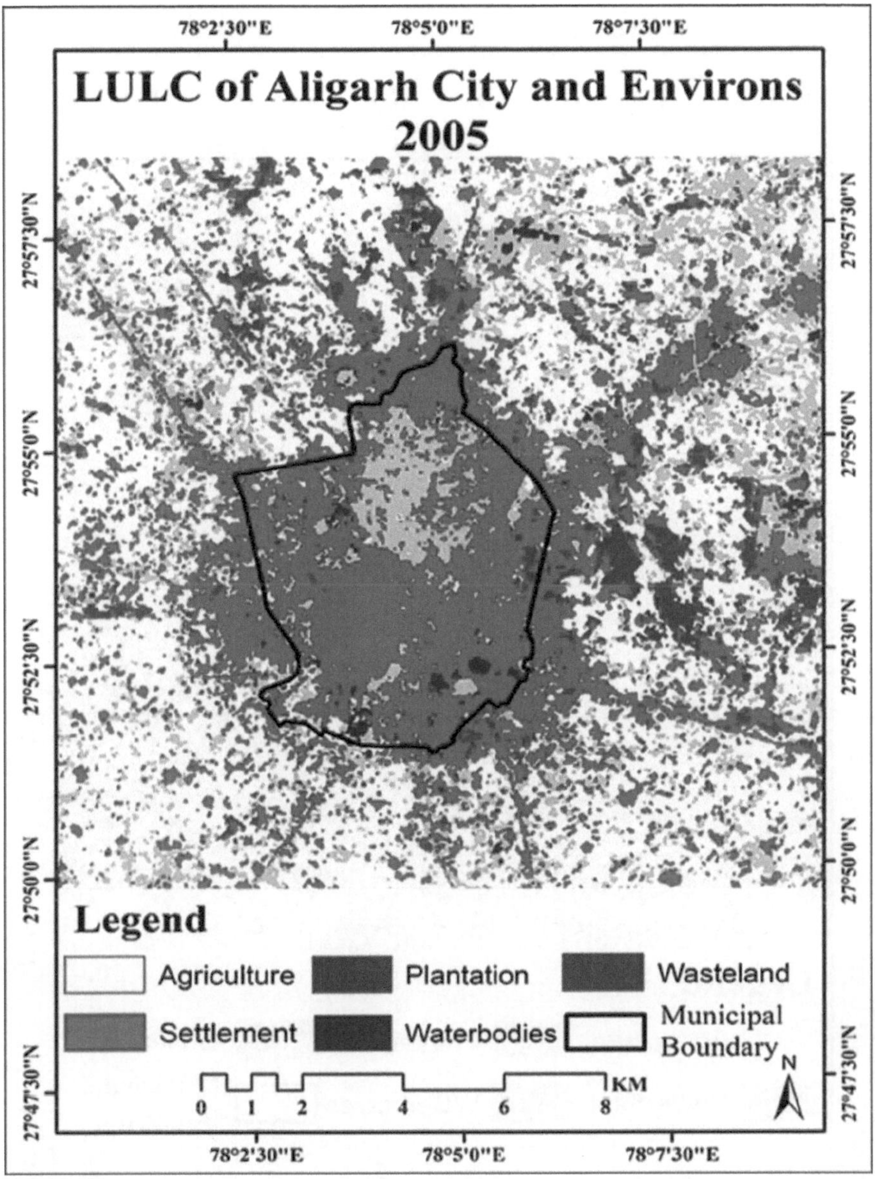

FIGURE 1.8 Land Use Land Change in Aligarh City and its Environs (2005)

change in Aligarh city and its environs in 2005 and 2018 respectively, i.e, with a time gap of 13 years. These figures show a remarkable change in the land use in Aligarh city because the places especially near Aligarh Muslim University have shown a tremendous increase in the number of settlements as many houses, restaurants and apartments have been built in the vicinity of the university to cope up with the demands of the students and faculty members and due to the neoliberal culture.

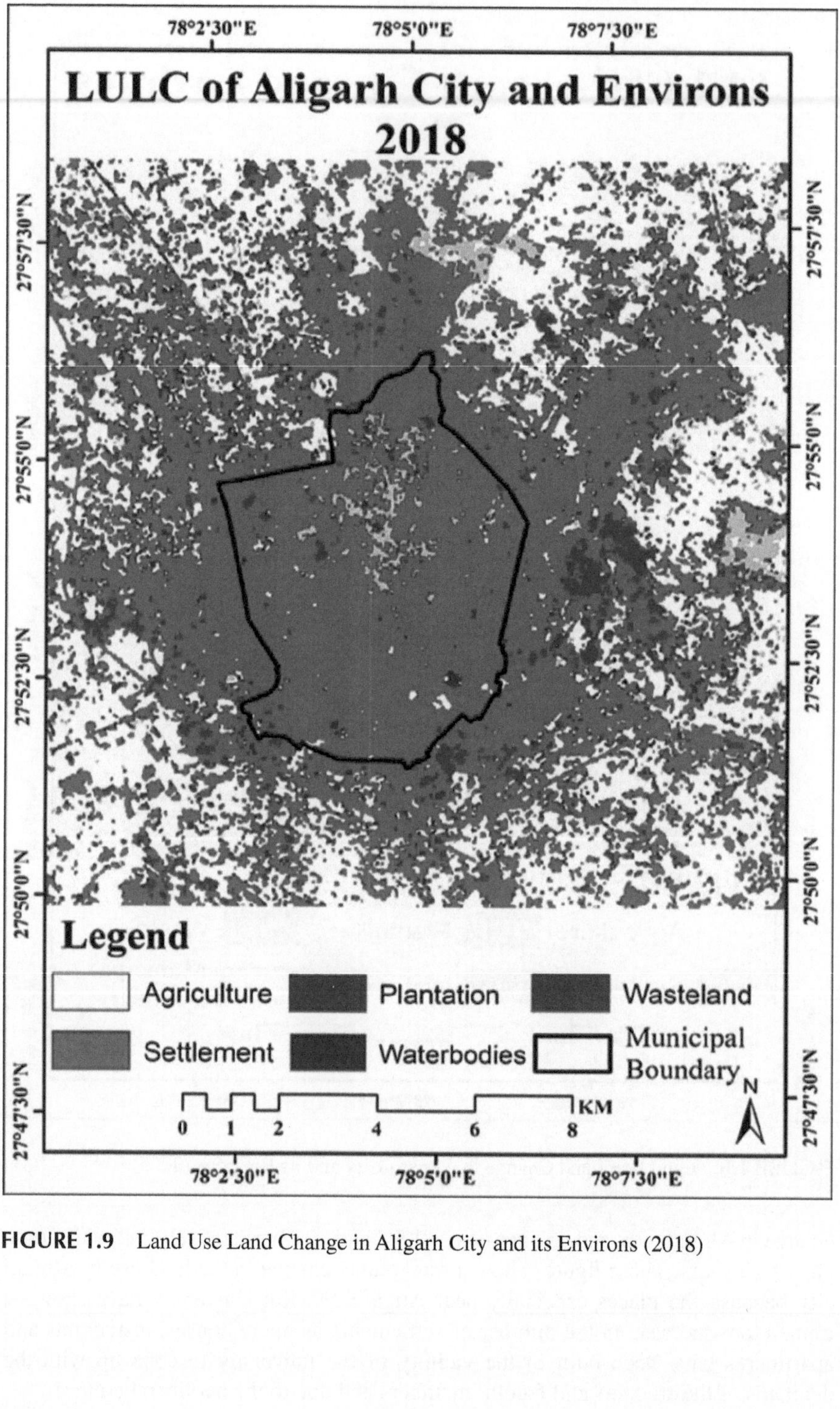

FIGURE 1.9 Land Use Land Change in Aligarh City and its Environs (2018)

1.3 CONCLUSION

Aligarh city has shown a remarkable change in land use and land change cover in 13 years, i.e, between 2005 and 2018, due to increase in population, increasing rate of urbanization and rising demands of people. The city has become a severe urban heat island which has increased the land surface temperature of the city, which in turn has affected the health of thousands of people. A significant portion of agricultural area has been transformed into built-up areas as the demand for more settlements has increased. The area under water bodies as well as plantation has shown a steep decline, due to various neoliberal activities. The need of the hour is to protect the water bodies from pollution as well as decrease in size. Agricultural land is in dire need of protection to maintain the ecological balance of the city.

REFERENCES

Arnfield, A. J. (2003). 'Two Decades of Urban Climate Research: A Review of Turbulence, Exchanges of Energy and Water, and the Urban Heat Island.' *International Journal of Climatology* 23(1):1–26.

Artis, D. A. and Carnahan, W. H. (1982). 'Survey of Emissivity Variability in Thermography of Urban Areas.' *Remote Sensing of Environment* 12(4):313–329.

EPA (US Environmental Protection Agency). (2008). *Reducing Urban Heat Islands: Compendium of Strategies*. US Environmental Protection Agency, Washington, D.C.

Estoque, R. C., Murayama, Y, and Myint, S. W. (2017). 'Effects of Landscape Composition and Pattern on Land Surface Temperature: An Urban Heat Island Study in the Megacities of Southeast Asia.' *Science of the Total Environment* 577:349–359.

Ganaie, T. A., Jamal, S., and Ahmad, W. S. (2020). Changing land use/land cover patterns and growing human population in Wular catchment of Kashmir Valley, India. *GeoJournal*, 1–18.

Grover, A., Singh, R., and Ram Singh, B. (2015). 'Analysis of Urban Heat Island (UHI) in Relation to Normalized Difference Vegetation Index (NDVI): A Comparative Study of Delhi and Mumbai.' *Environments* 2(2):125–138.

Jamal, S., Ahmad, W.S., Ali, M.A., & Sharma, A. A. (2019). 'Monitoring Land Use/Land Cover Change Detection and Urban Expansion with Remote Sensing and GIS techniques in Anantnag District of Kashmir Valley'. *The Geographer* 66(1):60–69.

Khan, T.A., Ahmad, W.S., and Tariq, S. (2020). 'A Multi-Temporal Land Use and Land Cover Change Analysis of Delhi-NCT, using Remote Sensing and GIS'. *Bhugol Swadesh Charcha, Multidisciplinary International Journal* 16(1):11.

Kikon, N., Singh, P., Singh, S. K., and Vyas, A. (2016). 'Assessment of Urban Heat Islands (UHI) of Noida City, India Using Multi-Temporal Satellite Data.' *Sustainable Cities and Society* 22:19–28.

Malik, I. H. (2021). Spatial Dimension of Impact, Relief, and Rescue of the 2014 Flood in Kashmir Valley. *Natural Hazards*, 1–19.

Malik, I. H., and Hashmi, S. N. I. (2020). Ethnographic Account of Flooding in North-Western Himalayas: A Study of Kashmir Valley. *GeoJournal*, 1–19.

Malik, I. H., and Hashmi, S. N. I. (2021). The Great Flood and its Aftermath in Kashmir Valley: Impact, Consequences and Vulnerability Assessment. *Journal of the Geological Society of India* 97(6):661–669.

Singh, R. and Kalota, D. (2019). Urban Sprawl and its Impact on Generation of Urban Heat Island: A Case Study of Ludhiana City. *Journal of the Indian Society of Remote Sensing* 47:1567–1576. https://doi.org/10.1007/s12524-019-00994-8

2 Use of a Smartphone to Map Noise Pollution

A Case Study of Ludhiana City

Harpinder Singh and Dheeraj Gambhir
Punjab Remote Sensing Centre, Ludhiana, India

CONTENTS

2.1 INTRODUCTION

Noise is any sound which discomforts the ears. It reduces one's quality of life by disrupting various normal activities like sleeping or having a conversation. Not all noises can be categorized as noise pollution. Noise pollution is increased levels of sound in the surrounding environment that are harmful and annoying to living beings. So, noise pollution is noise that is harmful to living beings' psychological and physical health. It is one of the major causes of deafness, impaired hearing and other health hazards. Research has also proved that not only human beings but also animals are affected by it – e.g. noise from ships and submarines disturbs fish and aquatic animals also. The unit used to measure the intensity of sounds is the decibel (dB). The pitch of sound can also be measured with frequency of sound vibrations per second. Sound becomes noise above a certain level. According to Noise (n.d.) more than 85dB sound is noise and is harmful. Ministry of Environment and Forest GOI (Government of India) have published the Environment (Protection) Rules, 1986. According to them, the recommended noise standards for automobiles, construction equipment and domestic appliances are as given in Table 2.1.

Today's smartphones are incredible little machines with lot of sensors packed into them. Some important sensors are:

DOI: 10.1201/9781003224624-4

TABLE 2.1
Recommended Noise Standards

		Limits in dB(A),Leq	
Area Code	**Category of Area**	**Day Time**	**Night Time**
A	Industrial area	75	70
B	Commercial area	65	55
C	Residential area	55	45
D	Silence Zone	50	40

Source: Data retrieved from http://www.envfor.nic.in/citizen/specinfo/noise.html

1. Proximity Sensor – Nearby objects can be detected using this sensor.
2. Accelerometer – This sensor is used to detect the change in velocity. Basically, it detects the movement in X, Y and Z manner.
3. Gyroscope – This sensor is used to track the rotations and twists on an object. It is also called a gyro.
4. Barometer – The barometer sensor is used to detect the altitude data.
5. GPS – Detects the location.
6. Microphone – Records the voice or sound.

In addition to the sensors, an app is required in the smartphone to collect the data recorded by the sensors. For this research the NoiseTube app has been used. Sony Computer Science Laboratory (Paris) started the NoiseTube research project in collaboration with the Vrije Universiteit (Brussels) in 2008. The project's objective was to turn a smartphone into a mobile noise level meter. Such an application helps citizens be aware and measure their exposure to noise in their everyday environment. They can also participate in this project by collecting the noise data of their neighborhoods and sending it to the project's server for further analysis. This project consists of two parts: a mobile application and a website (www.noisetube.net). These applications can be used by the users to submit, visualize and share the noise measurements. The user can also download the noise recordings from the smartphone itself.

According to Noise Map (n.d.) a noise map is a graphic representation of the sound level distribution existing in a given region, for a defined period. Doygun and Gurun (2008) measured and mapped the urban traffic noise pollution in a city of Turkey. Dursun, Ozdemir, Karabork and Koçak (2006) prepared the noise pollution map of Konya city using conventional techniques. Maisonneuve, Stevens, Niessen and Steels (2009) describe how the public can contribute and crowdsource data to create noise maps using GPS-enabled smartphones and the NoiseTube Android app. The main objective of the present research is to use a smartphone to map the noise levels in the industrial city of Ludhiana, which is getting heavily industrialized and urbanized, leading to increasing levels of noise pollution as well. Ludhiana is the largest city of Punjab and the commercial-industrial hub, with an increasing number of vehicles and haphazard industrial growth. This unchecked growth is also leading to conditions of urban sprawl and UHI in Ludhiana (Singh & Kalota, 2019). Considering the overcrowding and increasing noise pollution, this research aimed to utilize the sensors (microphone and GPS) on a simple Android smartphone to map the noise pollution in a part of Ludhiana city in Punjab.

2.2 METHODOLOGY

A smartphone is a pack of sensors and the sensors used in this research study are microphone and GPS (Global Positioning System). While the microphone records the sound, GPS records the location. An android app, 'NoiseTube', was installed on a smartphone. This smartphone was mounted on a vehicle and the vehicle was moved in a part of Ludhiana city to collect noise (along with the location) data. The parts of Ludhiana city where the vehicle moved were the commercial, commercial-cum-residential areas and main roads.

Once the data was collected it was extracted from the phone. The data was in the form of a table with each record giving information of location and noise at that point. This data was converted into the form of a map in the ArcGIS Online platform (Figure 2.1). The point map layer was symbolized according to the noise levels and density point cloud – i.e., the areas with yellow colour have higher noise as compared to red or grey areas.

2.3 RESULTS

From the map, it was clear that the road intersections and old city area were having high noise as compared to the other parts of the city. According to the collected data the maximum noise was 92.6 dB and the minimum was 43.9 dB. Finally, a story map application (Figure 2.2) was created on the storymaps.arcgis.com website which hosts the maps and the results of the research work. The link of the online application is: http://arcg.is/2wGDpBW

While collecting the noise data the vehicle stopped at various places due to the traffic. The map below (Figure 2.3) shows the traffic in different parts of Ludhiana city. Such data can help build maps which not only show us the shortest or traffic-free routes but also show us routes which have less noise. The result of the study to map the noise pollution amply demonstrates the power of sensors in a smartphone. Nowadays, costly and sophisticated equipment is required to record noise pollution by various agencies, but this method proved to be a simple and cost-effective solution.

2.4 FURTHER STUDY

An IoT (Internet of Things) network of such devices should be created by installing them in various parts of the city. The data should be sent to the cloud for real-time and low-cost monitoring of noise pollution. Further planners and policymakers can analyse the maps/data to find remedies to reduce it. GIS, crowdsourcing, data analytics, cloud and IoT when used together can provide effective solutions for noise and traffic management.

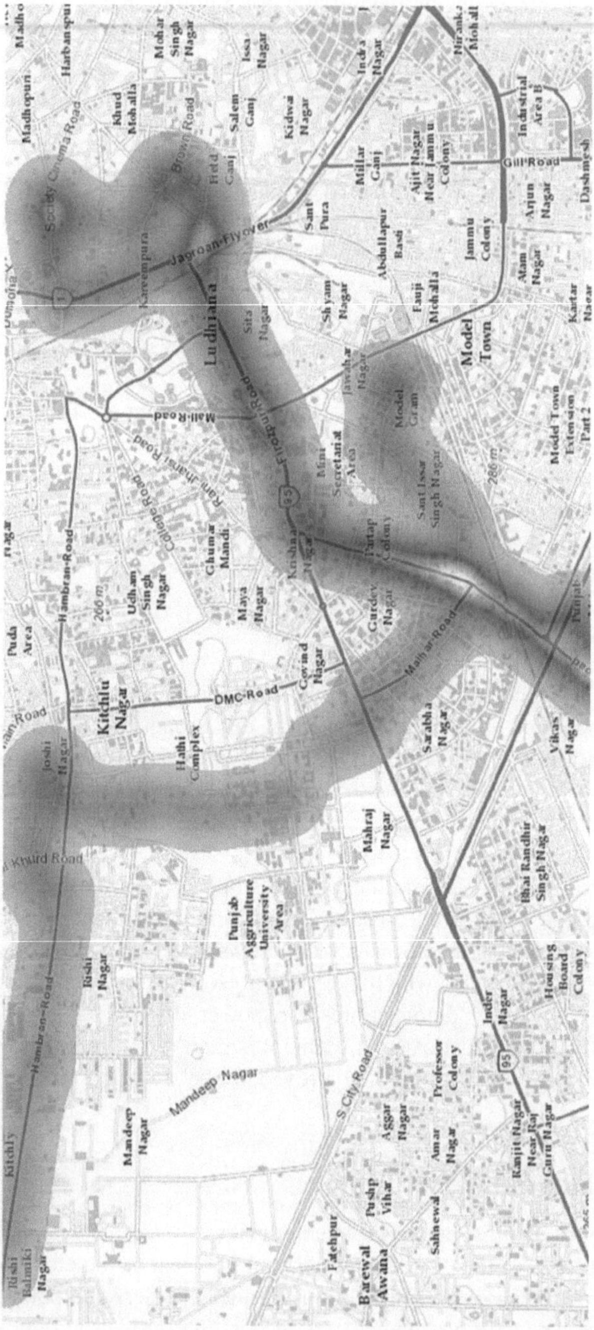

FIGURE 2.1 Noise Map of a part of Ludhiana city

FIGURE 2.2 Story Map Application

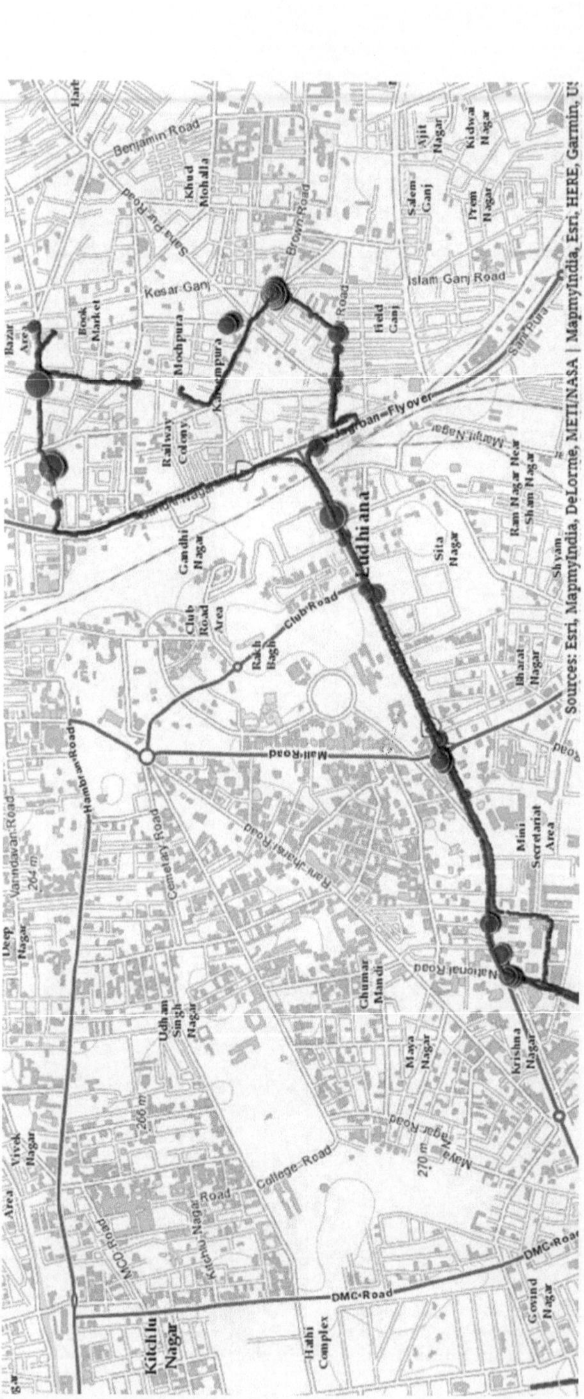

FIGURE 2.3 Traffic Map

REFERENCES

Dursun, S., Ozdemir, C., Karabork, H., & Koçak, S. (2006). Noise pollution and map of Konya city in Turkey. *Journal of International Environmental Application & Science*, *1*(1), 63–72.

Doygun, H., & Gurun, D. K. (2008). Analysing and mapping spatial and temporal dynamics of urban traffic noise pollution: A case study in Kahramanmaraş, Turkey. *Environmental Monitoring and Assessment*, *142*(1–3), 65–72.

Maisonneuve, N., Stevens, M., Niessen, M. E., & Steels, L. (2009). NoiseTube: Measuring and mapping noise pollution with mobile phones. *Information Technologies in Environmental Engineering*, 215–228.

Noise. (n.d.). In *http://dangerousdecibels.org*. Retrieved October 11, 2017, from http://dangerousdecibels.org/education/information-center/noise-induced-hearing-loss/

Noise Map. (n.d.). In *Wikipedia*. Retrieved October 11, 2017, from https://en.wikipedia.org/wiki/Noise_map

Singh, R. and Kalota, D. (2019). Urban Sprawl and its impact on generation of urban heat island: A case study of Ludhiana city. *Journal of the Indian Society of Remote Sensing*, *47*: 1567–1576. Retrieved from https://doi.org/10.1007/s12524-019-00994-8

3 An Assessment of Green Spaces Under the Smart Cities Mission

The Case of Qaiserbagh ABD Area of Lucknow City

Simran Purswani and Bhawna Bali
TERI School of Advanced Studies, New Delhi, India

CONTENTS

3.1 INTRODUCTION

The global agenda of Sustainable Development Goal 11 which focuses on 'inclusive, safe, resilient and sustainable' transformation of cities and human settlements critically hinges on the creation and improvement of green spaces for the teeming urban population and is one of several measures for improving urban planning and

DOI: 10.1201/9781003224624-5

managing urban spaces (United Nations Development Programme, 2016). Public green spaces embrace physical and psychological health, help in mitigating climate change, recharge ground water, prevent soil erosion, impact the economy of a city and provide space for all sections of society (Town and Country Planning Organization, 2014). The World Health Organization (WHO) also states that urban parks and gardens play a critical role in cooling cities and provide safe routes for walking and cycling for transport purposes as well as sites for physical activity, social interaction and recreation. Green spaces form the means to connect to nature, to our earth system (Wilenius & Jones, 2018). They may include public and private spaces available directly or indirectly to users. According to the European Commission (2014), there is a correlation between health impacts and better access to green spaces. Access to green spaces essentially translates into an improved urban environment which fosters social harmony. Overall, green spaces contribute to urban liveability which acts as one of the keys to urban systems (Simpson & Greg, 2018).

The rapid growth of cities in India has contributed to 44 per cent of carbon emissions originating from transport, industry, buildings and waste. In fact, according to the National Institute of Urban Affairs' estimates, 59 per cent of greenhouse gas (GHG) emissions in India is generated by megacities, metro cities and Class-I cities (Ministry of Housing and Urban Affairs, 2019). The biggest challenge to green spaces is their maintenance, preservation and strengthening, given the limited space available in urban areas. The designated green spaces across cities have borne the brunt of thoughtless urban development dominated by unpleasant buildings, vehicular traffic and noise, absence of dedicated space for walking, cycling or talking. More often, green spaces are covered with waste and exotic species of trees and plants, making such space useless and unsafe for people (Wilenius & Jones, 2018). This is indicative of a lack of integration of green spaces with urban planning. The immense benefits of green spaces remain untapped due to less quantity and deteriorating quality of green spaces. Also, information about both quantity and quality of green spaces is still questionable. The data are scarce and scanty on per capita availability of open spaces (TCPO, 2014). There is potential to improve the connectivity of green spaces and integrate it with socially inclusive planning (Rall, Hansen & Pauleit, 2018), which requires initiatives at city level to sustain natural resources. Interventions related to green spaces considered as long-term investments could help address public health issues. The integration of planning of green spaces should be with development strategies which include Master Plans, housing regulations, and policies related to transport, biodiversity and sustainability (WHO, 2017).

One of the ongoing programme interventions for cities in India, the Smart City Mission (SCM), launched in 2015, focused on enhancing and managing core elements of 100 potential cities with smart solutions by upgrading services and infrastructure (MoUD, 2015). Among the key features enshrined in the SCM is preservation and development of open spaces for augmenting quality of life as well as ensuring sustainable environment in urban areas. Public open spaces form an integral component of the physical pillar for comprehensive development of a city in the Liveability Standards codified by the Ministry of Housing and Urban Affairs (MoHUA, formerly Ministry of Urban Development) (NITI Aayog, 2018; MoHUA, 2019). An increasing focus on green spaces and open spaces in our urban policy and attendant programmes assumes significance in the light of our obligations to combat climate change, help in making cities resilient and sustainable and achieve the targets of SDG 11 (MoHUA, 2019).

3.2 AIM AND OBJECTIVES OF THE STUDY

This study is set against the backdrop of SCM's objective to preserve and develop open spaces for promoting ecological balance, reducing the urban heat island effect in areas and enhancing citizen engagement (MoUD, 2015). Accordingly, the study area for this research is Lucknow city – the administrative headquarters of the state of Uttar Pradesh. In view of the vision of the Lucknow Smart City proposal for strengthening quality of life, this study aims to analyse changes in land use/land cover and green spaces in Lucknow city in general and the Qaiserbagh Area-Based Development (ABD) area in particular. It further intends to assess the quality of parks in the Qaiserbagh ABD area with respect to facilities, security and inclusiveness. The study focuses on parameters of identity, location, typology, function and aesthetic values in order to identify the need for improvement and redesign the under-utilized and undeveloped green spaces. It suggests measures to activate parks within the city centre in order to enhance the liveability and attractiveness of surrounding areas, and thereby help in achieving the goal of quality of life for Lucknow Smart City.

3.3 STUDY AREA

Lucknow – the capital city of Uttar Pradesh (UP) – is situated in flat alluvial terrain with the River Gomti traversing the city. The municipal jurisdiction of Lucknow Municipal Corporation (LMC) extends over an area of 470 square kilometres, with a population of 2.8 million inhabitants in 2011 which is expected to grow to 3.6 million in 2021. Administratively, the area under LMC is organized into 8 zones comprising 110 wards in 2019 (Map 3.1). LMC has only 3 per cent of its land under recreational uses, out of which major open areas in the city are situated along River Gomti while the old city area lacks open and green spaces. The site analysed in detail for this study – Qaiserbagh – is a retrofit area under Lucknow Smart City. Located in Zone 1 of LMC, it is a densely built-up area spread over 329 hectares of land comprising 6 wards of old city of Lucknow and has major heritage complexes. It serves a resident population of 70,000 persons and a floating population of 22,000 persons. Besides amenities such as educational, medical and commercial, about 18 per cent of the total area of Qaiserbagh is devoted to green spaces (SCM 2016).

3.4 MATERIALS AND METHODS

3.4.1 Data Collection

The study has utilized data collected, collated and analysed from several sources such as satellite images, Google maps, site visits and questionnaire survey which helped in creating both data and compiling evidence. The data to assess green spaces was operationalized through land use classification. The satellite images of 2001 and 2019 were obtained from United States Geological Survey (USGS) Earth Explorer. Unsupervised classification in ERDAS IMAGINE and boundary of city received

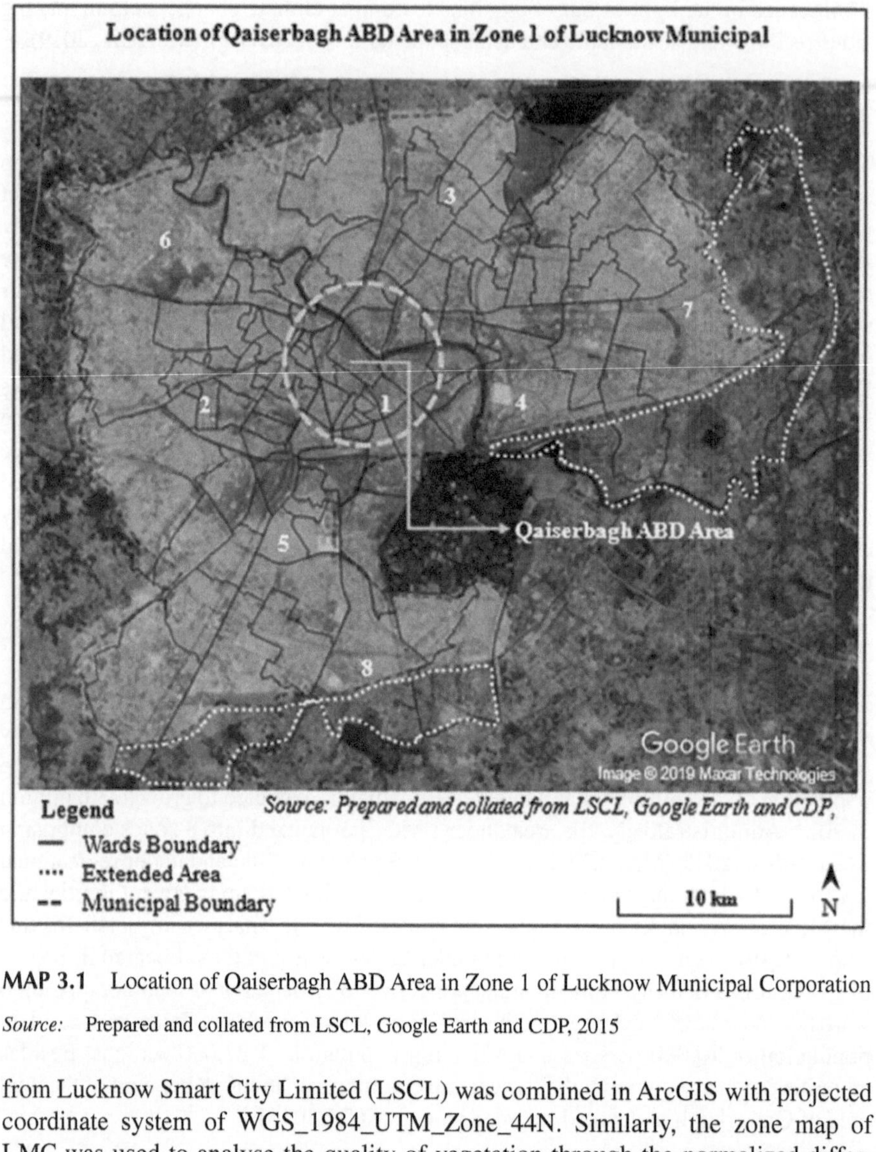

MAP 3.1 Location of Qaiserbagh ABD Area in Zone 1 of Lucknow Municipal Corporation

Source: Prepared and collated from LSCL, Google Earth and CDP, 2015

from Lucknow Smart City Limited (LSCL) was combined in ArcGIS with projected coordinate system of WGS_1984_UTM_Zone_44N. Similarly, the zone map of LMC was used to analyse the quality of vegetation through the normalized difference vegetation index (NDVI). The brief of data collected is mentioned in Table 3.1.

3.4.2 Methodology

Land Use/Land Cover Analysis using remote sensing data was carried out by acquiring satellite images and generating false colour composite (FCC) by layer stacking. With the help of the selection of area of interest (AOI) and unsupervised classification, images were categorized into four classes of land cover of Lucknow City and Qaiserbagh ABD (Table 3.2). The image interpretation keys were used to classify

TABLE 3.1
Data Used for Mapping and Assessing Green Spaces in Lucknow City and Qaiserbagh ABD Area

Data	Source	Description
Satellite Imagery	U. S. Geological Survey Earth Explorer Department of the Interior/USGS	Landsat 7 for year 2001 and Landsat 8 for year 2019
Zone Map	LSCL Google Earth Google Map	File of .kml extension 2019 Maxar Technologies 2019
City Information	Census of India 2011, Revised CDP, LSCL, Master Plan	Demographics of city and zones, details related to open and green spaces in city and ABD
Standards	URDPFI, NBC	Typology of green spaces, facilities in parks
Indicators	Climate Smart Cities Assessment Framework, Ease of Living Index	Per capita availability of green spaces, inclusiveness and safety
Field Data	Site visit and survey	Park user's perspective in Qaiserbagh ABD area
Good Practices	LSCL and online sources	National and international case studies with respect to design and management of parks

TABLE 3.2
Categories Used in Land Use Classification

Land Use Class	Description
Built-Up	All types of built structure like residential, commercial, industrial, institutional, paved surfaces
Vegetation	Forests, plantation, recreational spaces like parks, open spaces
Water Body	Lakes, reservoirs, drainages, tanks
Vacant Land	Unpaved roads, bare land, waste land

images. *NDVI classification* for LMC boundary was generated to analyse change in dense vegetation between 2001 and 2019. The reflectance value ranging from –1 to +1 represented no vegetation as –1, scarce vegetation as 0 and dense vegetation as +1. The area of different land uses through land use classification was used to analyse the per capita availability of green spaces in Qaiserbagh ABD as per standards given in the Urban and Regional Development Plans Formulation and Implementation (URDPFI) Guidelines, 2015.

The assessment of parks in Qaiserbagh ABD is based on ward-wise analysis of location and typology of 21 parks which are proposed to be rejuvenated and improved under SCM (Map 3.2), and a detailed site survey of three parks (Begum Hazrat Mahal Park, Dayanidhan Park, Suraj Kund Park) on the parameters of inclusiveness, safety and facilities. These parks were selected from the 21 parks based on random stratified sampling.

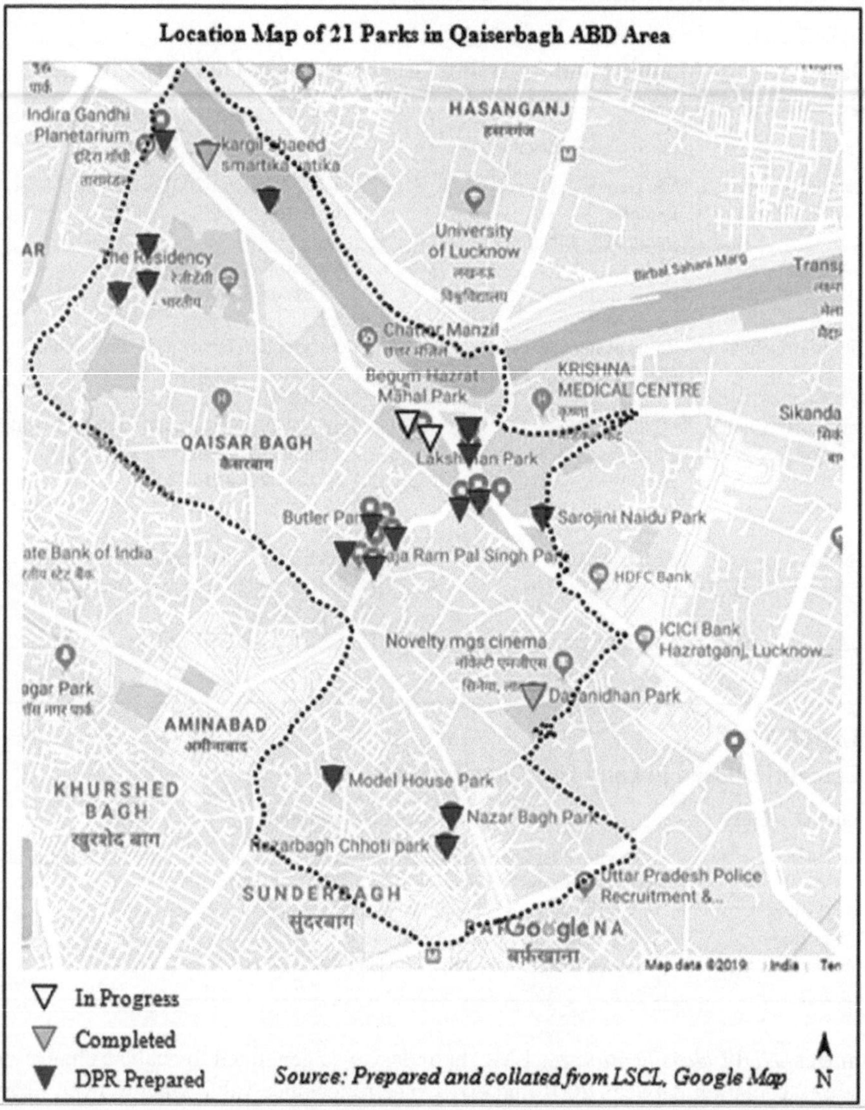

MAP 3.2 Location Map of 21 Parks in Qaiserbagh ABD Area

Source: Prepared and collated from LSCL, Google Map

A *questionnaire survey* was administered to park visitors who were chosen randomly during site visit. Among the 55 respondents in three parks, 60 per cent were male and 40 per cent female. Over two thirds of the respondents (68 per cent) were in 30–45 years age groups, followed by 45–60 years (18 per cent). In terms of employment status, 44 per cent of respondents were self-employed while a quarter constituted ‘at home’ category comprising home makers, retired people and unemployed persons and the remainder were students.

3.5 RESULTS AND DISCUSSION

3.5.1 Assessment of Existing Situation of Green Spaces in Lucknow City

As per Master Plan 2021 the developable area comprises 43,206 hectares of land, of which 9,469.4 hectares (21.9 per cent) is allocated to vegetation which includes recreational area, green and open spaces. However, of this total area under vegetation as much as 227 hectares have already been converted into residential, commercial, industrial, road development and other land uses. The revised Master Plan 2031 proposes the extension of developable area to 98,000 hectares and a healthy allocation of 22.3 per cent of total area or 15,894 hectares as green spaces including parks, open spaces, District Park, playground, botanical garden, Environment Park, green belt and forests (Mishra, 2011).

An assessment of green spaces in Lucknow using satellite imagery reveals an increase in the area under vegetation which includes recreational and open spaces (Table 3.3 and Map 3.3). However, the plan proposals for the year 2021 give an enhanced figure of area under these two categories. The proportion of total area devoted to green spaces/recreational uses consistently remained far below the standard enunciated in the URDPFI Guidelines (14 to 16 per cent).

An interesting fact to note is a decrease in quality of green spaces within a span of 35,000 hectares of municipal limits between 2001 and 2019 as revealed through NDVI classification (Map 3.4 and Map 3.5). The spaces with dense vegetation have been converted into barren land, built-up or less vegetation (LMC, 2015).

An assessment of green spaces in Qaiserbagh ABD area reveals a similar situation between 2001 and 2019. The land cover under built-up has nearly doubled and is the dominant land use category with nearly half the proportion to total area of Qaiserbagh (Table 3.4, Map 3.6 and Map 3.7). This has been largely at the cost of open spaces which occupied a third of the area and was the predominant land cover in 2001. It has shrunk by 66 per cent of its original area and forms 12 per cent of the area in 2019.

TABLE 3.3
Change in Land Cover of Lucknow City

Category	2004–2005		2010–2011		2019		2021 Proposed	
	Area (Ha)	%	Area (Ha)	%	Area (Ha)	%	Area (Ha)	%
Built-Up	13,505	83	22,711	92.8	35,070	81.2	33,736.7	78.1
Vegetation	2,455	15	998	4	7,186	16.6	9,469.4	21.9
Water Bodies	310	2	579	2.4	950	2.2	–	–
Others	–	–	194	0.8	–	–	–	–
Total Area	16,270		24,482		43,206		43,206	

Source: Prepared and Collated from Master Plan 2031, AMRUT Slip 2015, Satellite Imagery Analysis 2019

Note: Disaggregate data of area under category of open areas and recreational are not available as these are categorized as vegetation as per LULC Classification. Thus, vegetation land use includes open, green and recreational spaces; agricultural land use includes crop land, barren land and rocks.

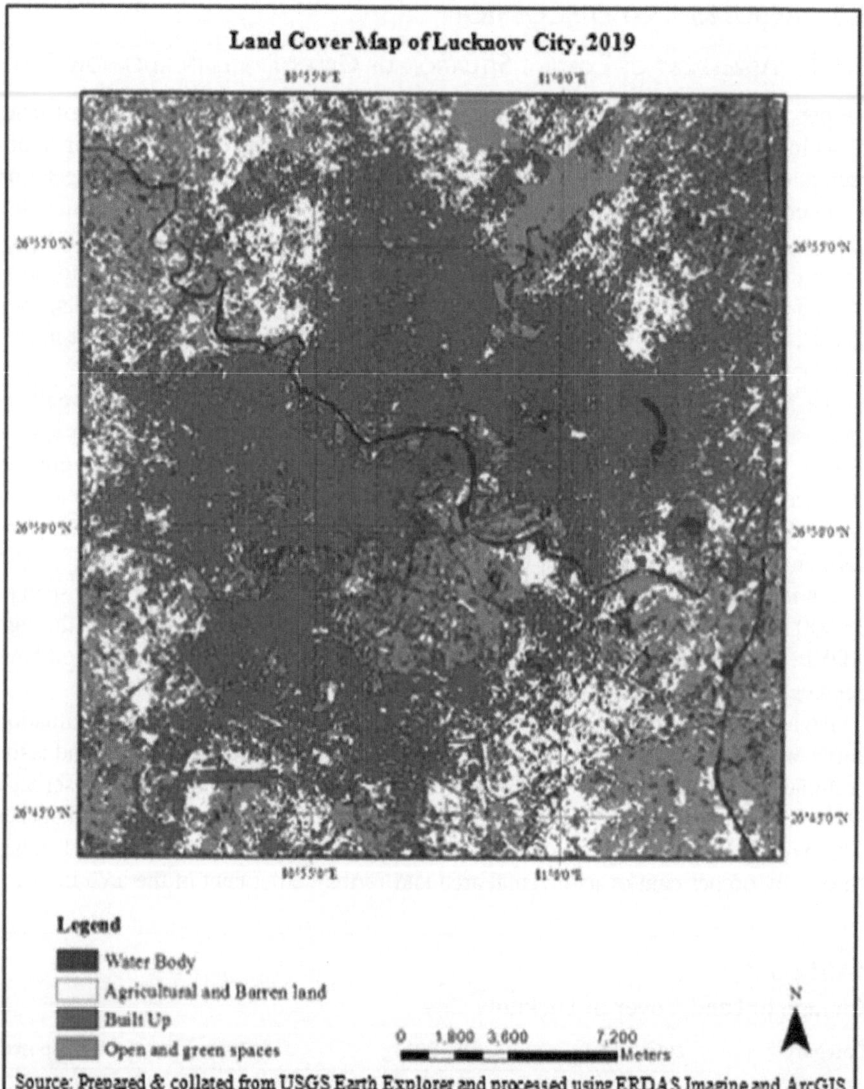

MAP 3.3 Land Cover Map of Lucknow City, 2019

Source: Prepared & collated from USGS Earth Explorer & processed using ERDAS Imagine and ArcGIS

Similarly, water body has diminished by 13 hectares and constitutes 7 per cent of the area in Qaiserbagh. On the other hand, vegetation cover has increased by 15 hectares making it the second predominant land cover category in 2019. Although vegetation has increased by 5 per cent, its per capita availability is far less than standards set by the URDPFI Guidelines (Table 3.5). This is related with increase in population of Qaiserbagh.

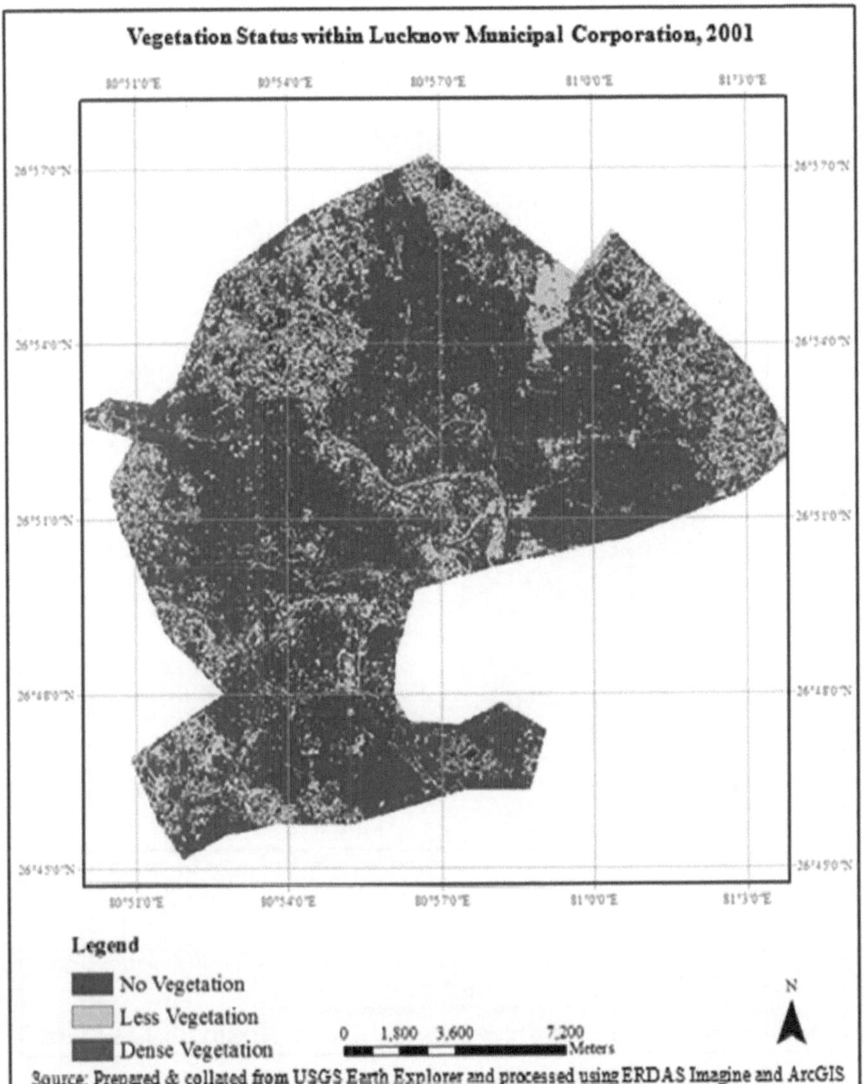

MAP 3.4 Vegetation Status within Lucknow Municipal Corporation, 2001

Source: Prepared & collated from USGS Earth Explorer and processed using ERDAS Imagine and ArcGIS

3.5.2 Green Spaces Managed by Lucknow Municipal Corporation

According to Urban Greening Guidelines 2014, the typology of green spaces varies with area, the surrounding population it serves and legal status. The mentioned types in national guidelines are reserved forest, protected forest, national park, district parks, community parks, neighbourhood parks, tot lots and playgrounds which vary in size and management authority. Another set of green spaces are the green belt

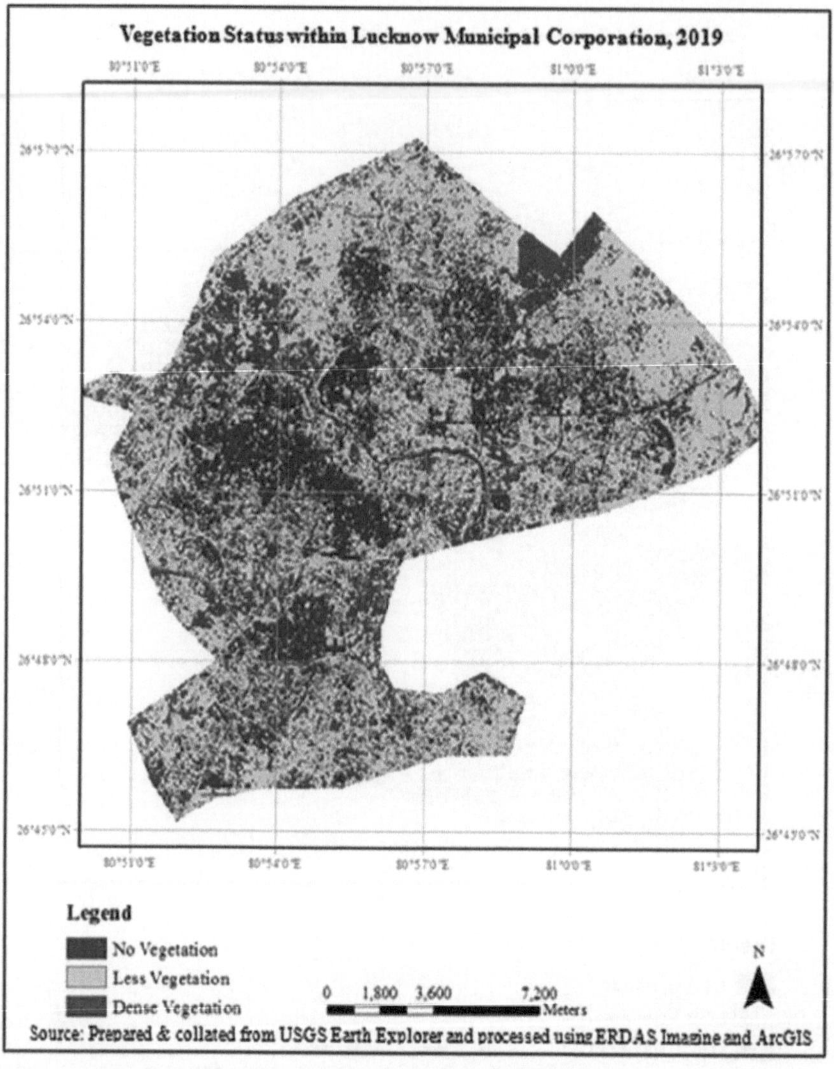

MAP 3.5 Vegetation Status within Lucknow Municipal Corporation, 2019

Source: Prepared & collated from USGS Earth Explorer and processed using ERDAS Imagine and ArcGIS

around the periphery of the city as parks, agriculture belt, and rural zone; the green strip used along roads or developed on vacant land; and tree cover considered as trees planted along roads or on the central verge. These are classified as recreational use along with water bodies and other natural features in the Master Plan (TCPO, 2014).

The UP Parks, Playgrounds and Open Spaces (Preservation and Regulation) Act, 1975, defines open space as any land (whether enclosed or not), belonging to the State Government or any local authority, on which there are no buildings or of which

TABLE 3.4
Land Cover Change in Qaiserbagh ABD

	2001		2019	
Land Cover	**Area (Ha)**	**Percentage**	**Area (Ha)**	**Percentage**
Built-Up	80	25%	157	49%
Water Body	35	11%	22	7%
Open Space	118	37%	39	12%
Vegetation	87	27%	102	32%
Total Area	329	100%	329	100%

Source: Survey Analysis of Parks and Park Visitors in ABD

not more than one-twentieth part is covered with buildings, and whole or the remainder of which is used for purposes of recreation, air or light parks as

> a piece of land on which there are no buildings of which not more than one-twentieth part is covered with or buildings, and the whole or the remainder of which is laid out as a garden with trees, plants or flower-beds or as a lawn or as a meadow and maintained as a place for the resort of the public for recreation, air or light

and playground as 'a piece of land adapted for the purpose of play, game or sport and used by any educational institution or club or other association' (UP Legislature, 1975).

A total of 1,684 parks and gardens with a combined area of 256 hectares is within the jurisdiction of LMC. These constitute a miniscule proportion of 0.82 per cent of the total area of LMC (Table 3.6). Out of this, only 488 parks are developed which include key parks like Begum Hazrat Mahal Park, Dayanidhan Park, Dr. Ram Manhar Lohia Park, Gautam Budha Park, Janeshwar Mishra Park, Shaheed Smarak Park, Suraj Kund Park. The native trees found are Shisham, Dhak, Mahua, Babul, Neem, Peepal, Ashok, Khajur, Mango, Gular, Sunflower, Rose and Marigold (LMC, 2015).

The zonal distribution of parks in the city (Table 3.6) clearly reflects an abysmally low proportion of area devoted to green spaces which is far less than the standards of 14 to 16 per cent of total area allocation given for million plus cities in the URDPFI Guidelines 2015. This existing poor situation of green spaces makes it an important part of infrastructure to be considered for enhancing liveability of the city. Zone 1, in which the study site is located, has the third lowest number (57) and proportion of parks (0.62 per cent) among the eight zones of LMC. Of these 57 parks, 21 are in the Qaiserbagh ABD area. Sixteen out of these twenty-one parks are functional while three are not in use, one is not developed, and one is semi-developed. The ward-level details of the 21 parks in the Qaiserbagh ABD area are mentioned in Table 3.7.

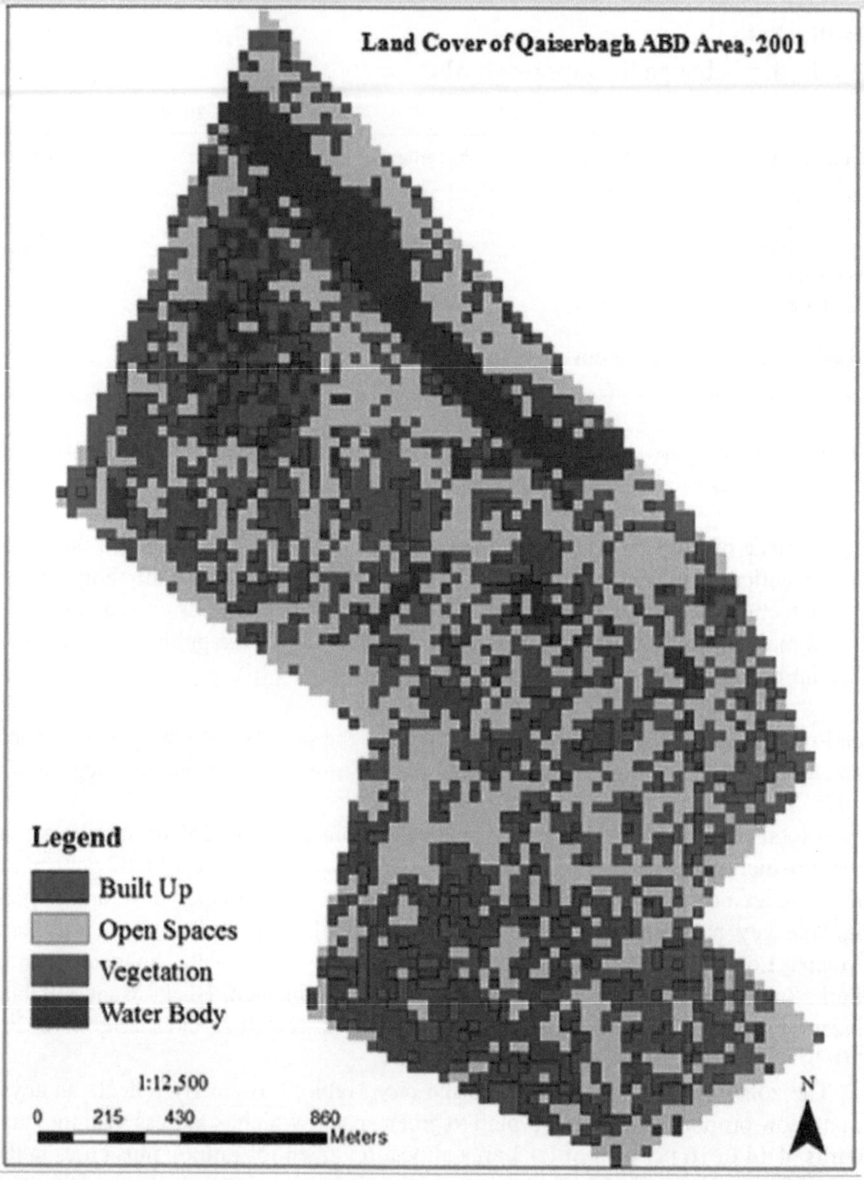

MAP 3.6 Land Cover Map of Qaiserbagh ABD, 2001

Source: Prepared & collated from USGS Earth Explorer and processed using ERDAS Imagine and ArcGIS

The Qaiserbagh ABD area is well served by nine community parks, seven neighbourhood parks and five tot lots. The largest among the parks – Begum Hazrat Mahal Park – receives a maximum footfall of about 500 persons per day (Table 3.7). Interestingly, this is the only park which has paid entry for visitors post-0730 hours.

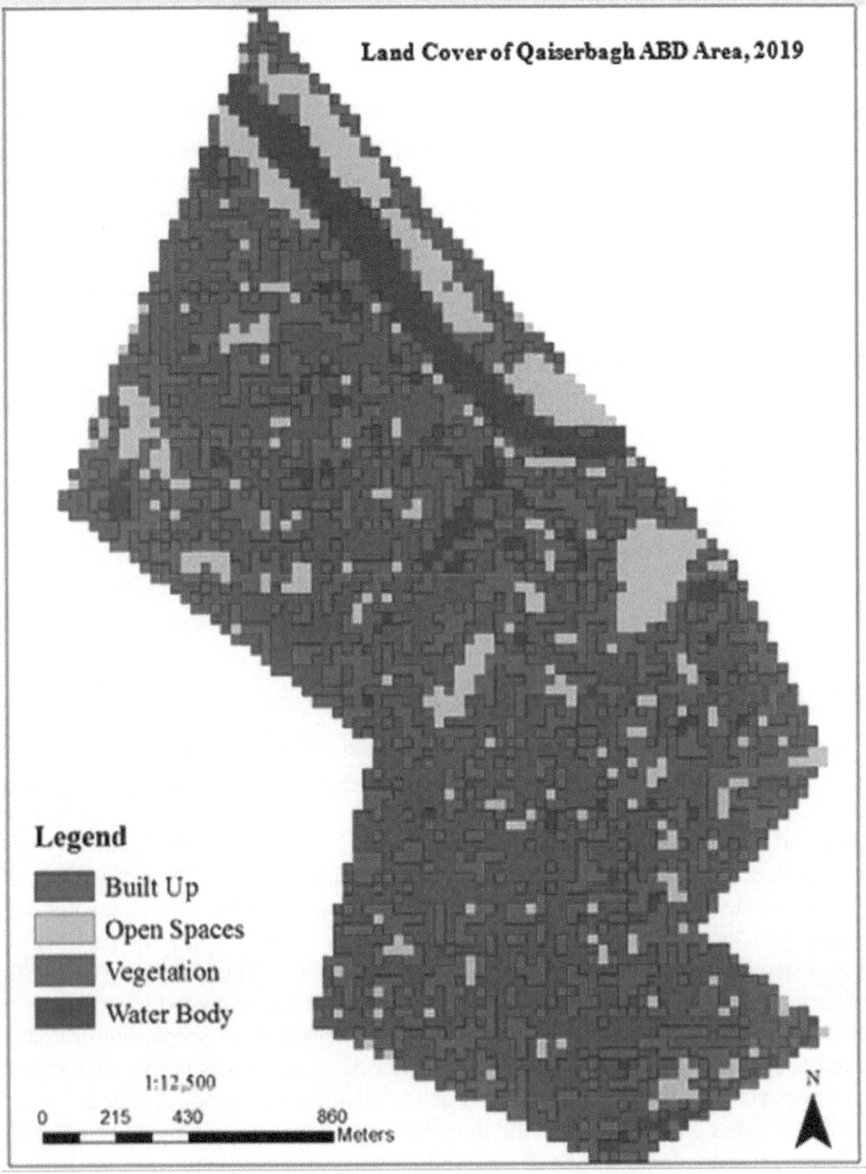

MAP 3.7 Land Cover Map of Qaiserbagh ABD, 2001

Source: Prepared & collated from USGS Earth Explorer and processed using ERDAS Imagine and ArcGIS

3.5.3 Assessment of Quality of Green Spaces in Qaiserbagh ABD Area

The three parks in the Qaiserbagh ABD area were assessed on norms regarding inclusiveness, facilities and safety (Table 3.8). As per the National Building Code (NBC), norms related with sustainability and inclusiveness are to be followed in every public

TABLE 3.5
Benchmark as per URDPFI Guidelines

Indicators	Benchmark (URDPFI)	2001	2019
Per person open space in built-up	3 sq. m.	0.00013	0.000035
Per person open space	10–12 sq. m.	11.04 sq. m.	3.7 sq. m.

TABLE 3.6
Zone-Wise Distribution of Parks in Lucknow Municipal Corporation

Zone	Area of Zone (Ha)	No. of Wards	No. of Parks	Area (Ha)	Percentage Area of Parks in Total Area of Zone
1	1,770	14	57	11	0.62
2	2,069	14	297	26	1.25
3	3,633	14	389	61	1.67
4	9,211	14	329	71	0.78
5	6,179	14	22	5.8	0.09
6	915	14	71	8.2	0.89
7	2,675	14	268	28	1.04
8	4,865	12	251	49	1.01
Total	31,317	110	1,684	256	0.82

Source: Collated from LSCL, Revised City Development Plan 2015, Google Earth, AMRUT SLIP 2015

place and buildings. Detailed assessment as per NBC norms and observational mapping of three parks of Qaiserbagh ABD Area reveal that none of the parks meet the criterion of sustainability because of an absence of compost beds and rainwater harvesting structures. It is noted that all the parks collect waste within the park which is removed by a contractor after two months of collection. The park lights are LED, adhering to the AMRUT Mission guidelines for energy efficiency fixtures. The parks are not universally designed rendering them inaccessible to specially/differently abled people. However, a noteworthy parameter of inclusiveness is space allocation for street vendors, which is available outside all the three parks. As far as safety parameters are concerned, while the three parks have guards on duty and LED park lights, there is an absence of CCTV surveillance and fire hydrants.

The analysis of questionnaire of 55 respondents across three parks – Suraj Kund Park, Begum Hazrat Mahal Park and Dayanidhan Park – revealed that an overwhelming majority of park visitors (82 per cent) were satisfied with the quality of park, followed by 14 per cent who stated park as good and 4 per cent people stated as in bad condition. However, nearly half the respondents stated the need for improvement in infrastructure provisions within parks including streetlights (Suraj Kund Park), toilets (Begum Hazrat Mahal Park), maintenance (Dayanidhan Park). Open air gym equipment which had been provided under the AMRUT Mission is only available in Dayanidhan Park. This facility is being sought for other parks as well. Overall, while

TABLE 3.7
Ward-level Details of 21 Parks in Qaiserbagh ABD Area

Ward No.	Ward Name	Area of Ward (Ha)	Ward Population	Name of Park (Under LMC)	Area of Park (Ha)	Typology of Green Space	Footfall per day	Footfall (6am-7am)
23	J C Bose	197	19,263	Model House Park	0.19	Neighbourhood Park		
34	Rani Laxmibai	122	15,832	Dayanidhan Park	1.65	Community Park	300	75
100	Nazar bagh	26	17,353	Nazarabagh Large Park	0.11	Totlot	40	-
				Nazarbag Small Park	0.05	Totlot	5	-
46	Golaganj-Peer Jaleel	25	24,264	Ram Temple park	0.07	Totlot	40	-
				D Block Park Riverbank Colony	0.06	Totlot	5	-
				Annant Kulkarni Park (River Bank Colony)	0.39	Neighbourhood Park	10	-
95	Wazirganj	171	13,468	Suraj Kund Park	1.19	Community Park	150	15
				Kargil Martyr Memorial Fort	0.65	Neighbourhood Park	100	10
				Shaheed Smarak Park	1.25	Community Park	300	50
95	Wazirganj	171	13,468	Globe Park	1.29	Community Park	70	20
				Begum Hazrat Mahal Park	3.47	Community Park	500	250
				Munshi Nawal Kishore Park	0.12	Totlot (not in use)		
95	Wazirganj	171	13,468	Awadhkunj Park	0.34	Neighbourhood Park	50	10
				Lakshman Park	1.19	Community Park	100	5
				Butlar Park	1.29	Community Park	300	100
95	Wazirganj	171	13,468	Near BSP Office	0.23	Neighbourhood Park (not developed)		
				Rajarampal Singh Park	1.09	Community Park	100	35
				Sanjay Singh and Gopal Singh	0.4	Neighbourhood Park (not in use)		
17	Hazratganj	1,203	16,702	Shubhas Churaha Park	0.17	Neighbourhood Park (not in use)		
				Sarojini Naidu Park	1.2	Community Park (semi developed)	10	-

Source: Ward information collated from LSCL, Revised CDP 2015, LMC; Typology reference as per Urban Greening Guidelines 2014, Footfall calculated from Site Visits

TABLE 3.8
Provisions of Inclusiveness, Facilities and Safety in Three Selected Parks of Qaiserbagh ABD Area

Details	Standards	Begum Hazrat Mahal Park	Dayanidhan Park	Suraj Kund Park
Area		34, 700 sq. m	16, 500 sq. m	11, 900 sq. m
Location		Mahatama Gandhi Marg	Dr. Souza Road	Mahatama Gandhi Marg
Visitors		Tourists, residential colonies	Office, residential	Hospitals, labour, workers
Important landmark nearby		Sadat Ali Khan Tomb,KDS Stadium	Novelty Cineplex	Indira Gandhi Planetarium
Nearby circle		Privartan Chowk	Lalbagh Chauraha	Daliganj crossing
Parks nearby		6 parks	No park within 100 m	3 parks
Transport mode available from park		KD Singh Metro Station, Rickshaw	Rickshaw	Auto, Bust Stand
Convergence with Smart City Project		Heritage park	Gym and Wi-Fi facility	Beautification of parks
Service Utilities				
Disposal of garden/park waste		Inside park, Collected in 2months by contractor	Inside park, Collected in 2months by contractor	Inside park, Collected in 2months by contractor
Availability of fire hydrant and hose reel		-	-	-
Authority to maintain the electricity of park lights		Electricity dept., LDA	Electricity dept., LMC	Electricity dept., LDA
CCTV surveillance on gates of park		-	-	-
Park Features and Facilities for Inclusiveness and Safety				
Walkways	Min 1m (one way)	3 m	2 m	2 m
Kerb Ramp	Min 1m	-	-	-
Footpath	2.5 m	2.5 m	2.5 m, but occupied	2.5 m, occupied by vendors
Handrails		-	-	-
Wheel Chair ramp	1.8 m	-	-	-
Tactile paving	0.3*0.3 m	-	-	-
Water feature-fountain, water body, other		6 nose water fountains	Well	30 feet deep water body
Monitoring of water quality of water feature		-	-	-
Number of Park Lights	LED	27 LED, 4 High Mast	56 LED	72 LED
Number of Dustbins		15	2	7
Spittoons		-	-	-
Rain Water Harvesting Structure		-	-	-

Compost beds		-	-	-
Number of Benches		16	10	8
Pucca or Semi Pucca Huts		-	1	-
Picnic Tables		-	1	-
Play area for Children	Swings, Sand Pit	Swings, Sand pit	10 play sets, no sand pit	-
Play area for adults	Courts, Gym Equipment	Badminton Court	Gym Equipment	-
Drinking water facility		One nose	Water Cooler	Pipe Water
Kiosks	Inside/Outside	-	-	-
Space for vendors		Outside	Outside	Outside
Toilets	Male, Female, Specially abled	1 M, 1F	Common	Common
Security Guards		2	1	1
Gardeners maintaining park		2	-	4 gardeners, 2 workers, 1 Incharge
Parking facility	Paid/Unpaid	Unpaid, On ground	Paid, Underground	-
Entry fee		Yes	-	-
Number of entry gates		4	3	1
Height of boundary wall	Less than Eye Level	Visible metallic fencing	Visible metallic fencing	Visible metallic fencing
Availability of signage boards		1 outside the park	1 outside park	1 outside park
Universally designed signage boards		-	-	-

Source: Data on norms collated from NBC 2016, LSC. Other information generated from observation mapping during site visit and interview with gardeners and security guards facilities like toilets, dustbins, benches, drinking water are available in these parks, their cleanliness and maintenance are questionable.

The survey also assessed environmental awareness among the park visitors. Nearly all the respondents (90 per cent) were aware of the benefits of green spaces. But only 27 per cent knew about the species of plants and 40 per cent about animals found in park. This lack of knowledge among park visitors about common flora and fauna in these parks could be attributable to an absence of signages indicating species and/or common name.

3.5.4 Solutions for Improving Existing Situation of Green Spaces in Lucknow City

The main objectives for designing and maintaining green spaces are playful and safe place, protected type of green space, inclusive and accessible to all (Bernard Van Leer Foundation 2019). Considering that new parks cannot be constructed in the densely populated old city of Lucknow, but these could be transformed to more inclusive and safer areas affording liveability to surrounding areas. A realistic approach towards preserving natural resources by increasing efficiency, liveability and environment quality of the city can be accomplished only if it is maintained, managed and integrated. Documentation of existing status of environment and biodiversity in the city involving schools, citizens and private parties and carrying out awareness activities could well be the beginning for a systematic improvement in green spaces within the city. Alongside this, a needs assessment of user group and local neighbourhood involving community participation at ward-level is imperative.

The ongoing Lucknow SCM has proposed projects on safety and security through CCTV surveillance, citizen participation through One Lucknow portal/app and upgrading the parks. The project of beautifying 21 parks under LSCL include installation of open gym and Wi-Fi connections. Wi-Fi connections in parks can connect to sites depicting park details, development, usage of gym equipment and other important information of that park. Renewable energy source through solar lights could be a replacement for LED. Although the current focus remains on the Qaiserbagh ABD area, this should become pan-city with the possibility of dovetailing it with other programmes like AMRUT Mission which aims at achieving minimum 15 per cent green coverage in cities. A comprehensive Green Action Plan could be prepared in order to focus on all green spaces throughout the city. While this is embedded in ongoing programmes like SCM and AMRUT Mission, planning for green spaces needs strict adherence through a Master Plan which is a statutory document.

The aspects of universal accessibility of public spaces is critical. To this effect, the design guidelines of the NBC need to be followed for revamping parks. In addition, provision of well-designed kiosks for street vendors operating outside these parks would enhance inclusiveness. As per survey, people are not aware of species of plants in the parks. This can be solved through adding bar code/signages on tree, plants and shrubs of all the parks which will help in enhancing knowledge of people regarding flora. The issue of lack of cleanliness can be tackled by adding sign boards with different quotes and message to prevent littering of waste and proper usage of parks.

Facilities like toilets could be made 'pay and use', given that respondents indicated willingness to pay with the expectation of getting clean facility.

3.6 CONCLUSIONS

The inevitability of the process of urbanization has led many green spaces to be degraded, neglected and finally converted into barren lands. The study reveals existing situation of green spaces within Lucknow city which have reduced considerably over past nearly two decades as well as there has been decrease in quality of dense vegetation. The per capita open space availability has been far below the standards. Adapting and transforming city centre lanes, bottom-up initiatives, site specific additions are some strategies which have been found effective in maximizing the environmental and equity benefits of green spaces, which in turn contributes to increasing liveability of cities.

ACKNOWLEDGEMENTS

We would like to thank Lucknow Smart City Limited for sharing their knowledge and time, U.S. Geological Survey for allowing access to extract satellite images and Google Earth and Google Maps for providing free access to maps.

REFERENCES

Bernard Van Leer Foundation, BDP. 2019. *ITCN Best Practices Compendium.* Ministry of Housing and Urban Affairs.

Bureau of Indian Standards. 2016. *National Building Code of India (Volume I).*

LMC. 2015. *Revised City Development Plan of Lucknow City-2040 (Volume I).* Lucknow: SENES Consultants India Private Limited. Accessed June 19, 2019.

Ministry of Housing and Urban Affairs. 2019. 'Smartnet, Climate Smart Cities'. Accessed July 10, 2019. https://smartnet.niua.org/csc/.

Mishra, A. K. 2011. *Lucknow Master Plan-2031.* Lucknow: Divisional Planning Block, Town and country planning department, Uttar Pradesh, LDA.

MoUD. 2015. 'Smart City Mission Statement and Guidelines'. June. Accessed June 16, 2019. http://smartcities.gov.in/upload/uploadfiles/files/SmartCityGuidelines(1).pdf.

NITI Aayog. 2018. 'SDG India Index, Baseline Report'. Accessed July 7, 2019. https://in.one.un.org/wp-content/uploads/2018/12/SDX-Index-India-21-12-2018.pdf.

Rall, E., Hansen, R., & Pauleit, S. 2018. 'The added value of public participation GIS (PPGIS) for urban green infrastructure planning'. Accessed June 19, 2019. https://www.sciencedirect.com/science/article/pii/S1618866717306453.

SCM. 2016. *Smart City Proposal.* Lucknow: Ministry of Urban Development. Accessed June 2019.

Simpson, J. P. and Greg, D. 2018. 'Multidisciplinary Digital Publishing Institute'. 18 December. Accessed June 20, 2019. https://www.mdpi.com/2073-445X/7/4/161.

Town and Country Planning Organization. 2014. 'Urban Greening Guidelines'. February. Accessed June 19, 2019. http://www.indiaenvironmentportal.org.in/files/file/urban%20green%20guidelines%202014.pdf.

Town and Country Planning Organization. 2015. *Urban and Regional Development Plans Formulation and Implementation (URDPFI) Guidelines (Volume I).*

United Nations Development Programme. 2016. *Goal 11: Sustainable cities and communities.* Accessed August 4, 2019. https://www.undp.org/content/undp/en/home/sustainable-development-goals/goal-11-sustainable-cities-and-communities.html.

UP Legislature. 1975. *Housing and Urban Planning Department, Uttar Pradesh.* Accessed July 23, 2019. http://awas.up.nic.in/acts_park1975.html.

WHO. 2017. 'Urban Green Space Interventions and Health'. Accessed June 19, 2019. https://www.cbd.int/health/who-euro-green-spaces-urbanhealth.pdf.

Wilenius, M. & Jones AM. 2018. *A green Urban Future Scaled-up prespectives in Urban Green for Human-Centred and Livable Urban Cores.* Turku: Finland Futures Research Centre. Accessed 2019. https://www.utu.fi/fi/yksikot/ffrc/julkaisut/e-tutu/Documents/FFRC_eBook_5-2018.pdf.

4 Impact on Land Surface Temperature and Urban Heat Island, Due to Land Use/Land Cover Change in Dhaka Metropolitan Area, Using Remote Sensing and GIS Techniques

Abdullah-Al- Faisal

Rajshahi University of Engineering and Technology (RUET), Rajshahi, Bangladesh

Operational Centre Amsterdam (OCA), Médecins Sans Frontières (MSF), Cox's Bazar, Bangladesh

Mohd. Shahinoor Rahman

New Jersey City University, Jersey City, USA

Abdulla-Al Kafy

Rajshahi University of Engineering and Technology (RUET), Rajshahi, Bangladesh

ICLEI South Asia, Rajshahi City Corporation, Rajshahi, Bangladesh

Sumita Roy

Rajshahi University of Engineering and Technology (RUET), Rajshahi, Bangladesh

DOI: 10.1201/9781003224624-6

Abdur Rouf Khan

Rajshahi University of Engineering and Technology (RUET), Rajshahi, Bangladesh

BRAC University, Dhaka, Bangladesh

Md Mostafizur Rahman

Rajshahi University of Engineering and Technology (RUET), Rajshahi, Bangladesh

University of Technology and Economics-1111, Budapest, Hungary

CONTENTS

4.1 INTRODUCTION

More than 70 per cent of the world's population is expected to live in urban areas in the next 30 years (Glazier et al., 2014). In 2100, global average surface temperatures will increase at 1.4–5.8°C and atmospheric carbon dioxide levels will double that of pre-industrial concentrations, according to the Intergovernmental Panel on Climate Change (IPCC) (Singh, Kikon, & Verma, 2017). The land use/land cover (LULC) has changed due to anthropogenic activities in both advanced and developing countries, which is the leading cause of environmental devastation, changes in urban hydrology, higher ambient temperatures, and climate change (Deng & Srinivasan, 2016). Although ecological degradation varies from place to place, urban areas are arguably one of the most responsible contributors because of the high concentration of population, economic activities, high use of metals, and energy consumption (Roy, 2009).

Urbanization refers to the process of the conversion of rural areas to urban areas, which is the result of population immigration, administrative services, construction of new infrastructure, and development of industry and the service sector (Bhatta, 2010). The urbanization process has been accelerating globally since the second half of the twentieth century, and the urbanization rate is faster in developing countries compared to developed nations (Singh et al., 2017). However, cities account for just 2 per cent of the Earth's surface, consume 75 per cent of all energy, and produce 75 per cent of all waste (Gallo & Owen, 1999). LULC changes have a significant effect on both local and regional biodiversity, surface temperature, precipitation, climate change, and global warming (Kafy, Islam, Khan, Ferdous, & Mamun, 2018; Singh et al., 2017). The process of urbanization adversely affects land, water, surface temperature, biodiversity and climate. For urban areas, the most significant problem is the change in surface temperature due to vegetation, wetland, and open space transformation into concrete surfaces. The impact of land degradation is more severe in areas with unplanned and haphazard urban development (Singh et al., 2017). An overall consequence of urbanization is the increase in the temperature of the surface, rainfall, evaporation rate, and hydrological areas in cities.

Urban areas appear higher on the surface than rural ones through a gradual shift in the surface, including the replacement of natural vegetation with buildings and roads, which are causing the phenomenon of the 'urban heat island' (Akbari, 2005). Additionally, the heat generated from motor vehicles, various manufacturing factories, air conditioners, aerosols, gas stations, and so on adds warmth to the nearby areas (Singh et al., 2017). Air generated by different refrigerants also contributes significantly to the urban heat island (Li, Zhou, Ouyang, Xu, & Zheng, 2012; Liu & Weng, 2012). The decrease in the airflow from high-rise buildings and high-density areas further aggravates the impact of the heat island. It is a significant case in which urban areas remain hotter than in rural areas. The urban heat island has been the most significant problem in the last 50 years because of the increase in urban areas and increased energy use. (Gutman, Huang, Chander, Noojipady, & Masek, 2013; Loyd, 2013). The urban heat island has a major impact on the quality of life of the increasing worldwide urban population (Singh et al., 2017). The urban heat island is among the most significant consequences (Landsberg, 1981), is a direct indication of the environmental degradation (Lu, Feng, Xiao, Shen, & Sun, 2009). Howard first discovered in London the phenomenon of the urban heat island (Mills, 2008). Luke Howard, a climate scientist, reported in 1818 higher temperature air in the city centre than in the suburban area, supporting the idea of the 'urban heat island' (Xu, Chen, Dan, & Qiu, 2011). Luke Howard proposed the concept of an urban heat island in early 1833, and after this work, the idea has gained considerable attention (Camilloni & Barros, 1997). After that, the phenomenon was reported in various cities of various sizes and observed worldwide.

The urban heat island issue has also become more relevant to the rapid urbanization cycle since it has a significant effect on society and the environment (Quanliang, Changjian, Zhan, & Jingxuan, 2009). The UHI is a known example of substantial anthropogenic climate change on Earth. It is a trend that indicates higher temperatures in urban areas compared with equally elevated non-urban areas (Kafy et al., 2019; Zhou et al., 2019). As with the rapid growth in urban development, urban and rural areas are no distinct borders, although the difference in the temperature between the central town and the city is observed. It is usually the most sustainable urban life-threatening phenomenon (Ngie, Abutaleb, Ahmed, Darwish, & Ahmed, 2014). The replacement of evaporative vegetation surfaces with impermeable surfaces primarily causes changes in energy exchange between land surfaces and the atmosphere. Anthropogenic heat release also contributes to this context (Ngie et al., 2014; Zhou et al., 2019). The transformation of vegetation into the region absorbs more heat by day and releases heat by night (Nuruzzaman, 2015). The city shape will affect the amplitude of UHI (Li et al., 2018). The urban design geometry with a narrow street and tall building decreases wind speed and increases the reflective surfaces, and this becomes a significant factor of UHI (Abutaleb et al., 2015). Increase in short-wave radiation absorption, increase in sensible heat storage, anthropogenic heat production, decrease in losses of long-wave radiation, decrease in evapotranspiration rates and decrease insensitive heat loss after urban warmth (Chow & Roth, 2006).

A plethora of global studies on the impacts on the spatial distribution and risk and urban vulnerability of heat islands based on remote sensing and earth surface temperature knowledge was recently published (Aminipouri, Knudby & Ho, 2016). UHI cases have recently been shown to cause significant health-related and environmental problems through deforestation and other anthropological activities in urban or suburban areas. Numerous studies conducted recently on the ecological impact of urbanization and land-cover changes in Bangladeshi cities revealed that land-cover changes had a considerable influence on surface temperatures (Ahmed, Kamruzzaman, Zhu, Rahman, & Choi, 2013; Dewan, Kabir, Nahar, & Rahman, 2012).

Artificial neural network (ANN) was first designed to understand the human brain's dynamic scenario and operation (Gatys, Ecker & Bethge, 2015). It does not need any prior knowledge about the elements, and its complicated structure reproduces the necessary process inside the framework for simulating the rise in potential LST (Dhamge, Atmapoojya, & Kadu, 2012; Maduako, Yun, & Patrick, 2016; Veronez, Thum, Luz, & Da Silva, 2006). ANN is based on a multi-layered perceptron (MLP) algorithm that produces automated network parameter decisions for improved modelling (Veronez et al., 2006). Once a trend is observed, it analyses the data and generates a performance with arbitrarily small precision. An automatic error function is conducted by calculating the difference between the random output and the goal output to get the necessary value. The ANN simulates the LST using LULC parameters as hidden layer and LST as output layer.

LULC parameters such as NDVI are weaker forecaster of LST. In contrast, NDBI and the normalized difference bare soil index (NDBSI), are comparatively stronger (Chen, Zhao, Li, & Yin. R. s. o. e., 2006b; Deng & Wu, 2013; Hua Li & Liu, 2008).

Prospects of different vegetation and anti-vegetation variables to predict LST confirmed that the urban index (UI) became the strongest predictor of LST. Since hypothesis-oriented indicators cannot accurately reflect historical patterns, the best-fitting model for LST simulation is the ANN (Civco, 1993; Mas & Flores, 2008; Mushore, Odindi, Dube, Mutanga, & Environment, 2017; Shatnawi & Abu Qdais, 2019).

Remote sensing data has already proved its applicability in studying the impacts of land-cover change and urbanization on the environment. There is an increasing trend for the utilization of remote sensing optical bands and remote sensing derived products from studying UHI (Y. Deng & Srinivasan, 2016). These remote sensing-derived products, such as NDVI, NDBI, LST and LSE, are frequently employed in conjunction with standard optical bands in UHI investigations. The purpose of this research was to determine the link between LULC and LST changes in DMA regions over a 28-year period using Landsat data. Landsat products' regular bands were used to monitor land-cover change. LST was detected using Landsat's thermal bands. Following that, a discussion was held about the spatial-temporal link between LST and LULC change, which can help with sustainable urban development. Finally, using the artificial neural network-based cellular automata (ANN-CA) technique, LST was predicted for the year 2029.

4.2 STUDY AREA

Dhaka city is situated between latitude 23.58°N and 23.90°N, 90.33°E and 90.50°E (Figure 4.1). The megacity is one of the world's most populous cities. The population

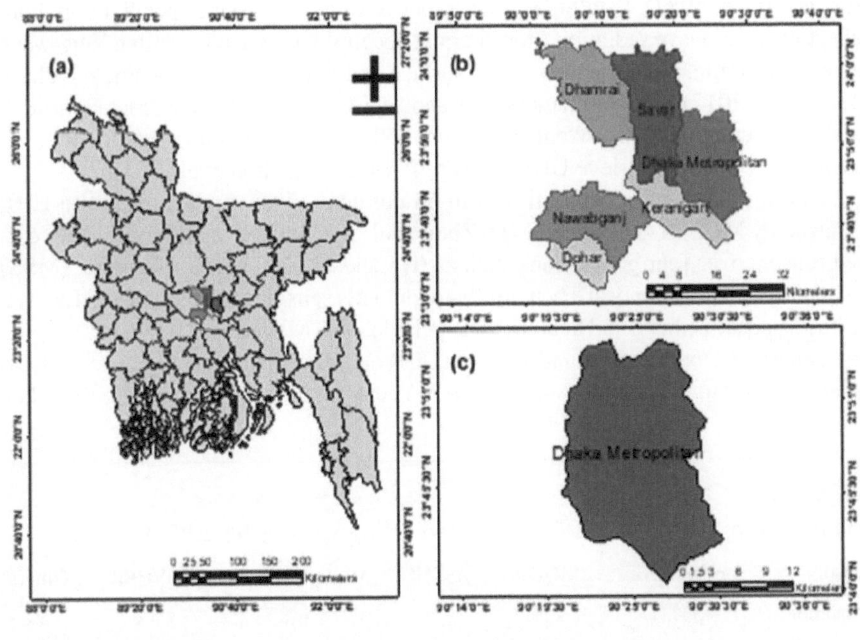

FIGURE 4.1 Location map of the study area a) In Bangladesh b) In Dhaka district c) Dhaka Metropolitan Area

is increasing in this rapidly growing megacity because of economic, physical, administrative, employment opportunities. The transition of the non-built-up environment to a built-up region changed the natural landscape to meet the need of the ever-growing population. Population growth also turns the natural and open land into the urban landscape. The DMA area is mostly situated on an alluvial terrace, which was formed during the Pleistocene period called Madhupur Terrace (Dewan et al., 2012).

The surface height of this area ranges between 1 and 14 m and almost average built-up areas range from 6 to 8 m elevations (Gutman et al., 2013; Khandelwal, Goyal, Kaul, & Mathew, 2018; Loyd, 2013). This town is surrounded by four major water systems; including the south, west, north and east areas, these are Buriganga, Turag, Tongi Khal, and Balu, respectively. This area receives 2,000 mm of rainfall annually per year because of its location in the subtropical monsoon climate, in humid weather. In the monsoon season (July to October), the monsoon rain contributes more than 80 percent of their annual rainfall. (Dewan et al., 2012). Fogs are seen from late December until late January. Rivers that originate from the plain are alluvial rivers fed by groundwater. The annual average rainfall is 114 mm (BBS, 2017). The city's average yearly temperature is 25°C, and the monthly mean ranges from 18°C in January to 29°C in August. (Banu, Hu, Guo, Hurst, & Tong, 2014).

4.3 DATA DESCRIPTION AND METHODS

4.3.1 Satellite Sensor for UHI Studies

UHI studies have been using satellite remote sensing since 1972 after Landsat was launched (Oke, 1982). Satellite sensors process a typical passive specification, both reflecting short-wave radiation (non-thermal spectral bands) and emitting long-wave radiation (thermal bands) from Earth's surface and atmosphere. (Mohamed, Odindi, & Mutanga, 2017). Two key factors LST and LULC change can be extracted through satellite image processing (Weng & Larson, 2005). Table 4.1 shows satellite imageries that were used to retrieve UHI all over the world (D. Zhou et al., 2019).

Landsat images have been utilized in 53 per cent of the studies related to UHI, followed by MODIS (25 per cent) (D. Zhou et al., 2019). The Landsat series has carried sensors over four generations such as (i) Landsat Multispectral Scanner (MSS) on Landsat 1–5, (ii) Landsat Thematic Mapper (TM) sensor on Landsat 4 and Landsat 5, (iii) Landsat Enhanced Thematic Mapper Plus (ETM+) sensor on Landsat 7 and (iv) Landsat 8 Operational Land Imager (OLI) and Thermal Infrared Sensor (TIRS) images (F. Chen, Yang, Yin, & Chan, 2017; Young et al., 2017). There are three key

TABLE 4.1
The Percentage of UHI Studies Analysed Using Various Satellite Images

Sensors	Landsat Series	MODIS	ASTER	Multiple Sensors	AVHRR	Others[1]
Proportion	53%	25%	7%	6%	4%	5%

SEVIRI, GOES, HCMM, HJ-1B, AATSR, ITOS-1, COMS, FY-2F, AMSR-E, AMSR2 reasons why Landsat images are used in UHI studies. Firstly, Landsat is Earth Observation's longest-running and stable system. (Wulder et al., 2016). Secondly, until 2008 there was a policy that scientists had to pay for Landsat image, but the plan was changed in 2008. A new policy was introduced where scientists can get the Landsat data freely from the USGS data repository (Popkin, 2018). Thirdly, the Landsat series 5,7 and 8 have swath coverage 185km × 185km, and they capture the surface of the Earth in a repeat period of 16 days. Moreover, Landsat provides sufficient spatial and temporal resolution to research land-cover and LST shifts at the city level (Mohamed et al., 2017). Thus, this study utilizes Landsat images to study UHI impact in Dhaka City Corporation (DMA) area.

4.3.2 Data Description

Table 4.1 shows that, in UHI studies, Landsat series are used mostly. Additionally, the Landsat data can be collected at free of cost. The data can be obtained from the USGS website (https://earthexplorer.usgs.gov). Therefore, the authors show interest in analysing the LST as well as UHI using Landsat data series. Before 2013, Landsat TM 5 data and after then Landsat OLI 8 data have been used for the study. The images and the acquisition date are given in Table 4.2.

LST data, derived from Landsat images, can significantly be affected by clouds (Mirzaei & Haghighat, 2010). That is why all the used Landsat data are collected as less than 10 per cent cloud coverage. ArcGIS 10.6.1, ERDAS 2014, and ENVI 5.3 were used for image processing and GIS procedures. For retrieving LST as well as UHI,

TABLE 4.2
Date Available and Specification

Satellite Data	Date of Acquisition	Sensor	Band No.	Spectral Range (Wavelength μm)	Spatial Resolution, m
Landsat 5	26-JAN-1991,	TM	1	0.45–0.52	30
	21-JAN-2001,		2	0.52–0.60	30
	01-JAN-2011		3	0.63–0.69	30
			4	0.76–0.90	30
			5	1.551.75	30
			6	10.4012.50	120 resampled to 30
			7	2.082.35	30
Landsat 8	07-JAN-2019	OLI	1	0.43–0.45	30
			2	0.45–0.51	30
			3	0.64–0.67	30
			4	0.53–0.59	30
			5	0.85–0.88	30
			6	1.57–1.65	30
			7	2.11–2.29	30
			8	1.36–1.38	15
			9	0.50–0.68	30
		TRIS 1	10	10.60–11.19	100 resampled to 30
		TRIS 2	11	11.50–12.51	100 resampled to 30

some Landsat calibration equation was needed, and some of the calibration equation was collected from 8handbooks (Chander & Markham, 2003; Gutman et al., 2013).

4.3.3 LST Estimation using Landsat Data

4.3.3.1 Sensor Brightness Temperature Estimation (T_i):

To begin, conversion from the infrared thermal band's spectral radiance to the actual radiance sensor brightness temperature is required. Plank's function equation provides an explanation for the preceding discourse, as seen in equation (4.1)

$$T_i = \frac{C_2}{\left(\lambda_i \times \ln\left(1 + \frac{C_1}{L_i \lambda_i^5}\right)\right)} \tag{4.1}$$

Here,

$C_1 = 1.19104356 \times 10^{-16}$ W.m^2,
$C_2 = 1.4387685 \times 10^4$ μm.K. Hence C1 and C2 both are constants.
T_i = sensor brightness temperature (unit kelvin),
L_i = spectral radiance
λ = the emitted radiance wavelength for optimal response
= 11.5 μm average limiting wavelengths for Landsat 5 band data, and
= 10.9 μm and 12 μm for Landsat 8 bands 10 and 11, respectively (Liu & Zhang, 2011; Stathopoulou et al., 2009).

To transform Landsat data, a simpler equation has been devised, which is shown in Equation (4.2).

$$T_i = \frac{K_2}{\ln\left(\frac{k_1}{L_\lambda} + 1\right)} \tag{4.2}$$

Here, K1 and K2 are calibration constants 1 in W/(m^2.sr. μm) and 2 in kelvin, respectively. The calibration constants for Landsat thermal bands are listed in Table 4.3.

TABLE 4.3
Calibration Constants for Landsat Thermal Bands

Constant	Unit K1 in W/(m^2.sr. μm)	Unit K2 in Kelvin
Landsat5 TM	607.76	1260.56
Landsat8 TIRS (band 10)	774.8853	1321.0789
Landsat8 TIRS (band 11)	480.8883	1201.1442

4.3.3.2 Retrieval of LST

Top of atmosphere (TOA) radiance is measured by the thermal infrared band. TOA consists of three parts of energy, such as the radiance emitted from the surface of the Earth, the radiance upwelling from the atmosphere and the radiance downwelling from the sky (Al Kuwari, Ahmed, & Kaiser, 2016; Buyantuyev & Wu, 2010; Weng & Larson, 2005). Both TOA and brightness temperature influence the atmospheric condition. Meanwhile, to obtain a suitable surface brightness temperature, atmospheric effects such as upward emission and downward surface irradiance reelection should be adjusted. Calculating land surface spectral emissivity (Ɛ) is a way to the above correction. Several variables govern the emissivity of the surface, such as chemical composition, built structure, water content, and roughness (Snyder, Wan, Zhang, & Feng, 1998). Many researchers defined that NDVI is closely related to surface emissivity, and therefore the emissivity can be estimated using NDVI (McCarthy, Best, & Betts, 2010; Speth, 2005; Thi Van & Duong Xuan Bao, 2010; Weng & Larson, 2005). The following Table 4.4 shows the land surface emissivity values according to their corresponding NDVI values (A. Kafy et al., 2019; J. Li et al., 2011; L. Liu & Zhang, 2011).

The NDVI value may be calculated using the reflectance values for the visible and near infrared (NIR) bands specified in Equation (4.3).

$$\mathrm{NDVI} = \frac{\mathrm{B_{NIR}} - \mathrm{B_{RED}}}{\mathrm{B_{NIR}} + \mathrm{B_{RED}}} \tag{4.3}$$

Where,

$\mathrm{B_{NIR}}$ = the pixel values of near infrared
$\mathrm{B_{RED}}$ = the pixel values red bands

Table 4.4 can be used to determine the land surface emissivity (Ɛ) using the NDVI value. The following Equation (4.4) calculates the terrestrial surface temperatures after adjusting for spectral emissivity (Ɛ) (Xiong et al., 2012; Yue, Liu, Fan, Ye, & Wu, 2012).

$$LST = \frac{T_i}{1 + \left(\lambda \times \frac{T_i}{\rho}\right) \times \ln(\varepsilon)} \tag{4.4}$$

TABLE 4.4
Values of Emissivity and Their Accompanying NDVI Values

NDVI	Spectral Emissivity (ε) of Land Surface
NDVI < −0.185	0.995
−0.185 ≤ NDVI < 0.157	0.970
0.157 ≤ NDVI ≤ 0.727	1.0094 + 0.047 ln (NDVI)
NDVI > 0.727	0.990

Here,

LST = land surface temperature,
T_i = the sensor's brightness temperature.

The emitted radiance's wavelength indicates as λ and ε indicates the spectral emissivity of the land surface.

In addition, $\rho = \frac{hc}{\sigma} = 1.438 \times 10^{-2}$ mk
Where,

h = Plank's constant = 6.626×10^{-34} Js,
c = velocity of light equal to 2.998×10^{8} ms^{-2}, and
σ = the Boltzmann constant (5.67×10^{-8} Wm^2 k^{-4} = 1.38×10^{-23} JK^{-1}).

4.3.4 Estimation of UHI

Urbanization influences the land-use change and this is the core driving factor of UHI (Li et al., 2018). The classic method of measuring the UHI effect is by differing temperatures in an urban area and the surrounding area (Huidong Li et al., 2018).

The impact of UHI has become a key issue in the field of urban climate and environmental change. To calculate the magnitude and scale of UHI, various methods have been taken in several places of the world (Ngie et al., 2014). The development of remote sensing (RS) technology has become a practical means in the UHI study. UHI can be continuously monitored and evaluated using multi-temporal infrared RS data (L. Liu & Zhang, 2011; Xu et al., 2011). These data can help in special mapping the large areas, not measuring the magnitude (Ngie et al., 2014). The LU/LC and the spatial distribution of vegetation intensity have a close relation with UHI (Mosammam, Nia, Khani, Teymouri, & Kazemi, 2017; Surawar & Kotharkar, 2017). The Surface Heat Island Intensity is studied separately in urban areas and rural areas using the remotely sensed data, and the selected pixels are analysed in this context. This estimation largely relies on the stations and pixels defined (Huidong Li et al., 2018).

Because of the seasonal variance within a year, LST is compared using a normalized technique. The normalized technique is shown as below Equation (4.5) (B. Ahmed et al., 2013; Ahmed, 2018).

$$\text{UHI} = \frac{\text{Ts} - \text{Tm}}{\text{SD}} \tag{4.5}$$

Where, Ts = LST

Tm = Mean of the LST
SD = Standard deviation.

The influence of UHI considers socioeconomic aspects and objectively calculates the urban thermal field variation index (UTFVI). The UTFVI was calculated using Equation (4.6) (Ahmed, 2018).

$$\text{UTFVI} = \frac{\text{Ts} - \text{Tm}}{\text{Ts}} \tag{4.6}$$

4.3.5 LULC Classification

Image classification is a procedure where land use/land cover (LULC) is categorized into different classes, whereas multispectral data are used for classification (Godschalk, 2004; Weng, Lu, & Schubring, 2004). Hence, several techniques and methods of classification process of supervised and unsupervised have been introducing by different researchers (Alavipanah, Wegmann, Qureshi, Weng, & Koellner, 2015; David R. Streutker, 2002).

4.3.5.1 Preprocessing

To define the spatial distribution of LULC classes and determining the surface temperature of the area, Landsat's satellite dataset for 26 January 1991, 21 January 2001, and 1 January 2011 and 7 January 2019 were used (Table 4.2). Data preprocessing with ENVI 5.3 Technology has been carried out. Landsat TM and Landsat 8 include separate band images that are stacked and then combined to create a multiband image. All these datasets were converted into 30 m of cell size and carried on the same spatial projection. By converting the digital number (DNs) into radiance, Landsat TM band 6 (thermal infrared band) and Landsat 8 band 10 and 11 (thermal infrared band) were used to measure terrestrial surface temperatures. This research also analysed the normalized vegetation difference index (NDVI) in which bands were used to derive vegetation indexes within the spectral spectrum of solar reflections. The satellite images were then used after the preprocessing phase to study the UHI and usage of ERDAS 2014 and Arc GIS 10.5 software for further process.

4.3.5.2 Image Classification

Support vector machine (SVM) algorithm was used to classify the image into four broader groups: water body, built-up area, vegetation land and bare land. Hence, the class water body contains all types of surface water features such as river, canals, lakes, ponds and so on. Built-up area contains all types of impervious infrastructures, including but not limited to buildings, roads, paved surfaces, commercial areas and residential impervious areas. Vegetation class includes pervious green surfaces, agricultural fields, grassland and homestead vegetations. Finally, bare land covers barren lands with no grass surface, riverbank soil and sandy regions. Due to easy handling and high accuracy of the SVM classifier, the classification process was done using ENVI 5.3 software. The pixels were assigned to the greatest probability class throughout the image classification process; the supervised classification technique was used to choose training units for the SVM classifier.

In SVM choices, the Radial Basis function was used as the kernel type and in the kernel function, Gamma was set to 0.07, the penalty parameter was set to 120.00, pyramid level was set by default to zero and the threshold of likelihood classification was set to 0.05. Additionally, a confusion matrix was created to ensure the categorized image and signature file were evaluated accurately.

4.3.6 Estimation of NDBI

NDBI is extremely built-up area sensitive. Recently, it was used as an instrument for calculating the extent of built areas (Chen, Zhao, Li, & Yin, 2006a; Zha, Gao, & Ni, 2003). NDBI has a strong correlation with LST; therefore, analysis among LST and NDBI creates great importance. Built-up area increases the impervious layers and results as higher LST. NDBI is concerned with built-up components; hence, the NDBI-LST relationship reveals the effect of the built-up region on the overall increase in surface temperature. Equation (4.7) below was used to analyse the NDBI of the study zone.

$$\text{NDBI} = \frac{B_{MIR} - B_{NIR}}{B_{MIR} + B_{NIR}} \tag{4.7}$$

Where,

B_{MIR} = middle infrared pixel values, and
B_{NIR} = near infrared pixel values.

The range of NDBI index is from –1 to +1. The data were utilized to establish a correlation between built-up areas and LST, and the analysis established the magnitude and contribution of built-up areas to LST.

4.3.7 LST Simulation using ANN-CA Algorithm

The LST is typically predicted with an artificial neural network (ANN) for 2029 with the help of MOLUSE Plugin in QGIS 2.18. ANN is an important tool for predicting future time series LST and LULC using previous years' datasets (Civco, 1993; Z.-L. Li et al., 2013; Maduako et al., 2016; Shatnawi & Abu Qdais, 2019). In this experiment, LST simulation was carried out with LULC images, NDVI, NDBI, latitude and longitude as input parameters (Gopal & Woodcock, 1996; Maduako et al., 2016). In addition, the ANN-CA model for future LULC was utilized to estimate road layer, NDBI, and NDVI data input parameters. The pixel value data of the images were transformed in ArcGIS software V10.6.1 to improve model performance (Patil, Deo, Ghosh, & Ravichandran, 2013). Figure 4.2 depicts the architecture of the LST prediction model.

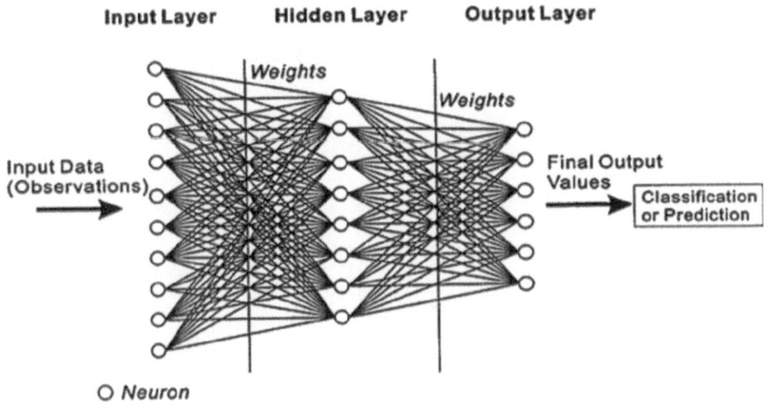

FIGURE 4.2 ANN-CA model architecture

4.4 RESULT AND DISCUSSION

4.4.1 Analysis of LULC

The LULC maps (1991, 2001, 2011 and 2019) are shown in Figure 4.4 and the area which were calculated from classified images is shown in Table 4.5. LULC maps demonstrate a rapid increase in built-up areas (33.95 per cent) and a significant decrease in the water body (–16.74 per cent) and vegetation land (–10.49 per cent) from 1991 to 2019. Moderate decrease in bare land (–6.72 per cent) was also noticeable. In 1991 the area of the water body was 71.62 km^2, which significantly was reduced to 20.66 km^2 in the year 2019. By 2019, the urbanized area was rapidly increased to (184.07 km^2) from 80.59 km^2. The vegetation and water body were the predominant land-use type in all the years but were faced a significant decrease in last 30 years. Maximum vegetation land initially was converted to bare land areas and finally was transformed into built-up areas. That is why bare land areas faced a moderate decrease from 50.99 km^2 to 30.50 km^2 in 1991 to 2019.

Table 4.6 demonstrates the overall classification accuracy, kappa coefficient, and validation of land-use classification. For each period, the overall accuracy is over 84

TABLE 4.5
Area and Percentage of LULC Changes in the Study Area from 1991 to 2019

Year	1991		2001		2011		2019		2019–1991
Land Use	Area (km^2)	%	Area (km^2)	%	Area (km^2)	%	Area (km^2)	%	%
Water body	71.67	23.52	71.80	23.56	30.69	10.07	20.66	6.78	−16.74
Urban area	80.59	26.44	122.70	40.25	129.99	42.65	184.07	60.40	33.95
Vegetation	101.51	33.31	72.58	23.81	102.05	33.49	69.54	22.82	−10.49
Bare land	50.99	16.73	37.71	12.37	42.03	13.79	30.50	10.01	−6.72

TABLE 4.6
Assessment of Classified Maps' Accuracy

Year	User Accuracy (%)				Producer Accuracy (%)				Overall Classification Accuracy	Overall Kappa Statistics
Waterbody	**Urban area**	**Vegetation**	**Bare land**	**Waterbody**	**Urban area**	**Vegetation**	**Bare land**	**1991**	**84.54**	**86.62**
88.51	84.26	83.36	85.86	86.51	83.53	84.04%	0.8356	2001	88.03	82.36
85.71	82.85	87.22	83.74	82.84	81.25	85.82%	0.8432	2011	86.53	83.56
80.64	91.75	88.12	86.43	85.82	89.85	86.23%	0.84235	2019	85.55	87.03
85.82	89.85	88.12	91.25	80.64	91.75	88.23%	0.8512			

per cent, and the overall accuracy in 2019 is the highest compared to other three years. For all images, the result of the kappa coefficient was over 0.82. When the coefficient of kappa is higher than 0.75, the degree of accuracy is categorized as excellent. In the same period 200 sampling points were compared with the corresponding point of the Google Earth images in order to validate the land-use classification. Validation rates are more than 80 per cent for all ages. In conclusion, the overall accuracy of the classification, the kappa coefficient statistics and validation are all excellent.

4.4.2 Changes in LST in the Research Zone

Climate is measured by temperature, humidity, atmospheric pressure, atmospheric particles and wind. Thus, climate is one of the most significant physical variables, which has an influence on the psychology, behaviours and physiology of human beings, as well as the economic activities of societies. It also controls the indoor and outdoor comforts of a person (Ogashawara & Bastos, 2012; Priyadarsini, 2012; Schwarz, Schlink, Franck, & Großmann, 2012). For measuring the climate change scenario, the estimation of LST distribution is necessary to identify the UHI effect in the urban environment. The term UHI become a hot topic in urban studies because of the severe adverse impact of UHI on the human and natural environment (Li et al., 2011; Liu & Zhang, 2011; Weng & Larson, 2005). Urbanization is considered one of the main contributors to climate change, in which humans modify the natural world (Qiao, Tian, & Xiao, 2013; Singh et al., 2017; Yu, Guo, Zeng, Koga, & Vejre, 2018). Residents frequently experience UHIs, particularly in major tropical and semi-arid cities. (Li, Wang, Peng, & Li, 2005; David R Streutker, 2003). UHI is a temperature anomaly that is higher in urban areas than in the surrounding suburban area. UHI occurs because of the alteration of the previous surface to the impermeable surface for buildings, industries, and other structures (Akbari, 2005; Weng & Larson, 2005). Urban areas are usually warmer than the surroundings because of the warm air between high buildings that is trapped in narrow streets.

Estimated LST data represents the temporal variation of the surface structure. A collection of images from Landsat is used to investigate seasonal, yearly, monthly and temporal variation of LST. The surface temperature varies season to season, as nature controls the changes. In the meantime, human-made activities can also affect natural changes adversely. This study examines the surface temperature from 1991 to 2019 time period.

The estimated LST values are categorized into five classes using 5°C interval. Figure 4.3 shows the pictorial representation of the surface temperature along with its temporal and spatial changes. The green (<22) and malachite green (22–27) colours indicate relatively lower to medium temperature, and dark umber (27–32), orange (32–37), and red (>37) colour tones indicate relatively high temperature, respectively.

The hierarchy of colour tone indicates the increasing temperature level from relatively lowest to highest. The percentage of each class represents the percentage of the total area according to its surface temperature class level.

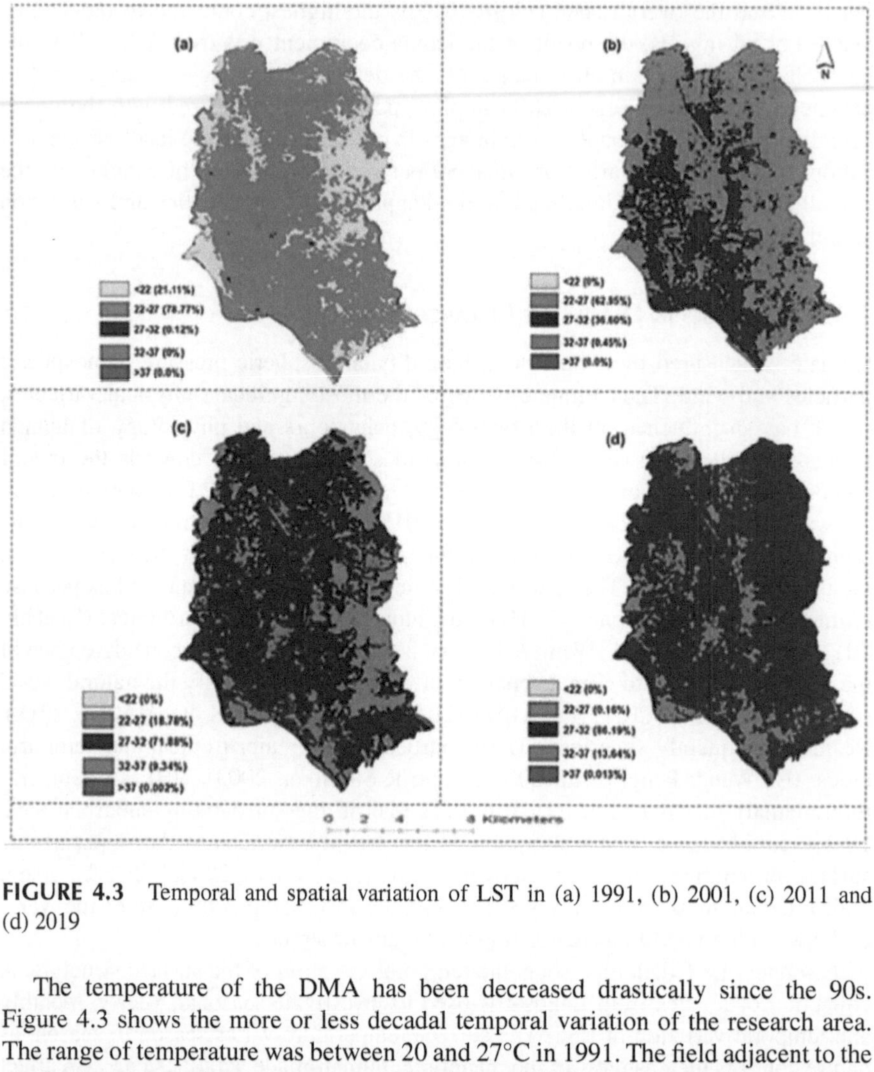

FIGURE 4.3 Temporal and spatial variation of LST in (a) 1991, (b) 2001, (c) 2011 and (d) 2019

The temperature of the DMA has been decreased drastically since the 90s. Figure 4.3 shows the more or less decadal temporal variation of the research area. The range of temperature was between 20 and 27°C in 1991. The field adjacent to the south-east region of the research area shows low LST due to the Buriganga River's open water body. However, the temperature of the south-east region of the area of interest is increased by urbanization along the riverbank. The urban built-up areas show higher temperatures compared to other land covers such as wetlands or vegetation. Although most parts of the area show the temperature above 22°C in 2002, the temperature was below 22°C in about 21 per cent areas in 1991. Similarly, the temperature of 36.6 per cent of the study area was in the range of 27–32°C in 2001, whereas only the temperature of 0.4 per cent of the total area was in the same range in 1991. Table 4.7 shows the change of urban land areas in different LST ranges. The negative change indicates small areas with a low temperature in 2001 compared to 1991. The positive change shows more land areas at a higher temperature in 2001 compared to 1991. Thus, the percentage of the land areas have been converted from

TABLE 4.7
Cross-Tabulation Summary of LST from 1991 to 2001 (Areas in km² and Temperatures in Degrees Celsius)

Range	Area (1991)	Area (%)	Area (2001)	Area (%)	Change (%) (2001-1991)
<22	64.322	21.105	0.000	0.000	–21.105
22–27	240.070	78.771	191.837	62.945	–15.826
27–32	0.379	0.124	111.552	36.602	36.478
32–37	0.000	0.000	1.382	0.453	0.453
>37	0.000	0.000	0.000	0.000	0.000
Total	304.771	100	304.771	100	

a low-temperature range to a high range. About 37 per cent of the study area, around 112.55 km², was converted from comparatively lower temperature class to higher class from in the year 1991 to 2001. In addition, the average temperature was around 26.45°C in the year of 2001, whereas the average temperature was 23°C in 1991. Thus, the average temperature increased by 3°C in 10 years period.

Most of the parts of the area appear radish in the Figure 4.3(c) because of the increment of surface temperature between 2001 and 2011, The temperature of around 78 per cent of the study area is in the range of 27–32°C. Table 4.8 shows 44.16 per cent of land converted from relatively lower temperatures in 2001 to a higher temperature in 2011. The winter temperature greater than 37°C is alarming for the study area. It is also noticeable that around 0.002 per cent of the land area has a temperature higher than 37°C, which is alarming for the winter season. In addition, the area in the class of 32–37°C was 0.45 per cent in 2001, which is converted to 9.43 per cent in 2011. The average temperature of the year 2011 was around 29.03°C. That means the average temperature increased by about 3°C and 6°C in a 10-year and 18-year period, respectively.

The current trend shows an almost similar trend of the change in LST in the research area, Fig. 4.3(d) indicates the high temperature in most of the portions of the study area in 2019 by red colour. Only a tiny percentage of the study region (0.16 per cent) shows the temperature in the rage of 22–27°C. The average temperature was

TABLE 4.8
Cross-Tabulation Summary of LST from 2001 to 2011 (Areas in km² and Temperatures in Degrees Celsius)

Range	Area (2001)	Area (%)	Area (2011)	Area (%)	Change (%) (2001–2011)
<22	0.000	0.000	0.000	0.000	0.000
22–27	191.837	62.945	57.249	18.784	-44.160
27–32	111.552	36.602	219.040	71.871	35.268
32–37	1.382	0.453	28.476	9.343	8.890
>37	0.000	0.000	0.005	0.002	0.002
Total	304.771	100	304.771	100	

TABLE 4.9
Cross-Tabulation Summary of LST from 2011 to 2019 (Areas in km² and Temperatures in Degrees Celsius)

Range	Area (2011)	Area (%)	Area (2019)	Area (%)	Change (%) (2019–2011)
<22	0.000	0.000	0.000	0.000	0.000
22–27	57.249	18.784	0.499	0.164	−18.621
27–32	219.040	71.871	262.680	86.187	14.316
32–37	28.476	9.343	41.561	13.636	4.293
>37	0.005	0.002	0.040	0.013	0.011
Total	304.771	100	304.771	100	

30.4°C in 2019, which is about 1°C higher compared to the temperature in 2011 (Table 4.9). Thus, the land areas in low temperature bound are decreased, whereas the land areas are increased in high temperature bound. The change in LST clearly indicates the increasing trend over time in the zone.

As UHI has a significant dependency on LST therefore, many researchers suggest using RS data for examining the most appropriate assessment of UHI (Abutaleb et al., 2015; Bhatta, 2010; Coutts et al., 2016; Mosammam et al., 2017). The relation between change in LST with LULC is investigated by many studies (Weng, 2009). The LST-LULC change comparative analysis allows scientists to explore the relation of the UHI impact on land-cover transition (Fabrizi, Bonafoni, & Biondi, 2010). To examine the relationship between LST and change in land cover, many forms of LULC indices have been introduced. The widely used indices are the normalized vegetation index (NDVI), the normalized difference built-up index (NDBI), the normalized difference water index (NDWI) and the normalized difference bareness index (NDBaI), which are highly associated with LST (Chen et al., 2006a; L. Liu & Zhang, 2011; Yue et al., 2012). The relationship between LULC and its indexs with LST will be discussed in the following section

4.4.3 Relationship between LST and LULC Indices

LULC changing patterns are the most influencing factors in the study of UHI. LULC, such as built-up environment, vegetation, body water, and bare land that effectively affect the LST (Deilami, Kamruzzaman, & Liu, 2018). The more the presence of impervious surface, results in the more the surface temperature (Estoque, Murayama, & Myint, 2017; D. Lu & Weng, 2006; Peng, Xie, Liu, & Ma, 2016; W. Zhou, Qian, Li, Li, & Han, 2014). If water body and vegetation present in the land cover, there would be less surface temperature as well (Coutts et al., 2016; A.-A. Kafy, 2018; A.-A. Kafy et al., 2018; A. Kafy et al., 2019; Liang & Weng, 2008; Naeem et al., 2018). The coolness of a city greatly depends on green coverage, which lies between 70 per cent and 80 per cent of the total land cover. Hence, the size and density of green coverage reduce the surface temperature effectively (Abutaleb et al., 2015; X. Li et al., 2012; Wang, Cheng, Xi, Yang, & Zhao, 2018; Yang et al., 2017).

The impact of bare land has both positive and negative correlations with the LST (Chen et al., 2006a; Peng et al., 2016). Additionally, analysing only the land-cover changes is not sufficient, and so land-use changes would be a better parameter for explaining the LST (Pan, 2015; Yue et al., 2012). Additionally, as mentioned earlier, considerable indicators such as NDBI and NDVI have so far received the most attention in the analysis of UHI (Berger et al., 2017; D. Lu & Weng, 2006; Qiao et al., 2013; Zhang, Murray, & Turner Ii, 2017). The classification of land-cover changes and its affecting indicators are described as follows.

Many types of literature show that LST has a positive NDBI correlation but a negative NDVI correlation. That means that the surface temperature is also increased with the increase in built-up areas. On the other hand, the more the NDVI value, the more will be higher dense vegetation, and so the surface temperature and the index have a negative relationship. In the study, several correlation analyses have been conducted to visualize the relation between LST with NDVI and NDBI. The Figure 4.4 below shows that LST has a positive relation with NDBI but has a negative relationship to NDVI. The Figure 4.4 shows four different years' correlation, the R^2 values of LST with NDVI and NDBI show that more or less 0.9 and that means both variables have a strong correlation with the LST.

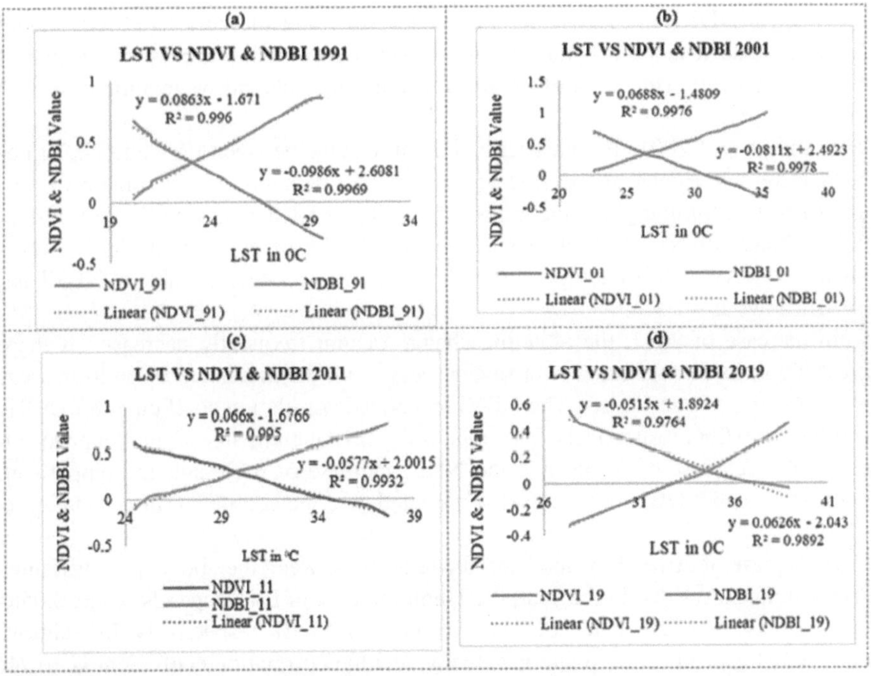

FIGURE 4.4 Correlation between LST with NDVI and NDBI in the year (a) 1991, (b) 2001, (c) 2011 and (d) 2019

TABLE 4.10
Correlation Coefficient Between LST, NDVI and NDBI of the Study Area

1991	LST	NDVI	NDBI	2011	LST	NDVI	NDBI
LST	1			LST	1		
NDVI	–0.099	1		NDVI	–0.058	1	
NDBI	0.083	–0.87	1	NDBI	0.066	–1.14	1
2001	**LST**	**NDVI**	**NDBI**	**2019**	**LST**	**NDVI**	**NDBI**
LST	1			LST	1		
NDVI	–0.08	1		NDVI	–0.052	1	
NDBI	0.07	–0.846	1	NDBI	0.063	–1.17	1

The Figure 4.4 shows linear regression analysis with the three variables LST, NDVI and NDBI. Meantime, the correlation coefficient indicates the extent of influence for one factor to another (Table 4.10). In the case of 1991, the correlation coefficient indicates that the 1-unit change of LST depends on 0.099-unit change of NDVI, negatively. Here intercept value for NDVI with LST is 2.61. In addition, LST has a positive relationship with NDBI, and the coefficient of correlation is 0.083. That means the 1-unit change of LST depends on 0.083-unit change of NDBI, positively. Both NDVI and NDBI compare negatively, and the correlation coefficient is –0.87. That means the 1-unit change of NDVI depends on 0.087-unit change of NDBI, negatively. Hence, built-up areas increase by reducing vegetation covers of the study area.

In the case of 2001, 10 years later, the unit change between LST and NDVI has been decreased. Hence, an increase of 1 unit of LST depends on reducing only 0.081 unit of NDVI. Similarly, the increment of 1 unit of LST depends on the rise of only 0.069 change of NDBI. That means, smaller change of NDVI and NDBI causes greater effects on surface temperature. Moreover, the unit change value of NDVI and NDBI shows as –0.846.

In the case of 2011, the affecting change of unit frequently decreases as there present a lacks of vegetation over several years with the increment of built-up areas as well as impervious layers. The NDVI and NDBI correlation coefficient for LST is –0.058 and 0.066, respectively. That means the LST increment of 1 unit depends on the NDVI decrease of 0.058 unit and NDBI increase of 0.07 unit. In comparison, both NDVI and NDBI are negatively correlated, and the coefficient of correlation is –1.14.

In the case of 2019, 28 years later, the relation between the above three variables shows frequent results. In the year, the 1-unit increase of LST depends on the 0.058-unit decrease of NDVI and 0.067-unit increase of NDBI, respectively. In addition, both NDVI and NDBI correlate negatively, and the correlation coefficient is –0.71. That means the 1-unit increase of NDVI depends on the 0.71-unit decrease of NDBI value. Each LULC represents a different temperature. The following section discusses the changes of LST with the changes in land cover/uses.

4.4.4 Association of Different LULC and LST

The radiative energy emitted from the surface of the Earth is traced by remote sensed LST including buildings, paved surfaces, wooded trees, bare soil and wetlands. To reflect LULC's association with LST, two cross-sections were made throughout the study area (Figure 4.5), and the average LST of each form of LULC was shown in Figure 4.5 (1–4). Because of the higher concentration of LST values in the southern and eastern zone, two cross-sections (AA/, BB/) has made in a south-east direction.

The profile reveals that urban core area, with a maximum of built-up environment, experienced LST > 27°C in 1991, LST > 30°C in 2001, LST > 35°C in 2011 and > 39°C in 2019 respectively. Also, high LST recorded > 25°C, > 29°C and > 33°C for Bare land in four different year cross-sections. Two other land uses (water bodies and vegetation land) reported the lowest temperature ranging from < 21°C to < 37°C in association with the built-up area and bare land. The maximum temperature recorded for the year 1991 in bare land (29°C) and from 2001 to 2019 in an urban area (Table 4.11).

4.4.5 Simulation of LST

To explore the optimum predictive model by applying the ANN-CA (artificial neural network-based cellular automata) algorithm; cases were characterized into two

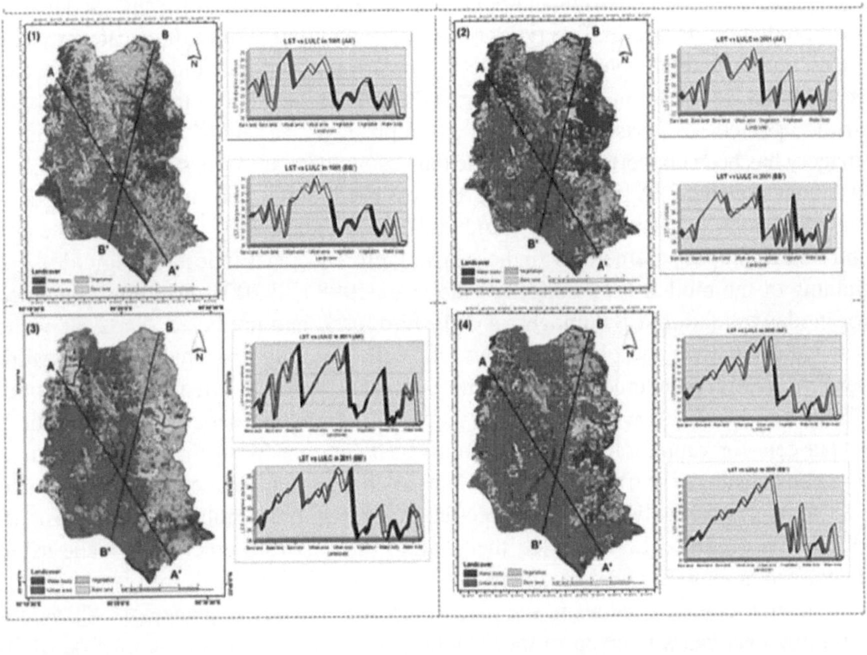

FIGURE 4.5 Association of LST with different land cover/use in the year of (1) 1991, (2) 2001, (3) 2011 and (4) 2019

TABLE 4.11
Temperature Variation in Different LULC

LULC	Year	MIN	MAX	MEAN	STD
Bare land	1991	21.50	29.59	23.59	0.80
	2001	23.91	34.91	27.15	1.44
	2011	24.97	37.61	30.06	1.55
	2019	28.00	33.99	29.95	0.84
Urban area	1991	21.06	28.76	23.92	0.91
	2001	27.07	35.33	27.96	0.94
	2011	24.97	37.22	30.43	1.44
	2019	33.00	38.87	33.61	0.63
Vegetation	1991	20.62	27.09	22.89	0.71
	2001	22.99	31.91	26.06	0.82
	2011	24.11	34.85	27.64	1.73
	2019	26.59	34.94	31.51	0.82
Water body	1991	20.17	26.25	21.93	0.63
	2001	22.53	30.60	25.06	1.01
	2011	24.11	34.85	26.50	1.27
	2019	26.51	31.99	28.34	0.66

categories in the form of input data. In case 1, exacted NDBI and NDVI set from satellite images as input variables and in case 2, extracted built-up area, bare soil and vegetation, water bodies from land-cover map as input variables were analysed to perform prediction. However, five types of cases were classified into three categories by the number of hidden layers. The node numbers composed by a hidden layer have to be modified from 1 to 5n, which corresponds to five times that of the input variables, so the updated LST has been applied as output layer variables. Temporal Landsat imagery has been converted into discrete data using the pixel values in ArcGIS 10.5 to use in the QGIS 3.2. A plugin (MOLUSE) was used to establish an ANN for predicting LST, using land-use imagery (temporal), NDVI and NDBI as input parameters, and LST as output parameter with their spatial references (latitude and longitude). For training of the model, LST images of the years 1998 and 2008 have used. Intended for model validation, LST images of 2019 have used and predicted. The LST value was validated with the observed value of 2019. Instantly, the analysis of network performance is considered acceptable, the proficient network is tied up for prediction. The best possible data was divided into the 80:10:10 ratio, 80 per cent for preparation, 10 per cent for validation, and the remaining 10 per cent for testing. The simulation function considers the trained network (Net) as an argument and returns the simulated LST to the respectively given input vectors. After that, the simulation result was parallel to the observed LST dataset; then, future time series prediction is made using the trained network's saved parameters. Using a model in Arc GIS software, the LST latitude and longitude values of the pixels were achieved. A sequence of LST values from previous years is given to the system as input vectors at a time interval of 10 years on a time scale of 20 years (2001–2019) so that the system can assess the hidden arrangement within the dataset and predict the time scale as shown in Figure 4.6.

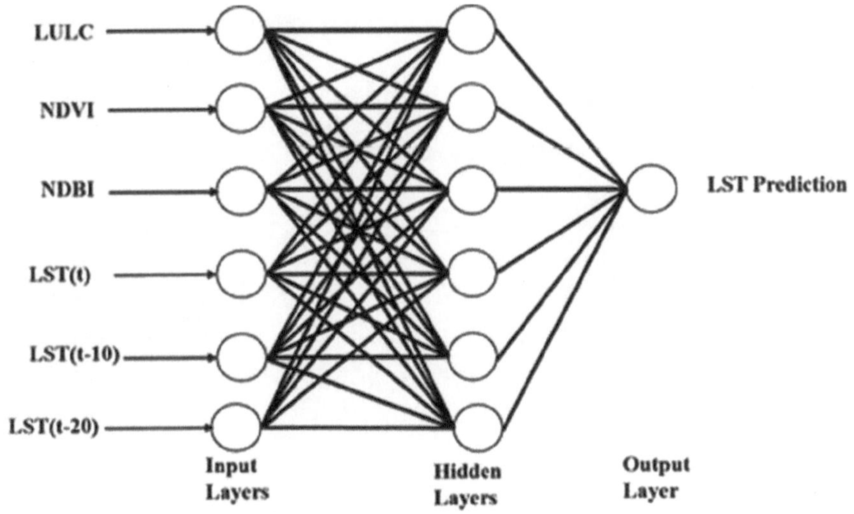

FIGURE 4.6 Model for LST prediction

The projected LST of a spatial unit for the next 10 years was used by the following equation, the t+10 feature of the earlier values of that spatial entity within the time scale.

$$LST(t+10) = f[LST(t), LST(t-10), LST(t-20)]$$

$$LST(t) = 2019.$$

For accurate estimation, 2019 LST prediction was conducted. Observed and predicted LST values have strong closeness. The maximum LST value was shown 39.87°C, and the minimum value was 25.51°C in the observed map of 2019, whereas the predicted map was recorded the maximum and minimum LST value as 40.03°C and 26.45°C, respectively. This predicted and observed LST values were demonstrated the precision, but the spatial distribution of LST within the region that was seen some variability. The predicted ANN-CA model was also demonstrated a high MSE (0.523) and R (0.892) value for future 2029 LST simulation.

Consequently, based on predicted LST 2019, the ANN model predicted the LST scenario for 2029 shown in Figure 4.7. In the LST of the predicted map for 2028, there is a slight variation because the earlier LST value used as an input parameter in the ANN model. As displayed in Figure 4.3, the area coverage of high LST value was gradually increased (150 km^2 area in the range more than 37°C), and the highest temperature was estimated in 2029 is 43°C and the lowest temperature is 31°C. The findings showed that the higher value of LST is gradually growing in the central part of Dhaka city and expanding to the neighbouring area (Table 4.12).

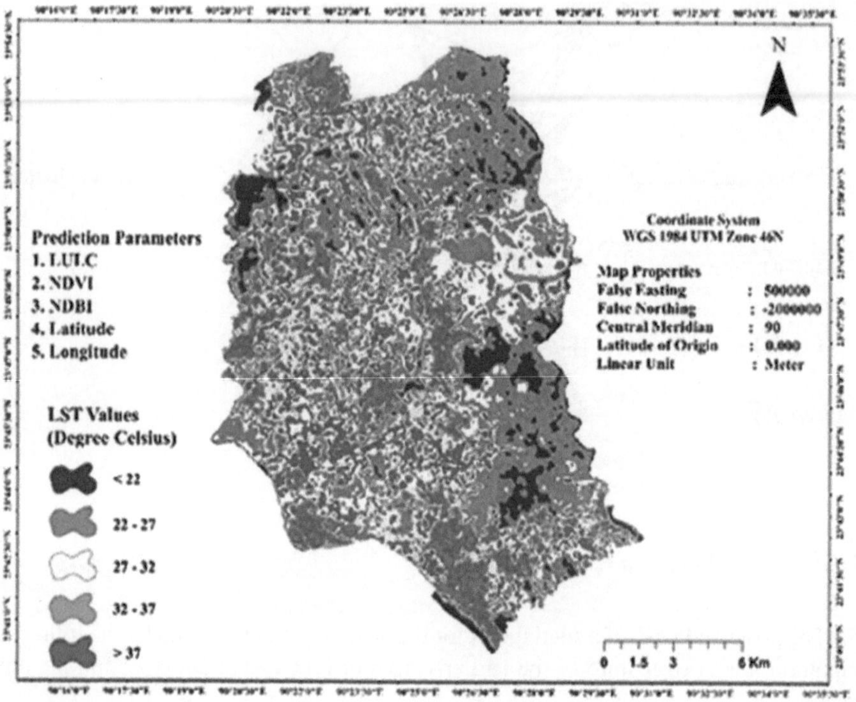

FIGURE 4.7 Prediction of LST

TABLE 4.12
LST Model Validation for the Year 2019

Prediction Year	ANN Model Validation for LST Prediction		
	No of Hidden Layer	Mean Square Error (MSE)	Correlation Coefficient-R
2019	4	0.523	0.892

During the analysis, it was observed that LST values have had variation in the study time epoch (1991–2019), but the maximum temperature value coverage was increased rapidly with the growth in the built-up sector. Likewise, in the forecast map of 2029, there was also a slight variation in the LST value compared to the earlier time epoch, but the coverage of the maximum temperature value in the city and its surrounding area has been increased efficiency with respect to the settlement area. The impact of urban heat islands is also rising day after day due to the continuous rise in temperature. The following section discusses the UHI effects and its extent to the study area.

4.4.6 UHI Effects in the Study Area

With substantial heterogeneity to the surface of urban land and the different ecosystems of the surrounding rural area, human activities strongly affect the metropolitan regions (A.-A. Kafy et al., 2018; Huidong Li et al., 2018). The UHI affects the environmental change factors e.g., regional climate, vegetation growth, and water and air quality. Such factors again impact the rising human well-being of mortality and morbidity, energy consumption and urban violence incidents. This impact is more significant in a warm climate, in a fast urbanizing world (D. Zhou et al., 2019). The buildings need more energy to cool down as a consequence of this phenomenon. To order to meet this increased energy demand, more energy is needed and the increases in greenhouse gases. Greenhouse gas emissions boost decreases the atmosphere and makes life more vulnerable to human (Nuruzzaman, 2015). Again, UHI causes increased cooling fossil fuel use, leading to higher CO_2 emissions causing global warming (Ngie et al., 2014).

The LST, presented in section 3.2, has been normalized in two steps using Equation (4.5) & (4.6). In the first step, the land has classified that experience the UHI where the most UHI experience gives a value 1, and the areas provide value 0 has no effect. The data of the year 2001, 2011 and 2019 will go to further analysis that is the evaluation of UHI changes over the year. The calculation of UTFVI will describe the ecological evaluation.

The comparison is made on a scale of UTFVI value 0 – 0.2 and in six categories where the 0 value is denoted with none and 0.2 as the strongest.

Table 4.13 clearly shows that there happened a change in UTFVI value, and the difference is from the none to the strongest. That means the area has become warmer, and as a consequence, the UHI is more. The change is so significant that 78.54 per cent more area experience the UHI effect in 2011, which had no UHI effect in 2001. 86.73 per cent of the city had the strongest UHI effect. This gives a red alert of the ecological endanger of the environment. Evaluation of further change in 2019 from 2011, shows an additional 12.99 per cent area in the strongest effect category (Table 4.14). In 2019, 99.72 per cent area experienced the strongest UHI effect and indicated a dangerous future-forward.

The maps (Figure 4.8) make the changes more visual with a colour range from dark green (no UHI effect) to dark red (strongest effect). The areas with the most substantial UHI effect are the most vulnerable to temperature rise. A change detection analysis has been taken place to analyse the changing effect more clearly. As there is no UHI effect in the year 1991, changes can only be possible from 2001 to 2011 and 2011 to 2019. The UHI change detection analysis is classified into five parameters such as decreased, some decrease, unchanged, some increase, and increased. Figure 4.9 shows the difference map where each colour tone indicates different UHI changes. The map indicates a significant increase in UHI effects from 2001 to 2019.

Table 4.15 indicates that there was a considerable change in the increase in the UHI effect. The table shows that 88.53 per cent of the whole area shows the increase

TABLE 4.13
Cross-Tabulation Summary of UTFVI from 2001 to 2011 (Area in km²)

UTFVI	Range	Area (2001)	Area in %	Area (2011)	Area in %	Change in % (2011–2001)
None	<0	259.580	85.173	20.212	6.632	−78.540
Weak	0.000–0.005	7.887	2.588	3.894	1.278	−1.310
Middle	0.005–0.010	6.592	2.163	4.646	1.524	−0.638
Strong	0.010–0.015	5.683	1.865	5.491	1.802	−0.063
Stronger	0.015–0.02	4.738	1.554	6.187	2.030	0.476
Strongest	>0.2	20.291	6.658	264.326	86.734	80.076
Total		304.769	100	304.756	100	

TABLE 4.14
Cross-Tabulation Summary of UTFVI from 2011 to 2019 (Area in km²)

UTFVI	Range	Area (2011)	Area in %	Area (2019)	Area in %	Change in % (2019–2011)
None	<0	20.212	6.632	0.399	0.131	−6.501
Weak	0.000–0.005	3.894	1.278	0.017	0.006	−1.272
Middle	0.005–0.010	4.646	1.524	0.079	0.026	−1.498
Strong	0.010–0.015	5.491	1.802	0.146	0.048	−1.754
Stronger	0.015–0.02	6.187	2.030	0.214	0.070	−1.960
Strongest	>0.2	264.326	86.730	303.924	99.719	12.989
Total		304.756	100	304.779	100	

in the UHI effect from the year 2001 to 2011 additionally, as most of the area indicates as the increase of the UHI effect and so the changes in UHI show unchanged UHI effect, which is 86.46 per cent of total space.

This high UHI effect has become an alarming concern for the city. The UHI effect depends on the geometry of the city, and Dhaka city's unplanned development causes this devastating impact of the temperature rise. The increase of vegetation through a planned growth of the town can minimize the urban heat island effect and ensure a healthy and liveable city life.

4.5 CONCLUSION

The temperature rise in urban areas is ubiquitous in recent times. UHI study became famous because of the adverse consequences of land surface temperature (LST) on urban residents. The effect was due to the increase of impermeable surface due to high-density built-up city areas. Dhaka being the most densely populated city, UHI study may be useful for future urban planning. This research aims at analysing Dhaka city's UHI effect using multitemporal Landsat images.

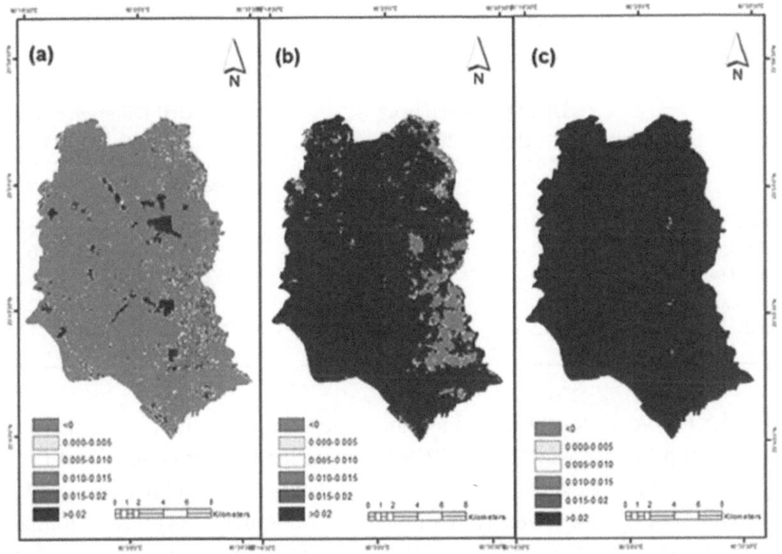

FIGURE 4.8 Temporal and spatial variation of UTFVI in (a) 2001, (b) 2011 and (c) 2019. The hierarchy of colour tone indicates the increasing temperature level from relatively lowest to highest

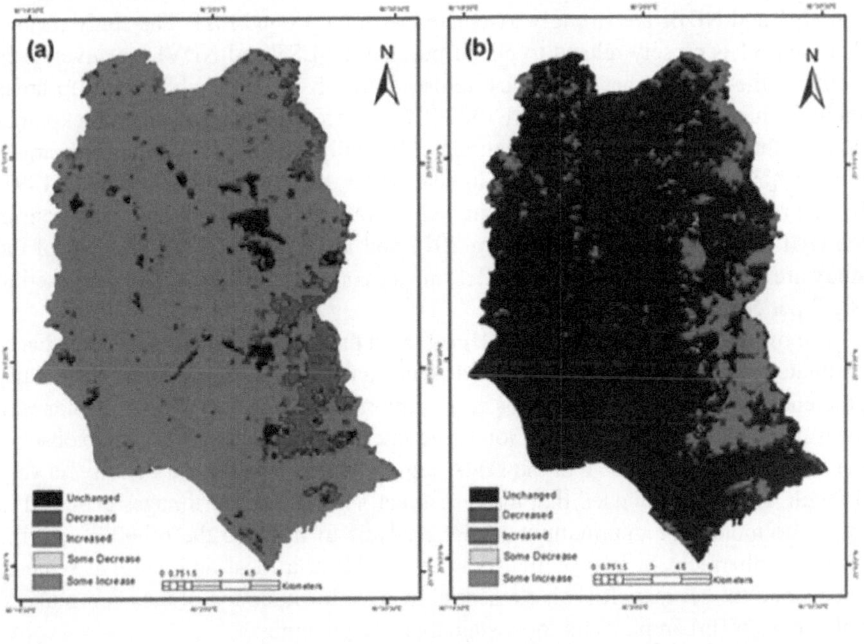

FIGURE 4.9 UHI change detection from (a) 2001 to 2011 and (b) 2011 to 2019

TABLE 4.15
UHI Difference from 2001 to 2011 and 2011 to 2019

	UHI Difference from 2001 to 2011		UHI Difference from 2001 to 2011	
Indicators	**Area (km²)**	**% of Change**	**Area (km²)**	**% of Change**
Decreased	18.0522	5.923209	0.2466	0.080913
Some Decrease	–	–	0.2142	0.070282
Unchanged	16.8912	5.542267	263.5038	86.45972
Some Increase	–	–	6.1776	2.026967
Increased	269.8272	88.53452	34.6284	11.36212

Since LST is very closely linked to land cover, this study monitors and examines the relationship between 1991 and 2019 between land-cover change and LST. The result indicates that LST increases in the past 20 years. About 37 per cent of the study area, around 112.55 km^2, was converted from a comparatively lower temperature class to a higher class from 1991 to 2001. Similarly, the average temperature was increased by about 3°C and 6°C in 10 years and 18 years' period, respectively. The changes in temperature were most significant between 1991 and 2011. Therefore, the most influencing factor for the surface temperature change in land use/cover change. This study also utilized RS derived products such as NDBI and NDVI to explore the relationship of these derived products with LST. The correlation coefficients suggest that LST and NDBI are strongly associated with NDVI and LST. The study reveals that the LST is closely related to NDBI but that the LST and NDVI are adverse. In addition, the wetland has the lowest temperature of 20.17°C, while built-up areas show the highest mean temperature (38.87°C). Therefore, the fundamental explanation for the regular rise in temperature can be said to be the rise in built-up areas. Furthermore, in the analysis of UHI indicates that there was no UHI effect in 1991, but the result was documented at an incredible rate in 2001. A gradual increment in the UHI effect is also continued from 2011 and 2019. About 86.73 per cent of the study areas appear in the strongest UHI impact zones in 2011, which is increased to 99.72 per cent in 2019.

The overall result shows a high UHI effect in the DMA area. As the selected area is situated in the Dhaka Metropolitan Area and acts as the most important financially efficient area for Bangladesh. Thus, it is important to take the necessary planning measures to reduce UHI effects. Cloud-free satellite images are necessary to observe the seasonal change in LST in the DMA region, which is a tropical country, is very difficult. It is recommended that the researcher should consider images acquired in the same month for multitemporal UHI analysis to increase the reliability of the result. Furthermore, it is essential to analyse UHI at a local and national scale to understand the severity. It is also essential to draw the attention of decision-makers, policymakers, urban planners for sustainable city planning.

ACKNOWLEDGEMENTS

The authors want to acknowledge the USGS Earth Explorer for the Landsat archives.

CONFLICTS OF INTEREST

No conflict of interest with the article is declared by the authors.

REFERENCES

Abutaleb, K., Ngie, A., Darwish, A., Ahmed, M., Arafat, S., & Ahmed, F. (2015). Assessment of urban heat island using remotely sensed imagery over Greater Cairo, Egypt. *Advances in Remote Sensing*, *4*(01), 35.

Ahmed, S. (2018). Assessment of urban heat islands and impact of climate change on socioeconomic over Suez Governorate using remote sensing and GIS techniques. *The Egyptian Journal of Remote Sensing and Space Science*, *21*(1), 15–25.

Ahmed, B., Kamruzzaman, M., Zhu, X., Rahman, M., & Choi, K. (2013). Simulating land cover changes and their impacts on land surface temperature in Dhaka, Bangladesh. *Remote Sensing*, *5*(11), 5969–5998.

Akbari, H. (2005). Energy saving potentials and air quality benefits of urban heat island mitigation: Ernest Orlando Lawrence Berkeley National Laboratory, Berkeley, CA (US).

Al Kuwari, N. Y., Ahmed, S., & Kaiser, M. F. (2016). Optimal satellite sensor selection utilized to monitor the impact of urban sprawl on the thermal environment in doha city, Qatar.

Alavipanah, S., Wegmann, M., Qureshi, S., Weng, Q., & Koellner, T. (2015). The role of vegetation in mitigating urban land surface temperatures: A case study of Munich, Germany during the warm season. *Sustainability*, *7*(4), 4689–4706.

Aminipouri, M., Knudby, A., & Ho, H. C. (2016). Using multiple disparate data sources to map heat vulnerability: Vancouver case study. *The Canadian Geographer/Le Géographe canadien*, *60*(3), 356–368.

Banu, S., Hu, W., Guo, Y., Hurst, C., & Tong, S. (2014). Projecting the impact of climate change on dengue transmission in Dhaka, Bangladesh. *Environment International*, *63*, 137–142.

Berger, C., Rosentreter, J., Voltersen, M., Baumgart, C., Schmullius, C., & Hese, S. (2017). Spatio-temporal analysis of the relationship between 2D/3D urban site characteristics and land surface temperature. *Remote Sensing of Environment*, *193*, 225–243.

Bhatta, B. (2010). *Analysis of urban growth and sprawl from remote sensing data*. Springer Science & Business Media.

Buyantuyev, A., & Wu, J. (2010). Urban heat islands and landscape heterogeneity: linking spatiotemporal variations in surface temperatures to land-cover and socioeconomic patterns. *Landscape Ecology*, *25*(1), 17–33.

Camilloni, I., & Barros, V. (1997). On the urban heat island effect dependence on temperature trends. *Climatic Change*, *37*(4), 665–681.

Chander, G., & Markham, B. (2003). Revised Landsat-5 TM radiometric calibration procedures and postcalibration dynamic ranges. *IEEE Transactions on Geoscience and Remote Sensing*, *41*(11), 2674–2677.

Chen, F., Yang, S., Yin, K., & Chan, P. (2017). Challenges to quantitative applications of Landsat observations for the urban thermal environment. *Journal of Environmental Sciences*, *59*, 80–88.

Chen, X.-L., Zhao, H.-M., Li, P.-X., & Yin, Z.-Y. (2006a). Remote sensing image-based analysis of the relationship between urban heat island and land use/cover changes. *Remote Sensing of Environment*, *104*(2), 133–146.

Chen, X.-L., Zhao, H.-M., Li, P.-X., & Yin, Z.-Y. s. o. e. (2006b). Remote sensing image-based analysis of the relationship between urban heat island and land use/cover changes. *Remote Sensing of Environment*, *104*(2), 133–146.

Chow, W. T. L., & Roth, M. (2006). Temporal dynamics of the urban heat island of Singapore. *International Journal of Climatology: A Journal of the Royal Meteorological Society*, *26*(15), 2243–2260.

Civco, D. L. (1993). Artificial neural networks for land-cover classification and mapping. *International Journal of Geographical Information Science*, *7*(2), 173–186.

Coutts, A. M., Harris, R. J., Phan, T., Livesley, S. J., Williams, N. S. G., & Tapper, N. J. (2016). Thermal infrared remote sensing of urban heat: Hotspots, vegetation, and an assessment of techniques for use in urban planning. *Remote Sensing of Environment*, *186*, 637–651.

Deilami, K., Kamruzzaman, M., & Liu, Y. (2018). Urban heat island effect: A systematic review of spatio-temporal factors, data, methods, and mitigation measures. *International Journal of Applied Earth Observation and Geoinformation*, *67*, 30–42.

Deng, Y., & Srinivasan, S. (2016). Urban land use change and regional access: A case study in Beijing, China. *Habitat international*, *51*, 103–113.

Deng, C., & Wu, C. J. R. S. o. E. (2013). Examining the impacts of urban biophysical compositions on surface urban heat island: A spectral unmixing and thermal mixing approach. *Remote Sensing of Environment*, *131*, 262–274.

Dewan, A. M., Kabir, M. H., Nahar, K., & Rahman, M. Z. (2012). Urbanisation and environmental degradation in Dhaka Metropolitan Area of Bangladesh. *International Journal of Environment and Sustainable Development*, *11*(2), 118–147.

Dhamge, N. R., Atmapoojya, S., & Kadu, M. S. J. P. T. (2012). Genetic algorithm driven ANN model for runoff estimation. *Procedia Technology*, *6*, 501–508.

Estoque, R. C., Murayama, Y., & Myint, S. W. (2017). Effects of landscape composition and pattern on land surface temperature: An urban heat island study in the megacities of Southeast Asia. *Science of the Total Environment*, *577*, 349–359.

Fabrizi, R., Bonafoni, S., & Biondi, R. (2010). Satellite and ground-based sensors for the urban heat island analysis in the city of Rome. *Remote Sensing*, *2*(5), 1400–1415.

Gallo, K. P., & Owen, T. W. (1999). Satellite-based adjustments for the urban heat island temperature bias. *Journal of Applied Meteorology*, *38*(6), 806–813.

Gatys, L. A., Ecker, A. S., & Bethge, M. J. (2015). A neural algorithm of artistic style. *Nature Communications*, *8*, 1–16.

Glazier, R. H., Creatore, M. I., Weyman, J. T., Fazli, G., Matheson, F. I., Gozdyra, P., … Booth, G. L. (2014). Density, destinations or both? A comparison of measures of walkability in relation to transportation behaviors, obesity and diabetes in Toronto, Canada. *PloS one*, *9*(1), e85295.

Godschalk, D. R. (2004). Land use planning challenges: Coping with conflicts in visions of sustainable development and livable communities. *Journal of the American Planning Association*, *70*(1), 5–13.

Gopal, S., & Woodcock, C. (1996). Remote sensing of forest change using artificial neural networks. *IEEE Transactions on Geoscience and Remote Sensing*, *34*(2), 398–404.

Gutman, G., Huang, C., Chander, G., Noojipady, P., & Masek, J. G. (2013). Assessment of the NASA–USGS global land survey (GLS) datasets. *Remote sensing of environment*, *134*, 249–265.

Kafy, A.-A. (2018). Importance of Surface Water Bodies for Sustainable Cities: A Case Study of Rajshahi City Corporation: Bangladesh Institute of Planners (BIP).

Kafy, A., Hossain, M., Prince, A. A. N., Kawshar, M., Shamim, M. A., Das, P., & Noyon, M. E. K. (2019). *Estimation of Land Use Change to Identify Urban Heat Island Effect on Climate change: A Remote Sensing Based Approach*. Paper presented at the *International Conference on Climate Change (ICCC-2019)*, Dhaka, Bangladesh.

Kafy, A.-A., Islam, M., Khan, A. R., Ferdous, L., & Mamun, M. (2018). Identifying Most Influential Land Use Parameters Contributing Reduction of Surface Water Bodies in Rajshahi City, Bangladesh: A Remote Sensing Approach.

Khandelwal, S., Goyal, R., Kaul, N., & Mathew, A. (2018). Assessment of land surface temperature variation due to change in elevation of area surrounding Jaipur, India. *The Egyptian Journal of Remote Sensing and Space Science*, *21*(1), 87–94.

Landsberg, H. E. (1981). *The urban climate* (Vol. 28): Academic Press.

Li, H., & Liu, Q. (2008). *Comparison of NDBI and NDVI as indicators of surface urban heat island effect in MODIS imagery*. Paper presented at the *International conference on earth observation data processing and analysis (ICEODPA)*.

Li, J., Song, C., Cao, L., Zhu, F., Meng, X., & Wu, J. (2011). Impacts of landscape structure on surface urban heat islands: A case study of Shanghai, China. *Remote Sensing of Environment*, *115*(12), 3249–3263.

Li, Z.-L., Tang, B.-H., Wu, H., Ren, H., Yan, G., Wan, Z., … Sobrino, J. A. (2013). Satellite-derived land surface temperature: Current status and perspectives. *Remote sensing of environment*, *131*, 14–37.

Li, W., Wang, Y., Peng, J., & Li, G. (2005). Landscape spatial changes associated with rapid urbanization in Shenzhen, China. *The International Journal of Sustainable Development & World Ecology*, *12*(3), 314–325.

Li, H., Zhou, Y., Li, X., Meng, L., Wang, X., Wu, S., & Sodoudi, S. (2018). A new method to quantify surface urban heat island intensity. *Science of the Total Environment*, *624*, 262–272.

Li, X., Zhou, W., Ouyang, Z., Xu, W., & Zheng, H. (2012). Spatial pattern of greenspace affects land surface temperature: evidence from the heavily urbanized Beijing metropolitan area, China. *Landscape ecology*, *27*(6), 887–898.

Liang, B., & Weng, Q. (2008). Multiscale analysis of census-based land surface temperature variations and determinants in Indianapolis, United States. *Journal of Urban Planning and Development*, *134*(3), 129–139.

Liu, H., & Weng, Q. (2012). Enhancing temporal resolution of satellite imagery for public health studies: A case study of West Nile Virus outbreak in Los Angeles in 2007. *Remote sensing of environment*, *117*, 57–71.

Liu, L., & Zhang, Y. (2011). Urban heat island analysis using the Landsat TM data and ASTER data: A case study in Hong Kong. *Remote Sensing*, *3*(7), 1535–1552.

Loyd, C. (2013). Putting Landsat 8's Bands to Work. *NASA, USA*.

Lu, Y., Feng, X., Xiao, P., Shen, C., & Sun, J. (2009). *Urban heat island in summer of Nanjing based on TM data*. Paper presented at the 2009 Joint Urban Remote Sensing Event.

Lu, D., & Weng, Q. (2006). Spectral mixture analysis of ASTER images for examining the relationship between urban thermal features and biophysical descriptors in Indianapolis, Indiana, USA. *Remote sensing of environment*, *104*(2), 157–167.

Maduako, I. D., Yun, Z., & Patrick, B. (2016). Simulation and prediction of land surface temperature (LST) dynamics within Ikom City in Nigeria using artificial neural network (ANN). *Journal of Remote Sensing & GIS*, *5*(1), 1–7.

Mas, J. F., & Flores, J. J. (2008). The application of artificial neural networks to the analysis of remotely sensed data. *International Journal of remote sensing*, *29*(3), 617–663.

McCarthy, M. P., Best, M. J., & Betts, R. A. (2010). Climate change in cities due to global warming and urban effects. *Geophysical Research Letters*, *37*(9), 1–5.

Mills, G. (2008). Luke Howard and the climate of London. *Weather*, *63*(6), 153–157.

Mirzaei, P. A., & Haghighat, F. (2010). Approaches to study urban heat island–abilities and limitations. *Building and environment, 45*(10), 2192–2201.

Mohamed, A. A., Odindi, J., & Mutanga, O. (2017). Land surface temperature and emissivity estimation for Urban Heat Island assessment using medium-and low-resolution space-borne sensors: A review. *Geocarto International, 32*(4), 455–470.

Mosammam, H. M., Nia, J. T., Khani, H., Teymouri, A., & Kazemi, M. (2017). Monitoring land use change and measuring urban sprawl based on its spatial forms: The case of Qom city. *The Egyptian Journal of Remote Sensing and Space Science, 20*(1), 103–116.

Mushore, T. D., Odindi, J., Dube, T., Mutanga, O. J. B., & Environment. (2017). Prediction of future urban surface temperatures using medium resolution satellite data in Harare metropolitan city, Zimbabwe. *Building and Environment, 122*, 397–410.

Naeem, S., Cao, C., Qazi, W., Zamani, M., Wei, C., Acharya, B., & Rehman, A. (2018). Studying the association between green space characteristics and land surface temperature for sustainable urban environments: An analysis of Beijing and Islamabad. *ISPRS International Journal of Geo-Information, 7*(2), 38.

Ngie, A., Abutaleb, K., Ahmed, F., Darwish, A., & Ahmed, M. (2014). Assessment of urban heat island using satellite remotely sensed imagery: a review. *South African Geographical Journal, 96*(2), 198–214.

Nuruzzaman, M. (2015). Urban heat island: causes, effects and mitigation measures-a review. *International Journal of Environmental Monitoring and Analysis, 3*(2), 67–73.

Ogashawara, I., & Bastos, V. (2012). A quantitative approach for analyzing the relationship between urban heat islands and land cover. *Remote Sensing, 4*(11), 3596–3618.

Oke, T. R. (1982). The energetic basis of the urban heat island. *Quarterly Journal of the Royal Meteorological Society, 108*(455), 1–24.

Pan, J. (2015). Analysis of human factors on urban heat island and simulation of urban thermal environment in Lanzhou city, China. *Journal of Applied Remote Sensing, 9*(1), 095999.

Patil, K., Deo, M. C., Ghosh, S., & Ravichandran, M. (2013). Predicting sea surface temperatures in the North Indian Ocean with nonlinear autoregressive neural networks. *International Journal of Oceanography, 2013*, 11–12.

Peng, J., Xie, P., Liu, Y., & Ma, J. (2016). Urban thermal environment dynamics and associated landscape pattern factors: A case study in the Beijing metropolitan region. *Remote sensing of environment, 173*, 145–155.

Popkin, G. (2018). US government considers charging for popular Earth-observing data. *Nature, 556*(7700), 417–419.

Priyadarsini, R. (2012). Urban heat island and its impact on building energy consumption. *Advances in building energy research* (pp. 267–276): Routledge.

Qiao, Z., Tian, G., & Xiao, L. (2013). Diurnal and seasonal impacts of urbanization on the urban thermal environment: A case study of Beijing using MODIS data. *ISPRS Journal of Photogrammetry and Remote Sensing, 85*, 93–101.

Quanliang, C., Changjian, N., Zhan, L., & Jingxuan, R. (2009). *Urban heat island effect research in Chengdu city based on MODIS data*. Paper presented at the 2009 *3rd International Conference on Bioinformatics and Biomedical Engineering*.

Roy, M. (2009). Planning for sustainable urbanisation in fast growing cities: Mitigation and adaptation issues addressed in Dhaka, Bangladesh. *Habitat international, 33*(3), 276–286.

Schwarz, N., Schlink, U., Franck, U., & Großmann, K. (2012). Relationship of land surface and air temperatures and its implications for quantifying urban heat island indicators—An application for the city of Leipzig (Germany). *Ecological Indicators, 18*, 693–704.

Shatnawi, N., & Abu Qdais, H. (2019). Mapping urban land surface temperature using remote sensing techniques and artificial neural network modelling. *International Journal of Remote Sensing, 40*(10), 1–16.

Singh, R. and Kalota, D. (2019) Urban sprawl and its impact on generation of urban heat island: A case study of Ludhiana city. *Journal of the Indian Society of Remote Sensing*, 47 (9): 1567–1576. https://doi.org/10.1007/s12524-019-00994-8

Singh, P., Kikon, N., & Verma, P. (2017). Impact of land use change and urbanization on urban heat island in Lucknow city, Central India. A remote sensing based estimate. *Sustainable Cities and Society*, *32*, 100–114.

Snyder, W. C., Wan, Z., Zhang, Y., & Feng, Y. Z. (1998). Classification-based emissivity for land surface temperature measurement from space. *International Journal of Remote Sensing*, *19*(14), 2753–2774.

Speth, J. G. (2005). *Global Environmental Challenges: Transitions to a Sustainable World.* Orient Blackswan.

Stathopoulou, M., Synnefa, A., Cartalis, C., Santamouris, M., Karlessi, T., & Akbari, H. (2009). A surface heat island study of Athens using high-resolution satellite imagery and measurements of the optical and thermal properties of commonly used building and paving materials. *International Journal of Sustainable Energy*, *28*(1–3), 59–76.

Streutker, D. R. (2002). A remote sensing study of the urban heat island of Houston, Texas. *International Journal of Remote Sensing*, *23*(13), 2595–2608.

Streutker, D. R. (2003). Satellite-measured growth of the urban heat island of Houston, Texas. *Remote sensing of Environment*, *85*(3), 282–289.

Surawar, M., & Kotharkar, R. (2017). Assessment of Urban Heat Island through Remote Sensing in Nagpur Urban Area Using Landsat 7 ETM+ Satellite Images. *International Journal of Urban and Civil Engineering*, *11*(7), 868–874.

Thi Van, T., & Duong Xuan Bao, H. A. (2010). Study of the impact of urban development on surface temperature using remote sensing in Ho Chi Minh City, Southern Vietnam. *Geographical Research*, *48*(1), 86–96.

Veronez, M. R., Thum, A. B., Luz, A. S., & Da Silva, D. (2006). *Artificial neural networks applied in the determination of soil surface temperature–SST.* Paper presented at the *I*nternational Symposium Accuracy Assessment in Natural Resources and Environmental Sciences, (Accuracy 2006).

Wang, X., Cheng, H., Xi, J., Yang, G., & Zhao, Y. (2018). Relationship between park composition, vegetation characteristics and cool island effect. *Sustainability*, *10*(3), 587.

Weng, Q. (2009). Thermal infrared remote sensing for urban climate and environmental studies: Methods, applications, and trends. *ISPRS Journal of Photogrammetry and Remote Sensing*, *64*(4), 335–344.

Weng, Q., & Larson, R. C. (2005). Satellite remote sensing of urban heat islands: current practice and prospects. *Geo-spatial technologies in urban environments* (pp. 91–111): Springer.

Weng, Q., Lu, D., & Schubring, J. (2004). Estimation of land surface temperature–vegetation abundance relationship for urban heat island studies. *Remote sensing of Environment*, *89*(4), 467–483.

Wulder, M. A., White, J. C., Loveland, T. R., Woodcock, C. E., Belward, A. S., Cohen, W. B., … Roy, D. P. (2016). The global Landsat archive: Status, consolidation, and direction. *Remote sensing of environment*, *185*, 271–283.

Xiong, Y., Huang, S., Chen, F., Ye, H., Wang, C., & Zhu, C. (2012). The impacts of rapid urbanization on the thermal environment: A remote sensing study of Guangzhou, South China. *Remote Sensing*, *4*(7), 2033–2056.

Xu, H., Chen, Y., Dan, S., & Qiu, W. (2011). Dynamical monitoring and evaluation methods to urban heat island effects based on RS&GIS. *Procedia Environmental Sciences*, *10*, 1228–1237.

Yang, C., He, X., Yu, L., Yang, J., Yan, F., Bu, K., … Zhang, S. (2017). The cooling effect of urban parks and its monthly variations in a snow climate city. *Remote Sensing*, *9*(10), 1066.

Young, N. E., Anderson, R. S., Chignell, S. M., Vorster, A. G., Lawrence, R., & Evangelista, P. H. (2017). A survival guide to Landsat preprocessing. *Ecology*, *98*(4), 920–932.

Yu, Z., Guo, X., Zeng, Y., Koga, M., & Vejre, H. (2018). Variations in land surface temperature and cooling efficiency of green space in rapid urbanization: The case of Fuzhou city, China. *Urban forestry & urban greening*, *29*, 113–121.

Yue, W., Liu, Y., Fan, P., Ye, X., & Wu, C. (2012). Assessing spatial pattern of urban thermal environment in Shanghai, China. *Stochastic environmental research and risk assessment*, *26*(7), 899–911.

Zha, Y., Gao, J., & Ni, S. (2003). Use of normalized difference built-up index in automatically mapping urban areas from TM imagery. *International Journal of Remote Sensing*, *24*(3), 583–594.

Zhang, Y., Murray, A. T., & Turner Ii, B. L. (2017). Optimizing green space locations to reduce daytime and nighttime urban heat island effects in Phoenix, Arizona. *Landscape and Urban Planning*, *165*, 162–171.

Zhou, W., Qian, Y., Li, X., Li, W., & Han, L. (2014). Relationships between land cover and the surface urban heat island: seasonal variability and effects of spatial and thematic resolution of land cover data on predicting land surface temperatures. *Landscape ecology*, *29*(1), 153–167.

Zhou, D., Xiao, J., Bonafoni, S., Berger, C., Deilami, K., Zhou, Y., … Sobrino, J. A. (2019). Satellite remote sensing of surface urban heat islands: progress, challenges, and perspectives. *Remote Sensing*, *11*(1), 48.

Part II

Geospatial Technology for Disaster Management

5 Spatio-Environmental Distribution of Drought Disaster Events

A Space-Based Approach Using Terra-MODIS Vegetation Index

Israel R. Orimoloye, Olusola O. Ololade, and Johannes A. Belle
University of the Free State, Bloemfontein, South Africa

CONTENTS

5.1 INTRODUCTION

Following the extreme global warming, the impacts of drought, a devastating natural disaster, has increased. Therefore, it is crucial to evaluate the spatio-environmental evolution and patterns of drought distribution. Drought, generally referred to as the water supply deficit (Wang et al., 2019), and this has serious environmental impacts and can be difficult to determine the extent, length and end of drought events. The phenomenon is a major factor of economic and social problems and causes damage to the environment and the ecosystems (Hao and Singh 2015; AghaKouchak et al., 2015). Climate-related disasters are responsible for 70 per cent of all natural disasters, and drought causes 50 per cent of these climate disasters (Wang et al., 2019). Continuous global warming has accelerated the global hydrological cycle and caused

DOI: 10.1201/9781003224624-8

extreme events such as drought, extreme heat and flooding (Huang et al. 2017; Wang et al., 2019). Although drought is slowly normalizing globally as seen by the continuing occurrence, frequency and severity of severe drought events, the effects of drought are becoming increasingly prominent and destructive (Field et al., 2012).

Drought disaster is a complicated and critical environmental problem. Several studies have shown that the mode of precipitation and changes in temperature are some of the factors that contribute to the severity and frequency of drought (Sinha et al., 2019). Many scientists have therefore researched drought, in particular as regards its monitoring and assessment (Kogan et al., 2017; Chen et al., 2017), spatial-temporal variability (Zeleke et al., 2017; Orimoloye et al., 2019), and prediction (AghaKouchak, 2015; Jiménez-Donaire et al., 2020). Drought indices can be estimated or measured from space-based or remote sensing parameters which can be determined by spatial size, length, frequency, and occurrence intensity (Krishnamurthy et al., 2020). Scientists have used various remote sensing data and related drought indices to measure the severity of the drought and enhance the management of drought disasters. Indices such as the Palmer severe drought index (Palmer, 1965), crop moisture index (Palmer 1968), standardized precipitation index (McKee et al., 1993), nonparametric multivariate standardized drought index (Huang et al., 2016), multivariate integrated drought index (Chang et al., 2016) and deficit soil moisture index were proposed. However, some of these indices have some drawbacks, including a fixed time scale, poor performance in events of short-term drought and lack of description of physical phenomena in spatial mode (Alley, 1984).

Space-based or remote sensing data have been employed to assess and evaluate drought disasters by various researchers globally (Swain et al., 2011; Abdi et al., 2019; Li and Wang, 2019; Singh, 2021). Remote sensing satellite systems based on providing spatially consistent and temporally repeated views of the surface of the earth, this is critical for monitoring vegetation dynamics over wide space both in short- and long-terms. In the last few decades, remotely sensed data have been used in thermal and climatic land surface studies including drought monitoring (Ji and Peters, 2003; Bayarjargal et al., 2006). Several satellite-derived drought conditions for vegetation indices (VIs) were developed to monitor vegetation conditions by quantifying the foliage density and canopy properties within the sensors field of view. VIs are useful in the assessment of active photosynthesis and transpiring vegetation and are therefore correlated with composite canopy properties such as leaf area, plant cover fraction and total chlorophyll content (Glenn et al., 2011).

Drought-based VIs have often been integrated with land surface thermal characteristics to enhance their effectiveness in detecting, quantifying and tracking the stress of drought on the vegetation. While the normalized vegetation difference index (NDVI) provides an estimate of the amount of vegetation present in a pixel, land surface temperature provides information on the moisture status of the soil. Studies however suggest that NDVI has a delayed response to deficiencies in precipitation or soil moisture, which means that the initial deficiency in soil moisture is not apparent in the NDVI signals. Consequently, this study will utilize a drought index (vegetation index) using Terra Moderate Resolution Imaging Spectroradiometer (MODIS) to assess drought episodes in the study area. MODIS data have been used for drought-related assessments over the years, for instance, agricultural drought (Patel et al.,

2012), hydrological drought (Abdi et al., 2019), meteorological drought (Li and Wang, 2019). MODIS vegetation indices, produced at intervals of 16 days and with multiple spatial resolutions, provide consistent spatial and temporal comparisons of vegetation canopy greeness, a composite property of the leaf area, chlorophyll and canopy structure. Two vegetation indices are derived from the atmospherically accurate reflection in the red, near-infrared and blue wavebands; the NDVI which provides consistency with NOAA's AVHRR NDVI time series record for historical and climatic applications; and the enhanced vegetation index (EVI), which minimizes canopy-soil differences and increases its sensitivity. MODIS sensors on board Terra and Aqua satellites are similar; the VI algorithm generates eight days apart (phased products) of each 16-day composite so that both data records can be combined to create a higher temporal resolution. The MODIS VI product suite is now widely used in all biodiversity, climate, and natural resource management and operational science studies, as demonstrated by the ever-increasing body of peer-reviewed publications (Patel et al., 2012; Abdi et al., 2019; Li & Wang, 2019).

The purpose of this research is to investigate the spatial-temporal distribution of drought events and to ascertain whether a significant antecedent relationship exists between the drought index and other environmental factors over the three years in the Free State Province of South Africa for 2016, 2017 and 2018. South Africa is part of a continental climate zone that is experiencing extreme climatic variability, including the study area. Social economy, food security, ecology and the environment are very much dependent on climate conditions (Tirado et al., 2010). The area of study faces numerous threats from extreme and regular drought events, and the combined effects of cold semi-arid and humid sub-tropical with dry winter climates have a major impact on the region. Global warming causes continuous changes in these climates, thereby influencing the spatial variations of the droughts and the droughts trend toward development. Thus, the assessment of the spatio-environmental characteristics of droughts is of great importance in water resource management, particularly in the area of research.

5.2 STUDY LOCATION AND METHODS

5.2.1 Study Location

The research area lies within the South African's Free State Province (Figure 5.1).

> The province is located between the equator's 26.6° South and 30.7° South latitudes, and between the Greenwich meridian's 24.3° East and 29.8° East longitude. The climate of the province is mainly semi-arid, except for the eastern and northeastern parts where, according to the Köppen climatic classification, the humid-subtropical climate is observed. The average annual rainfall ranges from 300 to more than 900 mm, with more than 70 per cent of the rainfall falling in September through April. The provincial topography is varied, with altitudes in the southern part < 1,200 m and in the eastern Free State > 1,800 m.
>
> (Moeletsi and Walker, 2012)

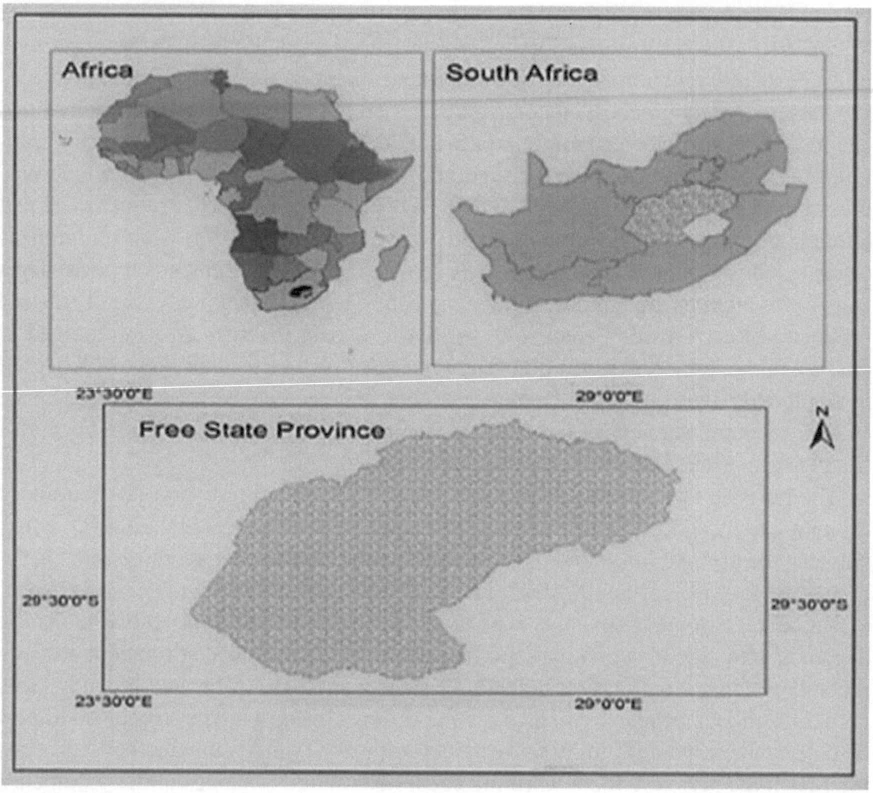

FIGURE 5.1 Study location

5.2.2 Data and Methods

Terra-MODIS 16-days composite EVI (MOD13Q1, collection 6) products for 2016, 2017 and 2018 in the Free State Province, South Africa (AREA_SQKM:129825.15644) were used in this study. The

> 16-day period provides a dense temporal coverage to monitor sufficiently detailed distributions of drought, eliminating short-term (daily) fluctuations and representing a period allowing vegetation to acclimatize to the surrounding surface air temperature and soil moisture conditions. MODIS vegetation indices (MOD13Q1) Version 6 data are generated as a Level 3 product every 16 days at a spatial resolution of 250 metres (m). The product MOD13Q1 features two primary layers of vegetation. The first is the National Oceanic and Atmospheric Administration-Advanced Very High-Resolution Radiometer (NOAA-AVHRR) derived NDVI, NDVI, which is referred to as the continuity index. The second layer of vegetation is the EVI, which has enhanced sensitivity to regions with high biomass. The algorithm selects the best available pixel value from the 16 days of all acquisitions.

The parameters used in the study are characterized by low clouds, low visual angle and maximum NDVI/EVI.

The VCI is calculated using R programming as shown below,

$$VCI_{\text{ijk}} = \frac{VI_{\text{ijk}} - VI_{\text{i, min}}}{VI_{\text{i, max}} - VI_{\text{i, min}}} * 100$$

where

> VCI_{ijk} is the VCI value for the pixel $_i$ during week/month/day of the years (DOY_j) for year $_k$, VI_{ijk} is the weekly/monthly/DOYs VI value for pixel $_i$ in week/month/DOY $_j$ for year $_k$ whereby both the NDVI or EVI can be utilized as VI, VI_i, min and $VI_{i,max}$ are the multiyear minimum and maximum VI, respectively, for pixel $_i$.

The UN-SPIDER recommended index of drought conditions index was used in this study. The drought values and categories are; 90 to 100 per cent (no drought), 80 to 90 per cent (no drought), 70 to 80 per cent (no drought), 60 to 70 per cent (no drought), 50 to 60 per cent (no drought), 40 to 50 per cent (no drought), 30 to 40 per cent (light drought), drought conditions between 20 and 30 per cent (moderate drought), 10 and 20 per cent (severe drought), 0 and 10 per cent (extreme drought).

5.3 RESULTS AND DISCUSSION

Figures 5.2–5.4 show the monthly space-environmental distribution and drought patterns in the field of study. The spatial distribution of monthly drought conditions and patterns dependent on VCI in the 2016 study region are illustrated in Figure 5.2. Dry episodes vary from month to month as shown in Figure 5.2. It was found in the study that 2016 is the year in which drought occurs in nearly all the regions, the driving conditions rise from northwest to southeast and the conditions of maximum drought in the southeast in January and February which lie within the summer months.

Findings from this study were obtained from the space-based information for drought characteristics assessment over the study area. Regions with high drought conditions index are identified as vulnerable environment or ecosystems. The proportion of drought-affected area with significant high impact trends was the area with 30 per cent drought conditions or below. There exist extreme drought episodes of less than 20 per cent drought condition in these months as presented in Figure 5.2. The potential spatio-environmental impacts due to drought conditions might have triggered more various environmental imbalances in the study area from water dearth, loss agricultural outputs, vegetation degradation, ecosystems extinction among others (John et al., 2019; Jordaan et al., 2019). Overgrazing perhaps another major environmental indicator of drought in the region under the survey, as described in the previous research (Jordaan et al., 2019; Vetter et al., 2020), is the lack of pasture for growth and the failure to match animal pasture to forage growth and development. Overgrazing happens as a result of over populated with many cattle or due to drought events in most cases not adequately managed grazing activities. This decreased soil

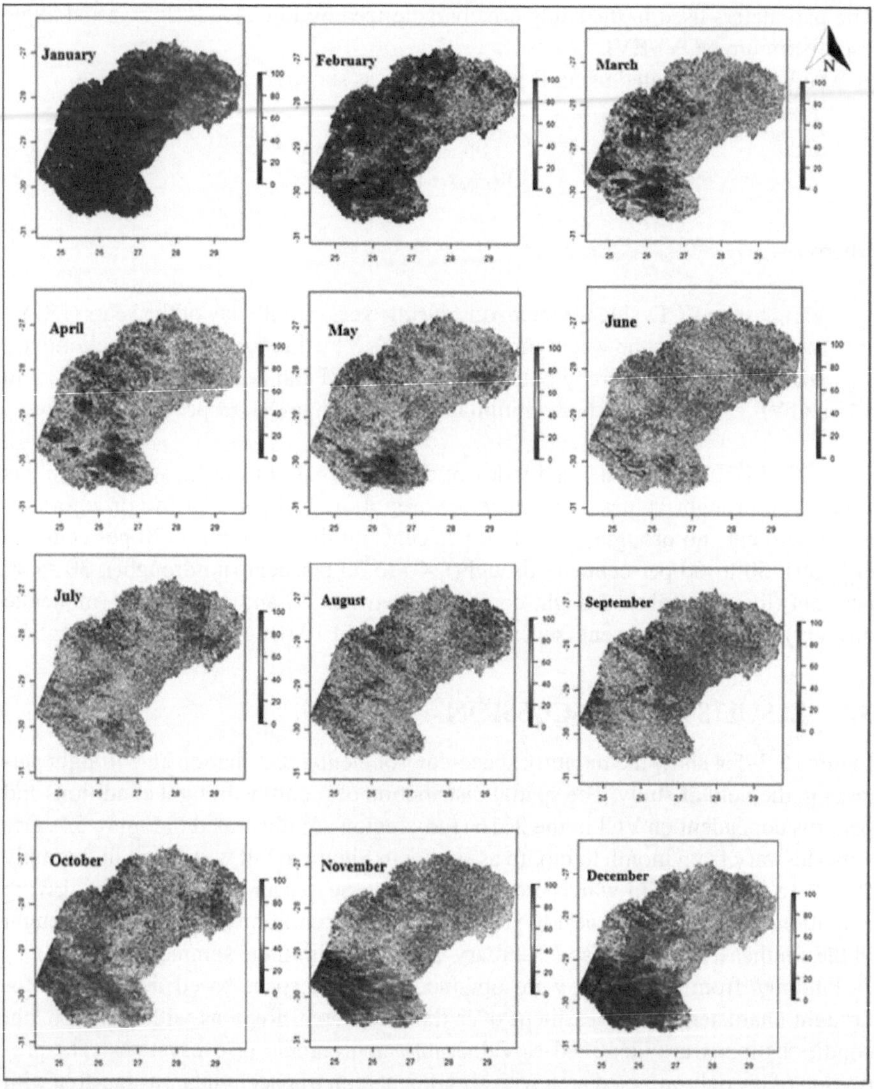

FIGURE 5.2 Spatiotemporal distribution of drought events in year 2016 (January to December)

vegetation involves and raises crushing chances during rainy periods. The crusting conditions limited water absorption and prolonged the plant recovery from the preceding or ongoing droughts (Frischen et al., 2020; Vetter et al., 2020).

The study further reveals more drought spatial distribution in year 2016 where March, October, November and December witnessed light drought to moderate drought condtions during the same period though the distribution varies from one place to another. The drought impacts were more felt in the south partiicularly in the southwestern region though not as severe as drought events in January and February of the same year. The drought condition in these four months ranges from no drought to severe drought and this condition is not spatially concordant, it is more evident in some areas than the other. More so, the study area experienced drought episodes in April through September with almost the same drought condition throughout the region in the same year.

The drought condition witnessed in these months ranges from no drought to moderate situtation during the period. It was identified that these months are winter, autumn and spring months where drought is less felt in region. The absence or light drought conditions in these months may not be directly connected to the season as other factors may be responsible for the situation for these months. Studies have however shown that climate conditions also cause extreme drought stress during winter, which can also be hypothesized to cause cavitation in the cover of vegetation (Mayr et al., 2002; Schwinning et al., 2005).

Furthermore, the information in Figure 5.3 shows the spatio-environmental distribution patterns of drought disasters in the study area for year 2017. Another spatial trend scenario of drought condition was identified in the study area where some months and regions were more affected than the other based on the drought index evaluated in this study. January, October, November and December witnessed more drought impact in year 2017, ranges from moderate drought to extreme drought conditions which were more felt in the southern region. Most of the regions witnessed drought conditions between 0 and 20 per cent (severe and extreme drought conditions) in these four months. The drought conditions in these months might have affected some of the environmental components in the area. For example, severe drought events have been reported in conjunction with decreased crop, agricultural land and forest production, increased fire risk, reduced water levels, increased mortality rates for animal and animal life and damages to wildlife and fish habitats (Nakagawa et al., 2000; Van Loon et al., 2014; Castro et al., 2020).

Drought conditions in the region can also affect many households and the climate in a devastating form. The threat depends in particular on the intensity and length of drought events in areas with a high drought index of less than 30 per cent and on the duration of drought disasters in the region (Ding and McCarl 2020; Hameed et al. 2020). It may have more effects on disadvantaged communities than more affluent societies with greater support from other resources. Nonetheless, drought in any type of society can be extremely serious (Cowell, 1984). The results also show that in February there is no drought except in the Northeast part of the province, where the same year there was moderate drought. March through September experienced mild drought to moderate drought conditions in the region in year 2017. The situation of drought in the area is different from location to location, as presented in Figure 5.3, which is spatially spread over the area. The spatial patterns of the dryness will differ as some areas are affected more than others. For instance, areas with no or light drought conditions will be less affected than the area with high moderate and severe drought conditions. More so, environmental components in the area will be spatially affected based on the level of severity of drought conditions in the affected environment (Jung & Chang, 2012; Porensky et al., 2013). Agricultural sectors or water-dependent sectors amy also be impacted, for instance, crops are affected whenever there are drought events esepcially in the area that received less water in volume.

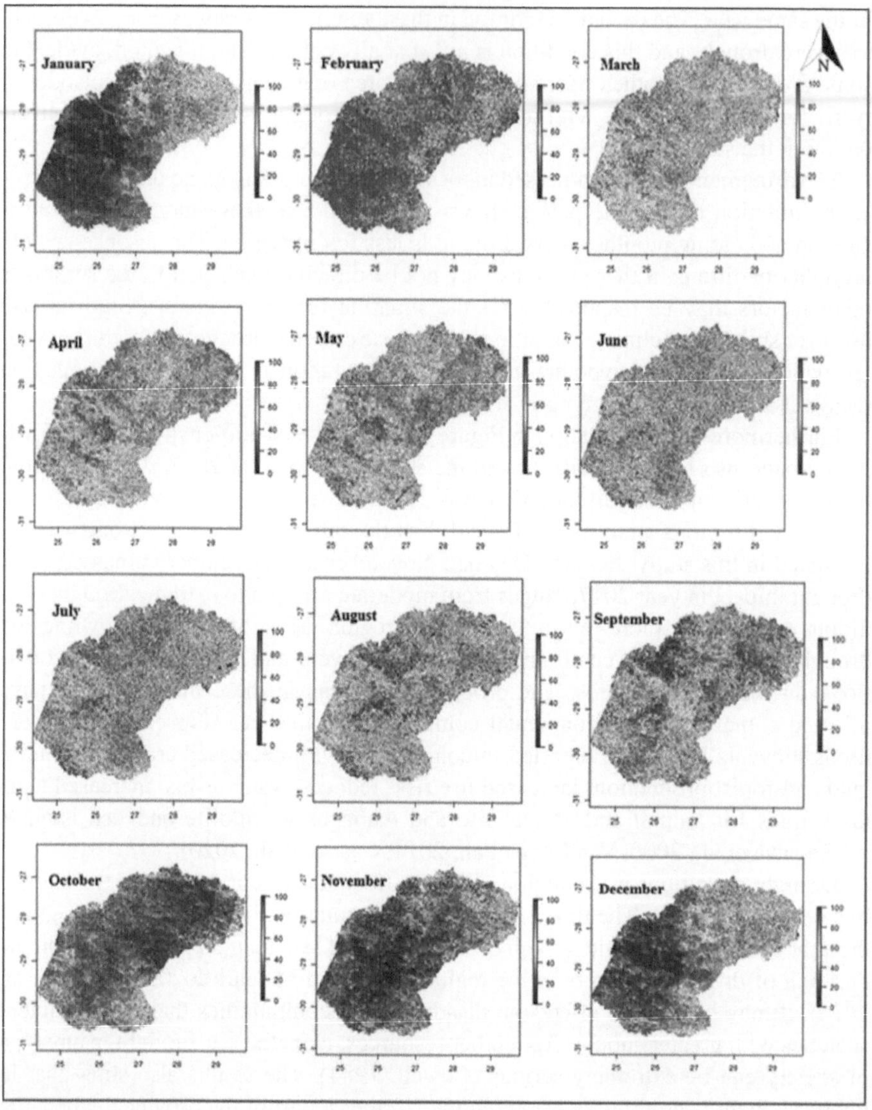

FIGURE 5.3 Spatiotemporal distribution of drought events in the year 2017 (January to December)

Farmers may not be able to irrigate using groundwater or surface waters such as agricultural ponds, especially in areas with moderate or serious drought (Peters et al., 2006; Maurer and Menzel, 2019; Wossenyeleh et al., 2019). When the drought happens, however, any other approaches or ways to alleviate drought might be lacking. Due to the drought conditions and potential water limits, farmers can not manage their fields in the extreme drought. Some farmers and communities may relovcate to other places that are less impacted by a drought condition to live (Jülich, 2011; Rain et al., 2011).

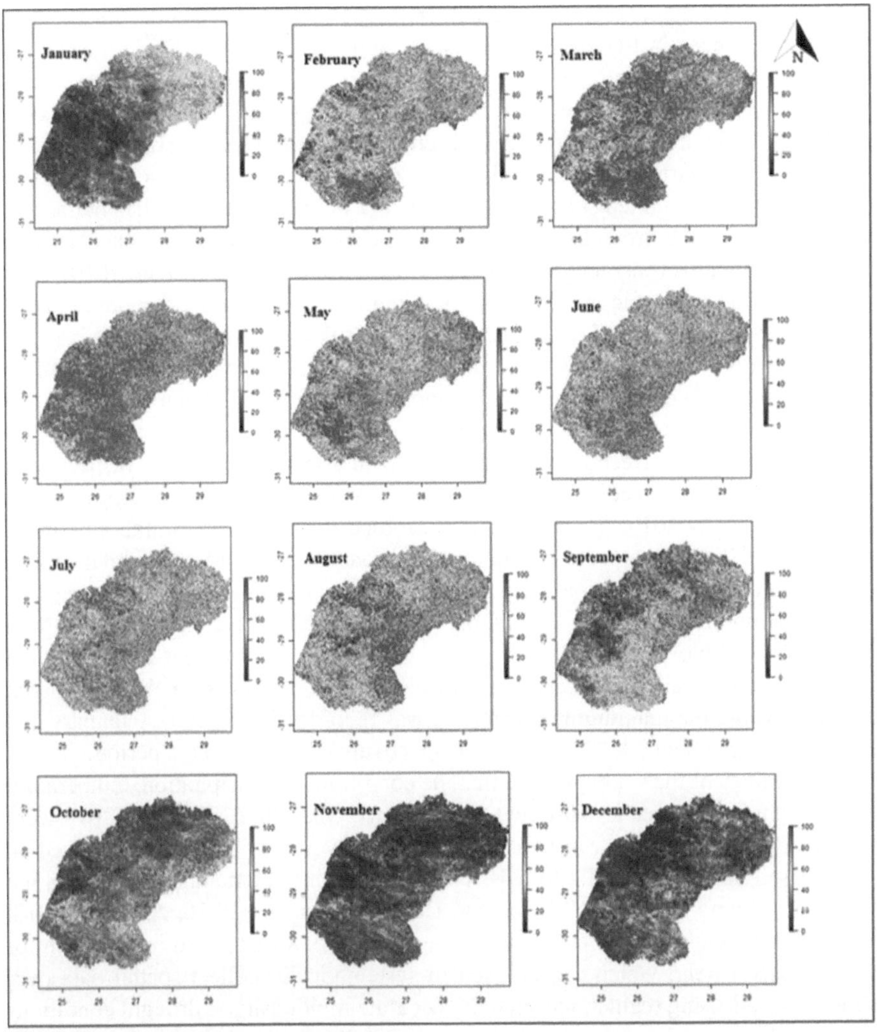

FIGURE 5.4 Spatiotemporal distribution of drought events in year 2018 (January to December)

The information in Figure 5.4 presents the spatial patterns of drought events for year 2018 over the study area. The analysis from the study reveals that year 2018 witnessed severe to extreme drought conditions in most of the area. The drought values and categories in January, October, November and December were less than 30 per cent (moderate drought), less than 20 per cent (severe drought) and between 10 and 0 per cent (extreme drought) in most of the region. These months were seriously affected by drought events as shown in Figure 5.4. The result further reveals that September witnessed light to moderate drought conditions in the same year range from 30 to 40 per cent and 20 to 30 per cent, respectively in the same year. No drought to light drought conditions were experienced in March through August in the study area. Drought conditions in year 2018 might have contributed to the environmental imbalances in the area especially in January, October to December which witnessed severe to extreme drought conditions. Relatively small storage reservoirs in the area could be vulnerable to the drought disaster especially in the most affected months of below-average rainfall or high drought conditions, triggering an agricultural drought scenario (Marengo et al., 2019; Zuo et al., 2019). More so, the extreme drought of 2018 exhibited substantial impacts not only on water supplies but across a broad range of the province's economic, social and leisure activities especially in the affected months (Farzanegan et al., 2019; Ribeiro, 2019; Zhao et al., 2019).

The information in Figure 5.5. presents rainfall and maximum temperature distribution for year 2016, 2017 and 2018. It was identified that year 2016 received the highest rainfall with about 54 mm in January while November and December received 46 and 39 mm, respectively. The months of August and September recorded the lowest rainfall amount of about 7 and 2 mm in the same year. Maximum temperature was recorded in December and January with about 34 and 33°C respectively, while June and July received the least maximum temperature with about 16 and 18°C, respectively 2016. August and July recorded the minimum rainfall amount of about 0.3 and 0.7 mm respectively, more so, February and January recorded the minimum rainfall amount of about 138 and 93 mm, respectively. The study area witnessed maximum temperature in the months of December with about 32°C, followed by January and November with about 30°C each in year 2017. Year 2018 recorded maximum rainfall amount of about 2 mm the months of January, February, March and April while the maximum temperature was recorded in December, January and November with about 34, 32 and 31°C, respectively during the same period.

The findings of the study indicate that the conditions of precipitation, temperature and drought in this area are different. For instance, year 2016, 2017 and 2018 were affected by drought disaster in January, October, November and December based on space-based data (Figures 5.2 to 5.4). The highest rainfall amounts recorded in the study were found in January, November, December for year 2016, 2017 and 2018. This development might be attributed to the potency of space-based information for drought assessment, which can be used to systematically collect spatial data over wide areas, allowing regular and spatially accurate monitoring of drought conditions and other land surface features (Abuzar et al., 2019; West et al., 2019; Orimoloye et al., 2019). However, the amount of rainfall and temperature received play an important controlling factor on how much water is available in the soil to sustain the environment and water-reliant sectors (Fiwa et al., 2014; Ayanlade et al., 2018).

5.4 CONCLUSION

Monitoring the spatio-environmental distribution, patterns of evolution and potential impacts of drought will help in managing the implications of drought on the environment, which is of critical importance in drought disaster management. Over the three-year cycles (2016, 2017 and 2018), the improvements in the sequence of drought

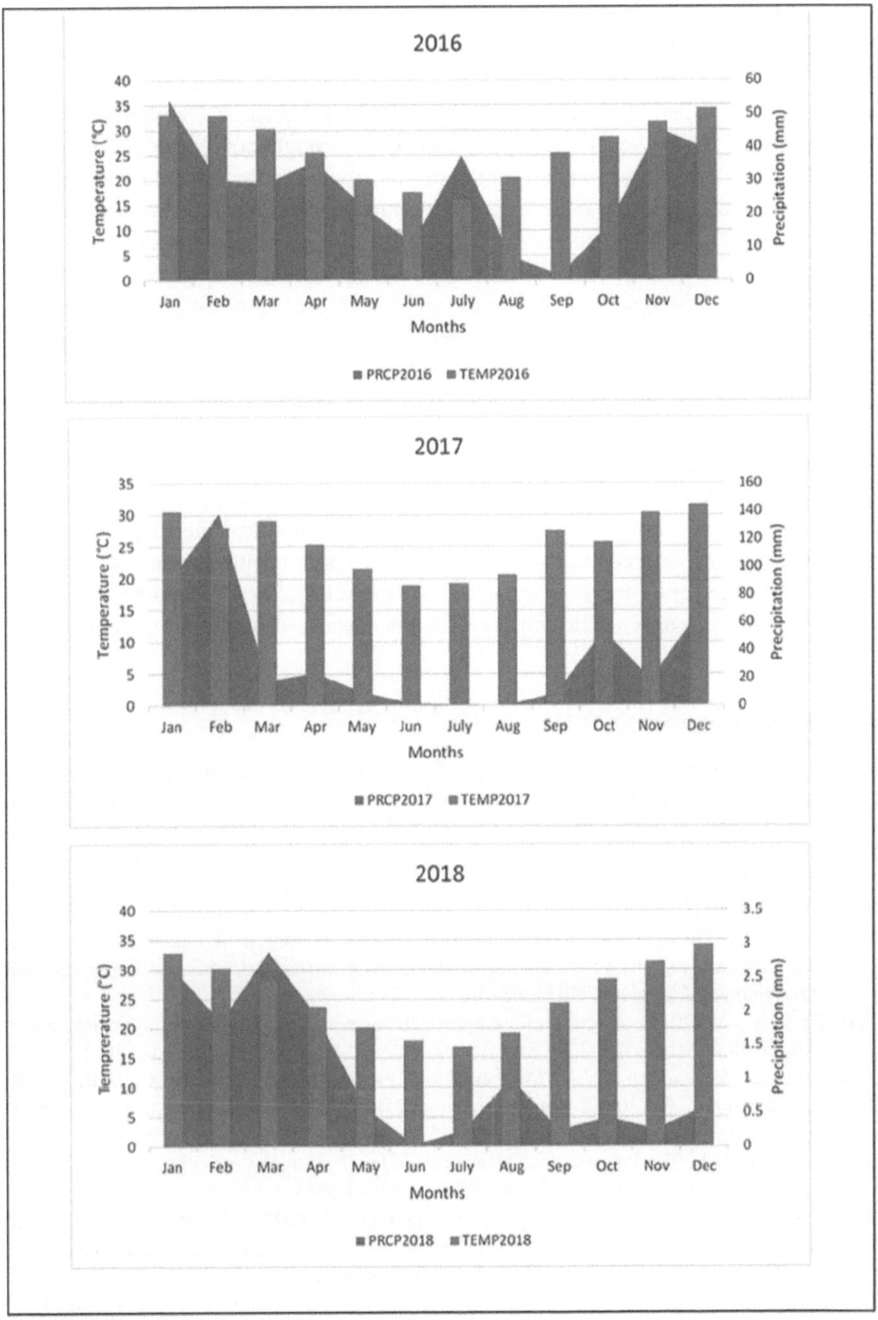

FIGURE 5.5 Rainfall and temperature trends for 2016, 2017 and 2018

indices (VCI), rainfall and temperatures in the region analysed were evaluated for the study. R programming and rainfall and environmental components were used for evaluating the distribution characteristics of the drought disaster. It was noted that the months most affected by drought disaster are January, February, October, November and December for the years 2016, 2017 and 2018. Drought conditions and its potential impacts in these three years might have contributed to the environmental imbalances in the area especially in the affected months which witnessed severe to extreme drought conditions during the period. In the time series of drought conditions, the major variations were essentially consistent and were especially well-converged in those months. Findings from this study provide information hotspots for environmental and ecosystem conservation under drought-related events and shift toward ecologically based environmental and disaster management.

More so, the outcomes from this study play vital roles in identifying the areas that are vulnerable to drought and to improve resilience by governments and relevant policymaking bodies. The analysis that is carried out in the study will help to prepare both for possible droughts and climate change, such as water conservation, water quality across the entire research area, water infrastructures, the identification of alternative water sources, drought preparation and drought-resistant planting. Additional steps to improve resiliency and energy efficiency in buildings to other environmental stressors would improve the resistance to drought in mutual benefit.

REFERENCES

Abdi, O., Shirvani, Z., & Buchroithner, M. F. (2019). Forest drought-induced diversity of Hyrcanian individual-tree mortality affected by meteorological and hydrological droughts by analyzing moderate resolution imaging spectroradiometer products and spatial autoregressive models over northeast Iran. *Agricultural and Forest Meteorology*, 275, 265–276.

Abuzar, M. K., Shafiq, M., Mahmood, S. A., Irfan, M., Khalil, T., Khubaib, N., ... & Shaista, S. (2019). Drought risk assessment in the Khushab region of Pakistan using satellite remote sensing and geospatial methods. *International Journal of Economic and Environmental Geology*, 10(1), 48–56.

AghaKouchak, A. (2015). A multivariate approach for persistence-based drought prediction: Application to the 2010–2011 East Africa drought. *Journal of Hydrology*, 526, 127–135.

AghaKouchak, A., Farahmand, A., Melton, F. S., Teixeira, J., Anderson, M. C., Wardlow, B. D., & Hain, C. R. (2015). Remote sensing of drought: Progress, challenges and opportunities. *Reviews of Geophysics*, 53(2), 452–480.

Alley, W. M. (1984). The Palmer drought severity index: limitations and assumptions. *Journal of climate and applied meteorology*, 23(7), 1100–1109.

Ayanlade, A., Radeny, M., Morton, J. F., & Muchaba, T. (2018). Rainfall variability and drought characteristics in two agro-climatic zones: An assessment of climate change challenges in Africa. *Science of the Total Environment*, 630, 728–737.

Bayarjargal, Y., Karnieli, A., Bayasgalan, M., Khudulmur, S., Gandush, C., & Tucker, C. J. (2006). A comparative study of NOAA-AVHRR derived drought indices using change vector analysis. *Remote Sensing of Environment*, 105(1), 9–22.

Castro, S. P., Esch, E. H., Eviner, V. T., Cleland, E. E., & Lipson, D. A. (2020). Exotic herbaceous species interact with severe drought to alter soil N cycling in a semi-arid shrubland. *Geoderma*, 361, 114111.

Chang, J., Li, Y., Wang, Y., & Yuan, M. (2016). Copula-based drought risk assessment combined with an integrated index in the Wei River Basin, China. *Journal of Hydrology*, 540, 824–834.

Chen, C. F., Son, N. T., Chen, C. R., Chiang, S. H., Chang, L. Y., & Valdez, M. (2017). Drought monitoring in cultivated areas of Central America using multi-temporal MODIS data. *Geomatics, Natural Hazards and Risk*, 8(2), 402–417.

Cowell, A. 1984. *No Respecter of Apartheid, Drought Scorches All* (https://www.nytimes.com/1984/03/16/world/no-respecter-of-apartheid-drought-scorches-all.html)

Ding, J., & McCarl, B. A. (2020). Economic and ecological impacts of increased drought frequency in the Edwards aquifer. *Climate*, 8(1), 2.

Farzanegan, M. R., Feizi, M., & Fereidouni, H. G. (2019). Drought and property prices: Empirical evidence from Iran (No. 16-2019). Joint Discussion Paper Series in Economics.

Field, C. B., Barros, V., Stocker, T. F., & Dahe, Q. (Eds.). (2012). *Managing the risks of extreme events and disasters to advance climate change adaptation: special report of the intergovernmental panel on climate change*. Cambridge University Press, Cambridge, UK.

Fiwa, L., Vanuytrecht, E., Wiyo, K. A., & Raes, D. (2014). Effect of rainfall variability on the length of the crop growing period over the past three decades in central Malawi. *Climate Research*, 62(1), 45–58.

Frischen, J., Meza, I., Rupp, D., Wietler, K., & Hagenlocher, M. (2020). Drought risk to agricultural systems in Zimbabwe: A spatial analysis of hazard, exposure, and vulnerability. *Sustainability*, 12(3), 752.

Glenn, E. P., Neale, C. M., Hunsaker, D. J., & Nagler, P. L. (2011). Vegetation index-based crop coefficients to estimate evapotranspiration by remote sensing in agricultural and natural ecosystems. *Hydrological Processes*, 25(26), 4050–4062.

Hameed, M., Ahmadalipour, A., & Moradkhani, H. (2020). Drought and food security in the middle east: An analytical framework. *Agricultural and Forest Meteorology*, 281, 107816.

Hao, Z., & Singh, V. P. (2015). Drought characterization from a multivariate perspective: A review. *Journal of Hydrology*, 527, 668–678.

Huang, S., Huang, Q., Leng, G., & Liu, S. (2016). A nonparametric multivariate standardized drought index for characterizing socioeconomic drought: A case study in the Heihe River Basin. *Journal of Hydrology*, 542, 875–883.

Huang, S., Leng, G., Huang, Q., Xie, Y., Liu, S., Meng, E., & Li, P. (2017). The asymmetric impact of global warming on US drought types and distributions in a large ensemble of 97 hydro-climatic simulations. *Scientific reports*, 7(1), 1–12.

Ji, L., & Peters, A. J. (2003). Assessing vegetation response to drought in the northern Great Plains using vegetation and drought indices. *Remote Sensing of Environment*, 87(1), 85–98.

Jiménez-Donaire, M. D. P., Tarquis, A., & Giráldez, J. V. (2020). Evaluation of a combined drought indicator and its potential for agricultural drought prediction in southern Spain. *Natural Hazards & Earth System Sciences*, 20(1), 21–33.

John, F., Toth, R., Frank, K., Groeneveld, J., & Müller, B. (2019). Ecological vulnerability through insurance? Potential unintended consequences of livestock drought insurance. *Ecological economics*, 157, 357–368.

Jordaan, A., Bahta, Y. T., & Phatudi-Mphahlele, B. (2019). Ecological vulnerability indicators to drought: Case of communal farmers in Eastern Cape, South Africa. *Jàmbá: Journal of Disaster Risk Studies*, 11(1), 1–11.

Jülich, S. (2011). Drought triggered temporary migration in an East Indian village. *International Migration*, 49, e189–e199.

Jung, I. W., & Chang, H. (2012). Climate change impacts on spatial patterns in drought risk in the Willamette River Basin, Oregon, USA. *Theoretical and Applied Climatology*, 108(3–4), 355–371.

Kogan, F., Guo, W., & Yang, W. (2017). SNPP/VIIRS vegetation health to assess 500 California drought. *Geomatics, Natural Hazards and Risk*, 8(2), 1383–1395.

Krishnamurthy R. P. K., Fisher, J. B., Schimel, D. S., & Kareiva, P. M. (2020). Applying tipping point theory to remote sensing science to improve early warning drought signals for food security. *Earth's Future*, 8(3), e2019EF001456.

Li, W., & Wang, Y. (2019, December). Long-term (2003–2017) Trends of Vegetation Condition Index (VCI) in Guangdong Using Modis Data and Implications for Drought Assessment. In 2019 Photonics & Electromagnetics Research Symposium-Fall (PIERS-Fall) (pp. 1944–1950). IEEE.

Marengo, J. A., Cunha, A. P., Soares, W. R., Torres, R. R., Alves, L. M., de Barros Brito, S. S., … & Magalhaes, A. R. (2019). Increase risk of drought in the semiarid lands of Northeast Brazil due to regional warming above 4 C. In *Climate change risks in Brazil* (pp. 181–200). Springer, Cham.

Maurer, V., & Menzel, L. (2019, January). The impact of groundwater recharge on the propagation of groundwater droughts in southwestern Germany. In Geophysical Research Abstracts (Vol. 21).

Mayr, S., Wolfschwenger, M., & Bauer, H. (2002). Winter-drought induced embolism in Norway spruce (Picea abies) at the Alpine timberline. *Physiologia Plantarum*, 115(1), 74–80.

McKee, T. B., Doesken, N. J., & Kleist, J. (1993, January). The relationship of drought frequency and duration to time scales. *Proceedings of the 8th Conference on Applied Climatology*, 17(22), 179–183.

Moeletsi, M. E., & Walker, S. (2012). Evaluation of NASA satellite and modelled temperature data for simulating maize water requirement satisfaction index in the Free State Province of South Africa. *Physics and Chemistry of the Earth, Parts A/B/C*, 50, 157–164.

Nakagawa, M., Tanaka, K., Nakashizuka, T., Ohkubo, T., Kato, T., Maeda, T., … & Teo, S. (2000). Impact of severe drought associated with the 1997–1998 El Nino in a tropical forest in Sarawak. *Journal of Tropical Ecology*, 16(3), 355–367.

Orimoloye, I. R., Ololade, O. O., Mazinyo, S. P., Kalumba, A. M., Ekundayo, O. Y., Busayo, E. T., … & Nel, W. (2019). Spatial assessment of drought severity in Cape Town area, South Africa. *Heliyon*, 5(7), e02148.

Palmer, W. C. (1965). Meteorological drought (Vol. 30). US Department of Commerce, Weather Bureau.

Palmer, W. C. (1968). Keeping track of crop moisture conditions, nationwide: The new crop moisture index.

Patel, N. R., Parida, B. R., Venus, V., Saha, S. K., & Dadhwal, V. K. (2012). Analysis of agricultural drought using vegetation temperature condition index (VTCI) from Terra/MODIS satellite data. *Environmental monitoring and assessment*, 184(12), 7153–7163.

Peters, E., Bier, G., Van Lanen, H. A., & Torfs, P. J. J. F. (2006). Propagation and spatial distribution of drought in a groundwater catchment. *Journal of Hydrology*, 321(1–4), 257–275.

Porensky, L. M., Wittman, S. E., Riginos, C., & Young, T. P. (2013). Herbivory and drought interact to enhance spatial patterning and diversity in a savanna understory. *Oecologia*, 173(2), 591–602.

Rain, D., Engstrom, R., Ludlow, C., & Antos, S. (2011). Accra Ghana: A city vulnerable to flooding and drought-induced migration. Case study prepared for cities and climate Change: Global Report on Human Settlements, 2011, 1–21.

Ribeiro, F. L. (2019). The Social Impact of Technology and Mega-infrastructures to Mitigate Drought: A Case Study of Changes in Social Capital Associated with the São Francisco Inter-basin Water Transfer in the Semiarid Region of Brazil (Doctoral dissertation, University of Delaware).

Schwinning, S., Starr, B. I., & Ehleringer, J. R. (2005). Summer and winter drought in a cold desert ecosystem (Colorado Plateau) part I: Effects on soil water and plant water uptake. *Journal of Arid Environments*, 60(4), 547–566.

Singh, R. (ed.) (2021) *Re-envisioning remote sensing applications: Perspectives from developing countries*. First Edition. CRC Press. https://doi.org/10.1201/9781003049210

Sinha, D., Syed, T. H., & Reager, J. T. (2019). Utilizing combined deviations of precipitation and GRACE-based terrestrial water storage as a metric for drought characterization: A case study over major Indian river basins. *Journal of Hydrology*, *572*, 294–307.

Swain, S., Wardlow, B. D., Narumalani, S., Tadesse, T., & Callahan, K. (2011). Assessment of vegetation response to drought in Nebraska using Terra-MODIS land surface temperature and normalized difference vegetation index. *GIScience & Remote Sensing*, 48(3), 432–455.

Tirado, M. C., Clarke, R., Jaykus, L. A., McQuatters-Gollop, A., & Frank, J. M. (2010). Climate change and food safety: A review. *Food Research International*, 43(7), 1745–1765.

Van Loon, A. F., Ploum, S. W., Parajka, J., Fleig, A. K., Garnier, E., Laaha, G., & Van Lanen, H. A. J. (2014). Hydrological drought typology: Temperature-related drought types and associated societal impacts. *Hydrology Earth System Sciences Discussions*, 11(9), 10465–10514.

Vetter, S., Goodall, V. L., & Alcock, R. (2020). Effect of drought on communal livestock farmers in KwaZulu-Natal, South Africa. *African Journal of Range & Forage Science*, 37(1), 93–106.

Wang, Y., Quan, Q., & Shen, B. (2019). Spatio-temporal variability of drought and effect of large scale climate in the source region of Yellow River. *Geomatics, Natural Hazards and Risk*, 10(1), 678–698.

West, H., Quinn, N., & Horswell, M. (2019). Remote sensing for drought monitoring & impact assessment: Progress, past challenges and future opportunities. *Remote Sensing of Environment*, 232, 111291.

Wossenyeleh, B. K., Verbeiren, B., Diels, J., & Huysmans, M. (2019, January). Vadose zone lag time effect on groundwater drought propagation in a temperate climate. In Geophysical Research Abstracts (Vol. 21).

Zeleke, T. T., Giorgi, F., Diro, G. T., & Zaitchik, B. F. (2017). Trend and periodicity of drought over Ethiopia. *International Journal of Climatology*, 37(13), 4733–4748.

Zhao, M., Huang, S., Huang, Q., Wang, H., Leng, G., & Xie, Y. (2019). Assessing socio-economic drought evolution characteristics and their possible meteorological driving force. *Geomatics, Natural Hazards and Risk*, 10(1), 1084–1101.

Zuo, D., Cai, S., Xu, Z., Peng, D., Kan, G., Sun, W., … & Yang, H. (2019). Assessment of meteorological and agricultural droughts using in-situ observations and remote sensing data. *Agricultural Water Management*, 222, 125–138.

6 Sustainable Management of Waterlogged and Salt-affected Areas Through Geospatial Technology

A Case Study of Central Haryana

Sandeep Kumar
Govt. College, Adampur, India

V. S. Arya
Haryana Space Applications Centre (HARASAC), Hisar, India

Ashok Beniwal
Government College, Adampur, Hisar, India

Manoj
CRM Jat College, Hisar, India

CONTENTS

DOI: 10.1201/9781003224624-9

6.1 INTRODUCTION

The problems of salt-affected soils are not new but their scale and intensity have been rising fast on a large scale. More areas have needed to be brought under irrigation in the last few years. It is usually decided that the future food needs of a rising population will be achieved by improving the management of soils that is previously under farming. Common properties and basic ideology are concerned in the identification, recovery and management of salt-affected soils throughout the world, while soil properties, climate, availability of water, farm management capacity, fiscal resources, accessible inputs and financial incentives differ from place to place, leading to differences in technique, extent and speed of soil recovery. The requirement for proper recognition and ensuing use of wrong retrieval methods result in losses of both capital and potential increases in crop production.

6.1.1 OBJECTIVES

i. The identification, delineation and mapping and classification of different types of waterlogged and salt-affected lands in Central Haryana using high resolution satellite data on a 1:10,000 scale.
ii. To suggest effective management strategies to counter the problem in the area in terms of improving the existing waterlogged and salt-affected soils and to suggest a proper drainage network to drain water from the waterlogged areas.

6.1.2 STUDY AREA

The study area is part of an inland drainage basin. The study area has a saucer-shaped topography. Due to the shape of the study area, rainwater creates a flooding problem in the monsoon season (Govt. of Haryana, 2015). To avoid the problem of floods, drains have been dug out in the study area. The length of the main canal is 182.83 km, while the branch canal is 297.36 km and the distributaries canal 943.69 km in length. Different drains in the study area occupy a length of 411.26 km. The location map of the study area is shown in Figure 6.1. The climate of the study area is dry with very hot summers and cold win. The study area is comprised of the vast Indo-Gangetic alluvial plains.

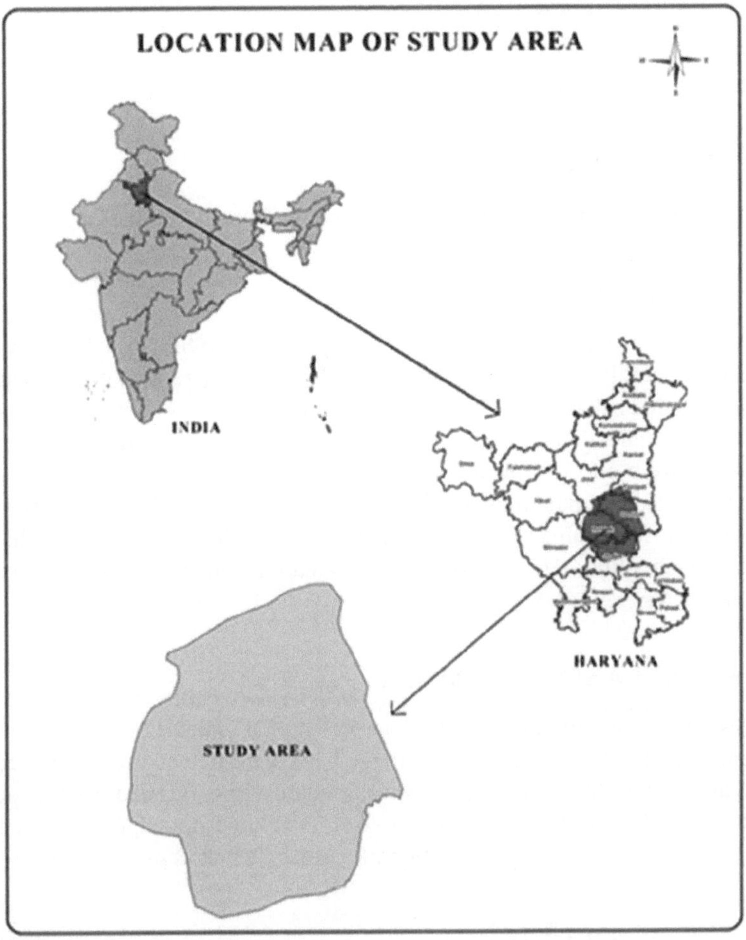

FIGURE 6.1 Location Map of Study Area

6.2 DATABASE

6.2.1 Satellite Data

IRS-IC/ID LISS-III, LISS-IV, Carto-Sat and Microwave digital data of different dates for the years 2009 to 2012 was used for the interpretation of the waterlogged and salt-affected areas. The projection system and datum of the satellite images were adopted as UTM and WGS 84, respectively. The data was received from the National Remote Sensing Centre (NRSC), Department of Space, and Government of India (GOI). The interpretation of satellite images was completed by using an on-screen interpretation technique using remote sensing (Singh, 2021).

6.2.2 Ancillary Data

Topographical maps of the Survey of India (SOI) were used for the identification of villages, transport network and different cultural features.

6.2.3 Software Used

Arc GIS Desktop 9.3, ERDAS Imagine 9.3, Arc Info Workstation and Geometica-10.3.

6.2.4 Methodology

Based on the standard image characteristics, an on-screen visual interpretation of LISS-IV satellite data was completed, adopting different interpretation keys. Waterlogged and salt-affected areas are mostly located in low-lying areas, impervious substratum and along the canals/river banks. The uncertain areas in the pre-field observation were checked during the field observation and afterwards these areas were customized by updating field observations. Water and soil samples from different sites of the study area were collected and analysed for their chemical characteristics in the laboratory. The methodology flowchart (Figure 6.2) shows the details of the methodology used.

6.3 RESULTS AND DISCUSSION

High resolution satellite images were interpreted on a scale of 1:10,000 to generate different thematic maps, such as waterlogged and salt-affected area maps, cropping pattern and crop rotation, underground water quality, underground water depth, soil texture, soil moisture, hydro geomorphology, canal and drainage, and settlements and infrastructure maps. The description of various themes is given below.

6.3.1 Salt-affected Area

Salt-affected land is identified as land having unfavourable effects on the growth of most plants due to the action or presence of excess soluble salts (saline soil), high exchangeable sodium (sodic soil) and both soluble salts and exchangeable sodium (saline-sodic) soil. A total of 60 soil samples from 30 different locations of salt-affected area were collected from various depths, i.e. from 0–15 cm and 15–30 cm. On the basis of their chemical characteristics (EC, pH and ESP), these soil samples were classified into the various classes of saline, sodic, and saline-sodic soil. It was observed that salt-affected areas are mostly found along the canals or drains. The spatial extent of these soils by class is shown in Table 6.1 and the salt-affected area by class is shown in Figure 6.3(A).

6.3.2 Waterlogged Area

Mostly the waterlogged patches in the study area are found along the canals or in depressions. Throughout the rainy period, the water accumulates in these depressions and canal seepage and creates the problem of waterlogging. Most of the seasonal

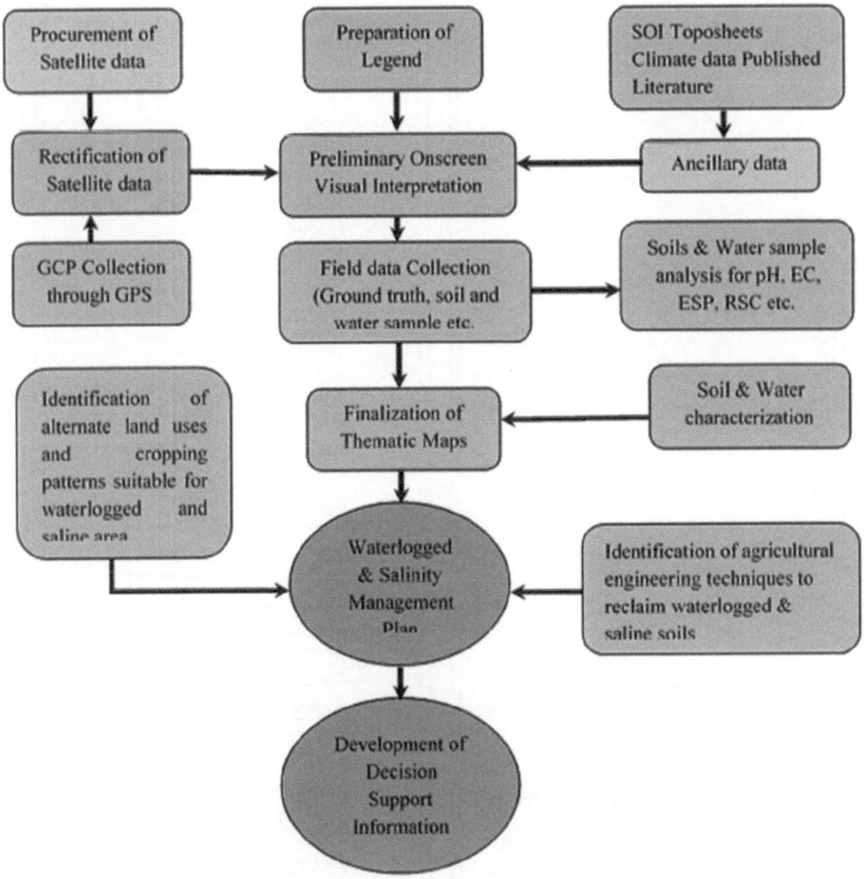

FIGURE 6.2 Methodology Flowchart

TABLE 6.1
Area Extent of Different Soil Salinity Classes

Salinity Classes	Area in km^2	Percentage of Total Salt-affected Area	Percentage of Total Geographical Area
Saline	50.14	79.32	1.12
Saline-Sodic	10.40	16.45	0.23
Sodic	2.67	4.22	0.06
Total Area	**63.21**	**100**	**1.41**

waterlogged or post-monsoon waterlogged area is distributed in the western and southern sides of the study area. The waterlogged area results in secondary salinization and salinity is developed in this area. The area under seasonal and permanent waterlogging is given in Table 6.2. The map of the waterlogged area is displayed in Figure 6.3(B).

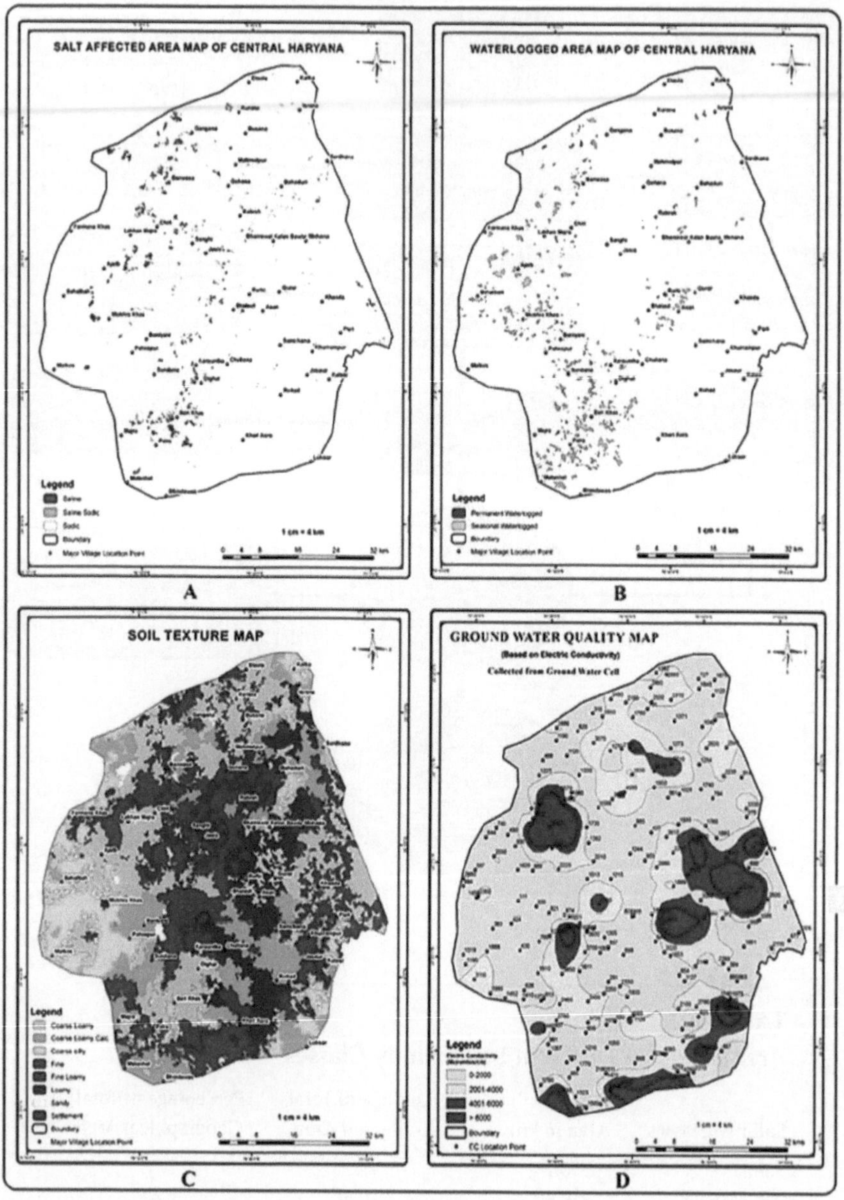

FIGURE 6.3 Spatial Attributes of Study Area

6.4 GROUNDWATER QUALITY AND WATER DEPTH

Underground water describes water available in the saturation zone. It is separated from the earth's surface by a permeable zone of aeration. A total of 207 water samples were collected from various locations of the region. These water samples from underground wells, tube wells and hand pumps were collected along with their GPS

TABLE 6.2
Area Extent of Waterlogged Areas

Waterlogged Classes	Area in km^2	Percentage of Total Waterlogged Area	Percentage of Total Geographical Area
Seasonal waterlogged	105.17	96.00	2.34
Permanent waterlogged	4.38	4.00	0.10
Total Area	**109.55**	**100.00**	**2.44**

location. The chemical properties of these samples like EC (dS/m), pH, Ca+Mg (me/L), CO_3, HCO_3 and residual sodium carbonate (RSC) were examined in the laboratory. It was observed that the sample with EC of 0–2000 μS/m seems fresh but its RSC value is >3 which make it unfit for irrigation. So, these samples were also classified based on their RSC values. Based on these EC values, a water quality map was also generated that is presented in Figure 6.3 (D).

6.5 SOIL STUDIES

Texture is a significant characteristic of soil which drives field management and crop productivity. The textural category of soil is decided by the percentage of silt, sand and clay. A soil texture map of the region was prepared by analysing the soil samples collected from 0–15 cm and 15–30 cm depth from different physiographic units. It was found that various soil texture family classes like coarse loamy, fine loamy, coarse loamy calcareous, coarse silty, fine, loamy and sandy are present in the study area. The spatial distribution of different soils is presented in Figure 6.3 (C).

6.6 LAND AND WATER RESOURCES DEVELOPMENT PLAN

The integration of various thematic maps like soil, slope, land use/land cover, geomorphology, water quality and groundwater depth coupled with agroclimatic data was done and a suitable combination of agricultural practices has been suggested. Several series of packages of practices have been recommended for overall sustainable development in the area on a long-term basis. On the basis of these developmental packages, various attributes have been grouped in two parts as follows:

Environmental Factors: landforms/geomorphology, groundwater depth and quality, types of soil and rainfall characteristics.

Human-induced Factors: land use/land cover, irrigation practice system, cropping pattern and crop rotation, socio-economic status, drainage, settlements and infrastucture system.

6.6.1 Management of Waterlogged Area

Waterlogged land in the area could be reclaimed by selecting a proper combination drainage method. The possibilities for the disposal of drainage water should also be chalked out. In the study area, the following three methods depending upon the existing conditions are suggested, as presented in Figure 6.4.

Horizontal drainage system: A horizontal drainage system is one which collects surplus water directly from the field and transfers it to the main

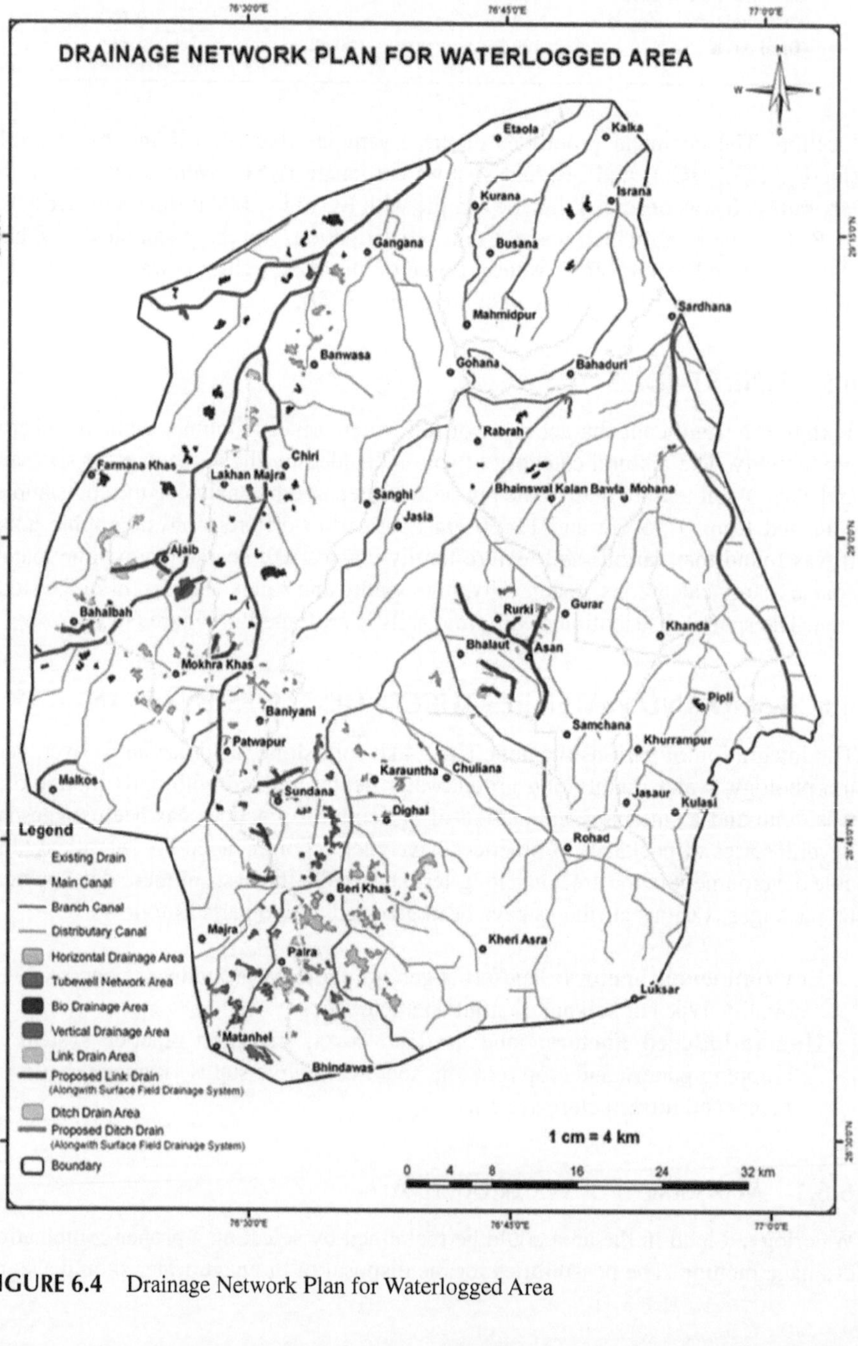

FIGURE 6.4 Drainage Network Plan for Waterlogged Area

drainage system through a network of field laterals, sub-collectors/collectors and main drains. The sub-surface horizontal drainage system is adopted in several locations around Mokhara Khas and Beri villages in the study area for reclaiming waterlogged and salinity problems where the stagnation of rainwater during the monsoon, saline groundwater and secondary salinization problems are being faced.

Vertical drainage system: The areas in the state experiencing a rise in the water table are generally underlain by poor underground water quality. With the excessive recharge from fresh canal water, water seepage, losses in the field, losses in water courses, minors and distributaries and so on, a relatively less saline water layer has developed over the main groundwater body.

Bio-drainage system: Bio drainage refers to the planting of tree species which have been found effective in controlling the water table. Some of the studies regarding bio drainage were conducted at HAU, Hisar and found that eucalyptus tereticomis is the most effective species in reducing the water table depth, due to its huge quantity of evapotranspiration.

Network of tube wells: The artificial removal of groundwater can be effected by pumping groundwater through tube wells in the areas along irrigation channels under groundwater mound conditions. Tube well drainage is an effective method of lowering shallow water tables and reducing the waterlogging hazard in the irrigated area that was suggested in the south of the study area.

6.7 MANAGEMENT OF SALT-AFFECTED AREAS

All types of soils contain some amount of water-soluble salts. Even non-saline soils in arid and semi-arid regions contain a much larger amount and variety of salts. Given this extensive occurrence, special management is required and is considered below. These are saline soils and alkali soils.

6.7.1 Reclamation of Saline Soil

- Exclusion of salts by scraping the surface layer and dumping these into a wasteland. Later on flood the field to leach down the remaining salts.
- Level the field as meticulously as possible and divide the field into small plots.
- Add sufficient organic wastes, compost or waste straw in these plots, and plough these plots and mix organic wastes very well into the layer.
- Leach the field with 30 cm of water in two consecutive instalments.
- After the soil is at the friable stage, sow recommended salt-tolerant crops and add 25 per cent extra fertilizer over the recommended doses.
- Keep the water table low (below 1.5 m) by pumping out sub-soil water. If the quality of this water is reasonably good, this water may be used for leaching and irrigating the crops.
- Do not allow the field to dry hard.

6.7.2 Reclamation of Saline-Sodic Soils

- Get the soil samples tested at the nearest soil-testing laboratory to find out the exact requirement of gypsum.
- Bund the field; the bunds should be 30 m high.
- Give two irrigations of 15 cm deep water to remove excess of sodium carbonate/bicarbonate from the surface layer.
- Grow the green manuring crop of 'dhaincha' during April and May. Green manure this crop at the end of June and transplant rice nursery of 40–45 days old by keeping 15 cm distance between rows and 15 cm between plants. At least three seedings may be planted in one hole. Add 25 per cent extra dose of nitrogen and 50 kg zinc sulphate per hectare.
- The field should not be allowed to go dry. Keep on giving light but frequent irrigations.
- After two years of paddy-wheat/barley rotation, practise the paddy-berseem rotation.
- Keep on getting the soil samples tested to find out the extent of reclamation and problems, if any.
- Keep water table below 2.5 m, and as far as possible mulching should be done on hot summer days.

6.8 CONCLUSIONS

The current study mainly explains the problems of waterlogging and soil salinity in Central Haryana, adopting the latest techniques of geoinformatics on a 1:10,000 scale with the assistance of LISS-IV sensor data. The brackish groundwater table is constantly rising because of leakage of water through canals, and irrigation channels with irrigated fields in the internal drainage basin in the saucer-shaped landscape of Central Haryana, creating the problems of waterlogged and secondary soil salinization. So, the utility of the present study is to increase the foodgrain production for the increasing population. Land and water resources have been subjected to great stress and inappropriate land use which has resulted in soil degradation in the study area. Sustainable and environmentally sound recommendations have been suggested for the waterlogged and salt-affected areas separately.

REFERENCES

Arya, V. S., Kumar Sandeep, Singh Hardev, Kumar Anil, Hooda R. S. and Arya Sandeep. 2014. Wastelands change analysis using multi-temporal satellite data in arid zone of Haryana. *Current Trends in Technology and Science*. 3 (1):60–64.

Arya, V. S., Kumar Sandeep, Arya Sandeep, Khatri S. S. and Singh Hardev. (2012). Wastelands Atlas of Haryana. (Change analysis based on temporal satellite data of 2005–06 & 2008–09). HRSAC/TR/03/2012.

Arya, V. S., Hooda R. S., Singh Sultan, Kaur Amarjeet and T. R. Nayak (1999a). Integrated land resources action plan development for Matenhail block of Jhajjar district. HARSAC/TR/06/99.

Arya, V. S., Singh Sultan, Kumar Ashok, Rao B. V. M. and Rao G. S. (1999b). Mapping of soil and water resources of Mewat area. Problems and their management using remote sensing techniques. HARSAC/TR/25/99.

Arya, V. S., Singh Sultan, Kumar Ashok, Rao B.V.M. and Rao G.S. (1999c). Integrated study through space applications for sustainable development in Gurgaon district. HARSAC/TR/26/99.

Barrett, E. C. and Curtis L. F. (1976). *Introduction to Environmental Remote Sensing*. Chapman and Hall, London. 336.

Choubey, V. K. (1997). Detection and delineation of waterlogging by remote sensing techniques. *Journal of the Indian Society of Remote Sensing*. 25(2):123–135.

Goossens, R. E. A., Badawi M. El., Ghabour T. K. and Dapper M. (1993). A simulation model to monitor the soil-salinity in irrigated arable land in arid areas based upon remote sensing and GIS. *EARSEL Advances in Remote Sensing*. 2(3):165–171.

Government of Haryana (2015). Statistical Abstract of Haryana (2013–2014). Department of Economic and Statistical Analysis, Govt. of Haryana, Panchkula.

Hooda, R. S., Parsad J., Saroha G. P., Arya V. S. and Sultan S. (2003). Wastelands Atlas of Haryana. Haryana Space Applications centre (HARSAC), Dept. of Science and Technology, Govt. of Haryana.

Jenson, S. K. and Dominque, J. O. (1988). Extracting topographic structure from digital elevation data for GIS analysis. *Photogrammatery and Remote Sensing*. 54(11):1593–1600.

Johnstone, R. M. and Barson, M. M. (1990). An assessment of the use of remote sensing techniques in land degradation studies. Australian Department of Primary Industries and Energy, Bureau of Rural Resources, Canberra, Australia. 5: 64.

Joshi, M. D., and Sahai, B. (1993). Mapping of salt-affected land in Saurashtra coast using Landsat satellite data. *International Journal of Remote Sensing*. 14(10):1919–1929.

Kalra, N. K. and Joshi, D. C. (1997). Evaluation of multi-sensor data for delineating salt-affected soils in arid-Rajasthan. *Journal of the Indian Society of Remote Sensing*. 25(2):79–91.

Manchanda, M. L. (1984). Use of remote sensing techniques in the study of distribution of salt-affected soils in northwest India. *Journal of Indian Soil Science*. 32:701–706.

Metternicht, G. I. and Zinck J. A. (1996). Modelling salinity-sodicity classes for mapping salt affected top soils in the semi-arid valleys of Cochabamba (Bolivia). *ITC Journal*. 11: 125–135.

Mougenot, B., Pouget M. and Epema G. F. (1993). Remote sensing of salt-affected soils. *Remote Sensing Review*. 7:241–259.

Moulders, M. A. (1987). *Remote sensing in soil science. Developments in soil science*. Elsevier Publication, Amsterdam. 15.

Narayan, L. R. A., Rao D. P. and Gautam N. C. (1989). Wasteland identification in India using satellite remote sensing. *International Journal of Remote Sensing*, 10(1): 93–106.

Rao, B. R. M. and Venkataratnam L. (1991). Monitoring of salt-affected soils – a case study using aerial photographs. *Geo-carto International Journal*, 1: 5–11.

Sethi, Madhurama, Gupta S. K. and Dubey D. D. (1996). Assessment of soil salinity and waterlogging in the Ukai-Kakrapar command area using remotely sensed data. Proc Workshop on Ukai-Kakrapar Command Area, Anand, Gujrat. 21–33.

Siderius, W. (1991). The use of remote sensing for irrigation management with emphasis on ILMI research concerning salinity, waterlogging and cropping patterns. Mission Report Enschede, The Netherlands: International Institute for Aerospace Survey and Earth Sciences.

Singh, A.N. (1994). Monitoring change in extent of salt-affected soils in northern India. *International Journal of Remote Sensing*, 15(16): 3173–3182.

Singh, R. (ed.) (2021) *Re-envisioning remote sensing applications: Perspectives from developing countries*. First Edition. CRC Press. https://doi.org/10.1201/9781003049210

Sreenivas, K., Venkatratnam L. and Narasimha Rav P.V. (1995) Dielectric properties of salt affected soils. *International Journal of Remote Sensing*. 16(4):641–649.

Steven, M. D., Malthus T. J., Jaggard F. M. and Andrieu B. (1992). *Monitoring responses of vegetation to stress. Remote Sensing from Research to Operation: Proceedings of the 18th Annual Conference of the Remote Sensing Society*. University of Dundee, 15–17, September, 1992. Nottingham, UK.

Venkatratnam, L. (1983). Monitoring of soil salinity in Indo-Gangetic plain of North West India using multi-date Landsat data. In *Proc. 17th International Symposium on Remote Sensing of Environment*. Ann Arbor, Michigan, USA. 1:369–377.

Wildman, W. E. (1982). Detection and management of soil, irrigation and drainage problems. In: *Remote sensing for resource management*, C.J. Johannsen, J.L. Sanders, eds., Soil Conservation Society of America, USA. 387–401.

7 Flood Risk Assessment and Analysis of Kashmir Valley Floor

Ishfaq Hussain Malik
Aligarh Muslim University, Aligarh, India

CONTENTS

7.1 INTRODUCTION

Disasters are events of destruction and enormous loss. The human response to disasters seems to be generating or at least permitting an increase in property losses, especially in countries where economic growth is rapid and modern technology is spreading fast. Some hazards are created by persistent habitation of dangerous areas or by alteration of land or water, while others are exacerbated by efforts to reduce the risk. Risk has several connotations in the discourses of different fields of knowledge. In general, risk is defined as 'the combination of the probability of an event and its negative consequences' (UNISDR, 2009). Von Kotze (1999) opines,

> The term risk is thus multidisciplinary and is used in a variety of contexts. Risk is usually associated with the degree to which humans cannot cope (lack of capacity) with a particular situation (e.g., natural hazard). The term disaster risk therefore refers to the potential (not actual and realised) disaster losses, in lives, health status, livelihoods, assets and services, which could occur in a particular community or society over some specified future time period. Disaster risk is the product of the possible damage caused by a hazard due to the vulnerability within a community. It should be noted that the effect of a hazard (of a particular magnitude) would affect communities differently.

These views make sense because the coping mechanism of different sections and classes of people varies depending on their social and economic conditions. The poorer sections of society and communities are at greater risk than the communities who have greater capacity to cope with disasters such as floods. Risk exists or

DOI: 10.1201/9781003224624-10

is created within social systems by the conditions of means and modes of production. The social and political factors underlying the identification of risk are crucial aspects of flood risk. The perception of vulnerability and risk varies among people because of prevailing political, social and economic conditions. When determining disaster risk, three important elements are taken into consideration: vulnerability to the hazard, a hazard, and some form of coping mechanism or capacity.

In order to explain the dynamics of disaster risk, the United Nations International Strategy for Disaster Reduction (2002) states: 'Disaster risk is the product of the combination of three elements – vulnerability, coping capacity and hazard.' The following notation illustrates this interaction:

$$R = f(h,v)$$

where, H = Hazard, V = Vulnerability

Oliver-Smith (1999) opines, 'disasters are clearly periods in which people experience a vast spectrum of intense emotions – anxiety, fear, terror, loss, grief, gratitude, anger, frustration, relief, and resignation – in all their shadings and intensities'. Birkland (1996) points out, 'there is a human "dread" of disaster. What is viewed as a disaster to one group may not be a disaster to the entire society.' In order to explain the historicity of disasters and their complex relationship with human beings and their way of life and settlements, Coppola (2011) says that human settlements have always been motivated by the needs of societies and individuals, such as the need for food to survive, water, defence, and access to trade and commerce. Increased natural hazard risk has been assumed in favour of these needs of humankind, almost without exception, often as a result of confidence that hazard risk can be accepted as 'part of life or can be effectively managed'. Evidence of such behaviour is apparently seen in so many examples of previous human settlements: communities along rivers build levees; those located along the sea coasts construct sea walls and jetties; farmers place their houses and sow their crops upon the fertile slopes of active volcanoes. Britton (1998) presents the relationship between population growth and disasters and says,

> As the population and size of these settlements grow, the assumed risk becomes more and more concentrated. The overall rates by which people have relocated from rural areas into cities (urbanization) have continued to increase over time. Rising populations in almost all countries of the world amplify the urbanization effect. In 1950, less than 30% of the world's 2.5 billion people lived in an urban setting. By 1998, the number of people on earth had grown to 5.7 billion and 45% of them lived in cities. UN estimates state that by 2025 there will be 8.3 billion people on earth, and over 60% of them will live in cities.

The damage to crops, infrastructure and housing, and the negative impacts on health and sanitation caused by floods and cyclones are particularly severe in the populous floodplains and coasts of many Third World states. For example, one in 20 people in India is vulnerable to flooding (Alexander, 1993).

Disasters like floods affect the social, economic and environmental aspects in the regions they occur. Floods are experienced every year in several corners of the world especially in the Himalayan region. Kashmir Valley, being part of the Himalayan region, has experienced many disastrous floods in the last century but the most disastrous flood was experienced in 2014, adversely affecting the socio-economic and environmental aspects. The data, reconstruction and analysis of flood chronologies and information from documentaries reveal that floods have been affecting people and their property in the Kashmir Valley since 635 AD. The most recent devastating flood, i.e. the 2014 flood in Kashmir Valley, had a devastating impact on the socio-economic, environmental and political conditions in Kashmir. It was the worst disaster that the Valley had experienced in the last 109 years and resulted in the death of 277 people in the whole state. The dynamics of relief and rescue were varied in different parts of the Valley. The flood gave rise to an interesting aspect of rescue operations and distribution of relief. As per the analysis of disasters in Kashmir, floods are not exactly an uncommon phenomenon in the region, which has witnessed 30 major floods. The region witnessed disastrous floods in the years 1950, 1959, 1992 and 2010 and most recently in 2014, 2015, 2017 and 2019. All these floods had different impacts but the flood of 2014 is the most devastating flood in the living memory of Kashmir; it affected all socio-economic and environmental aspects and created a political rupture in the Valley.

Kashmir is located in the northwestern part of the Himalayas and consists of several beautiful and picturesque valleys, mountains and gardens. Most of the people are involved in agricultural and horticultural activities to earn their daily living, so floods have a profound impact on farmers. Due to climate change in the last few decades, the weather conditions in Kashmir are showing great variability. According to Walter Lawrence (1895),

> several devastating floods are experienced in the vernacular histories of Kashmir, but the greatest was the dangerous inundation which followed the slipping of the Khadanyar mountain below Baramula in 879 A.D. The channel of the Jhelum River was blocked and a large part of the valley was submerged. In 1841 there was a serious flood, which caused much damage to life and property. A century later, in 1893, the state was affected by a flood of devastating proportions when 52 hours of continuous rainfall, beginning 18 July, caused what Lawrence describes as 'a great calamity' in which 25,426 acres of cropland was submerged, 2225 houses were wrecked and 329 cattle killed. There were major floods at the turn of the century, including the flood classified as the 'greatest flood ever known,' which came down the valley and inundated Srinagar on 23 July 1903, converting the city into 'a whole lake.'

Several floods followed in subsequent years (Malik & Hashmi, 2021) and more recently in 2014, 2016, 2017 and 2019 which had a great impact on Kashmir and its ecology and environmental balance (Malik & Hashmi, 2020) but they were not as devastating as the 2014 flood (Malik, 2021).

Meraj et al. (2013) while discussing the historical geography of flooding in Kashmir say,

> Jammu and Kashmir has had a long history of flooding, especially in the Kashmir Valley. The Kashmir Valley is one of the most flood hazard-prone Himalayan regions. The unique geomorphologic set-up of the Jhelum Basin, with heterogeneous lithology and varied hydrological condition, renders the basin particularly vulnerable to flooding.

Due to its geographic, climatic and geological setup, the Kashmir Valley is vulnerable to all types of hazards (Meraj et al., 2015; Romshoo et al., 2012 and Ray et al., 2009). The historical records reveal that the Kashmir Himalayan region has suffered heavy casualties and loss of property due to the recurrent floods, earthquakes, avalanches and other hydro-meteorological disasters (Mohammed et al., 2015). The study conducted by Rashid and Naseem (2007), while using geospatial technology regarding the wetland loss in Kashmir, revealed

> the impact of the loss of water bodies comes in the form of problems relating to drainage. The Srinagar city is facing acute problems of drainage since those wetlands and lakes acted as sponges during floods. Over the years, it has been observed that with a continuous rain for two to three days in Kashmir valley, the city is threatened with floods in river Jhelum while nothing would happen with this much of precipitation two to three decades back, further it has also been observed during the last decades that residential areas which never had floods in the past are getting inundated during floods in river Jhelum, this is because, there are hardly any wetlands to hold the excess water and to act as sponges during floods.

From Figure 7.1, it can be seen that the area from Pulwama to Srinagar and Banipora and Baramulla districts falls in the category of high-risk zone index, i.e. 3.6 and above. This area lies along the Jhelum River floodplain and lies in the zone of high flood risk. Anantnag, Shupian and Kulgam lie in the low flood risk zone while Ganderbal and Budgam districts lie in the medium flood risk zone. The area is vulnerable to floods because of the location of settlements along the Jhelum River. The encroachment on the banks of the Jhelum River along with urbanization has made it highly vulnerable to floods. The floodplain from the origin of the Jhelum River, i.e. Verinag Spring in district Anantnag in South Kashmir, to the north of Kashmir is inhabited by a huge population. The livelihood of all these people depends on the Jhelum floodplain which is susceptible to high risk of floods.

Urbanization is an important feature of land use and land cover in Kashmir Valley. In the last two decades there has been a substantial rise in the urban population and urbanized areas. The highly urbanized Srinagar city has been reshaped in the last two decades which saw both horizontal and vertical rises in urbanization in the Jhelum floodplain. The Jhelum floodplain in Srinagar city is occupied by human settlements, hotels and restaurants. Rajbagh, Natipora, Lalchowk and some other areas in the Jhelum floodplain in Srinagar city are highly urbanized, which has increased the flood risk in these areas. Urbanization has also been on the rise in the districts of Anantnag, Kulgam and Pulwama along the Jhelum floodplain. Urban places like Khanabal, Bijbehara, Awantipora and Pampore fall in the high flood risk zone of the

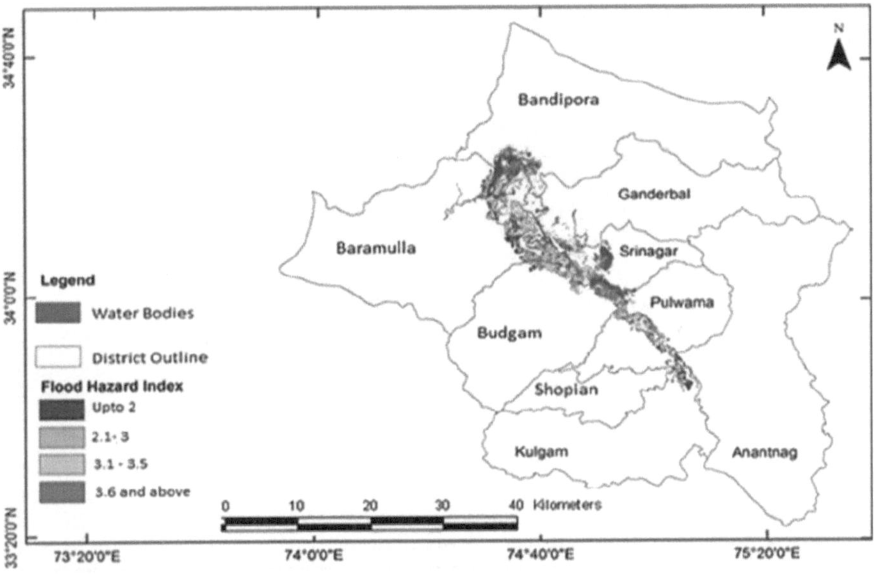

FIGURE 7.1 Flood Risk Map of Kashmir Valley Floor

Jhelum floodplain. Urbanization has also risen in the districts of Baramulla and Bandipora in the north of Kashmir along the Jhelum floodplain, which has also increased the flood risk of these districts.

From Figure 7.2, it can be seen that the Kashmir Valley floodplain comprises a diversity of land use. Most of the floodplain is covered with agricultural land, as the floodplain is highly fertile and is comprised of sedimentary soil brought from the upper reaches of the Kashmir Valley by the tributaries of the Jhelum River, such as the Lidder River from Pahalgam, Sandran and Vishav. The agricultural land lies in the zone of high flood risk as it is greatly affected during the floods. During recent floods, thousands of acres of agricultural land were affected and huge damage to crops was experienced. Since most of the people in Kashmir Valley are engaged in agricultural activities, the occurrence of floods in the Kashmir Valley floodplain greatly affects the economy of the region and leads to poverty and social anxiety. The Kashmir Valley floodplain also comprises forests, horticultural land and plantation land. Several water bodies are found in the Jhelum floodplain. The two most important lakes of the Kashmir Valley floor in terms of geography and economy are Dal Lake and Wular Lake. Both these lakes possess high flood risk. During recent floods, these two lakes have been greatly damaged. The ecology of these lakes has been profoundly hampered due to the floods. The risk of flooding in these two lakes makes thousands of people vulnerable to different types of loss due to flooding, as hundreds of families live on the banks of these lakes. The houseboat population known locally as *Hanjis* live on the Dal Lake and thus are at higher risk of floods. During the 2014 flood, the houseboat population evacuated the Dal Lake and shifted to higher areas for their safety. Some of them lived on boats for several days without proper food and clothes. Thus these people lie in the zone of high

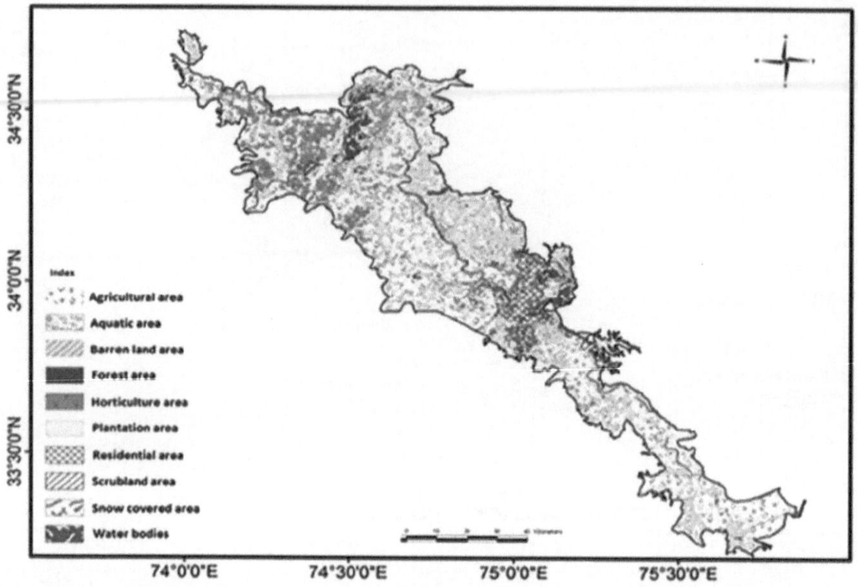

FIGURE 7.2 Land Use/Land Cover Map of Kashmir Valley Floodplain (2020)

flood risk. Several respondents, expressing the condition of flooding in the Jhelum River, said:

> During floods, Jhelum River overtops its natural banks, while as at the time of high intensity of flood the water passes over the artificial embankments which have been constructed on either side of Jhelum. Development initiatives are not risk neutral. They can exacerbate vulnerabilities and increase the number of hazards if they are poorly formulated or implemented, or if they cause serious environmental damages. Encroachment on the banks of Jhelum especially in some parts of Srinagar like Raj Bagh has resulted into lowering down the water holding capacity of Jhelum River and the structures along its banks were greatly damaged by the flood. The state has turned a blind eye towards the encroachment on the banks of Jhelum and the subsequent governments have failed to stop the encroachment. The Jhelum floodplain is occupied by roads and railway line which has led to encroachment on its banks and reduction in the carrying capacity of the river during recent floods. Houses, huts and other building have also been constructed along the banks of Jhelum River which have made the floodplain a high flood risk zone.

Tali (2011) reveals,

> The marshy areas (wetlands) in the upper Jhelum basin decreased by 50% between 1961 and 2001. In Kashmir, swamps covered an area of 209 km^2 during 1911–1946. These swampy areas shrank by 14.8% between 1946 and August

25, 2014 due to expansion of built-up area and cropland. The floodplain along Jhelum has been converted into agricultural lands and built-up areas at many places especially in Anantnag and Srinagar city. This has resulted into lowering down of water holding capacity of Jhelum. The construction of elevated railway track and newly developed highway from South Kashmir to North Kashmir resulted into shrinking of Jhelum flood plain and acted as a physical barrier during 2014 flood.

The wetlands in Kashmir Valley have been greatly damaged by the unregulated and unchecked growth of urban centres. Faulty developmental practices have changed the ecology of the wetlands and increasing pollution has resulted in the degradation of the wetlands. Several wetlands have been used for rice cultivation. The depletion of wetlands has resulted in a heightened flood risk in the Kashmir Valley as their water-carrying capacity has been drastically reduced.

Analysis of the historical records and documentation reveals that about 30 major floods have been witnessed in the recent history of Kashmir. The Valley of Kashmir has a long history of devastating floods. The first major flood was recorded in 1360 AD in which 20,000 houses were damaged in the low-lying plains of Sonawari and Srinagar. In 1970, Kashmir witnessed a flood as well as a famine which lasted for three years and led to extreme poverty and malnutrition of thousands of people in the Valley. In the last 127 years, the Valley of Kashmir has witnessed six disastrous floods, i.e in the years 1893, 1950, 1959, 1992, 2010 and 2014, which have resulted in overwhelming loss in terms of environmental, political, economic and social aspects. One of these disastrous floods was the flood of 18 July 1893 which resulted in the deaths of 32 people, economic loss of Rs. 64,804 in land revenue, submergence of 25,426 acres of crops, destruction of 2,225 houses and the loss of 329 cattle.

The flood of September 1950 in Kashmir resulted in the deaths of 100 people in different parts of the Valley and in the destruction of 15,000 houses. One of the most disastrous floods in the entire flood history of Kashmir was witnessed in July 1959 which caused the deaths of 477 people. On 5 July 1959, the flood waters of the Jhelum River reached 30.25 feet which was 6 points above the prescribed danger level. The 1992 flood caused the deaths of 200 people, while 60,000 people were marooned in the Valley. Seventy-one towns and villages were damaged by the flood of 6 August 2010. This resulted in the deaths of 255 people and the flood directly affected 9,000 people.

The most disastrous flood in the recent history of Kashmir Valley was witnessed in September 2014 when the whole Valley was submerged. The flood affected the socio-economic, political and ecological aspects in an adverse way which was felt for several years. It resulted in the deaths of 277 people in Jammu and Kashmir. Among the 277 fatalities, 87 were from the Kashmir Valley. About 2,600 villages were flooded in all the districts of the Valley and 390 villages were submerged in Kashmir. An economic loss of Rs. 1.0 trillion was encountered as 5 million people were affected and almost 2 lakh people were rescued from the flooded areas of Srinagar, Anantnag, Pulwama, Shopian, Baramulla and Kulgam.

The most hit by the 2014 flood was Srinagar city as the Jhelum River spilled its waters throughout the city. Srinagar experienced more deaths – 44 – than other districts of Kashmir because the water from South Kashmir accumulated in the city of Srinagar and there was no way for the people of Srinagar to save themselves in a short span of time. Within hours of rainfall, the district of Srinagar was inundated which made it difficult for the people to reach safer places and save themselves. A respondent from Srinagar, recollecting his memories of the flood, said that he had never seen such a disastrous flood in his entire life. The flood water entered his house within the first three days of the flood and the first storey of their house was completely inundated and their belongings were washed away by the water. They had no food for several days.

The floods in Kashmir Valley are caused by the combination of anthropogenic and natural factors, which combined together to cause disastrous floods and devastation. The high rainfall caused by the convergence of western disturbances and monsoonal winds accentuates the problem. The anthropogenic causes such as unplanned urbanization, encroachment on the banks of the Jhelum River, depletion of wetlands and administrative negligence aggravate the condition of the flood and results in widespread devastation in the floodplain of Kashmir and increases the risk of flood.

Analysis of the 2011 population census reveals that an area of 1,505.06 km^2 of the Kashmir Valley lies in the flood risk zone. The data reveals that 796 villages, 260,479 households and 1,574,997 people fall in the flood risk zone and are affected greatly by devastating floods. These villages and households mostly lie in the Jhelum floodplain and thus are highly vulnerable to floods every year. But it is worth mentioning that different sections of society are affected differently and are at different levels of risk. The economically lower-class people are at the highest risk of flooding because of their housing conditions and the avenues to survive the impact of floods, while the upper-class people are at lower risk of flooding because of their better housing conditions and the avenues to survive and cope during and after the floods.

7.2 CONCLUSION

The Kashmir Valley floor is a zone of high flood risk, with the region experiencing floods almost every year. Kashmir has experienced several disastrous floods throughout its history which have caused the deaths of thousands of people over the years and a tremendous economic loss. These floods have affected all sectors of society. The need of the hour is to protect the Jhelum River from encroachment and desiltation should be done regularly so that it can contain the extra water during the floods. Wetlands should be conserved so that more water could be accumulated during floods. A comprehensive disaster management plan is needed in Kashmir so that the intensity and frequency of floods could be reduced. The sustainable management of water bodies throughout the Kashmir Valley is key to reducing the intensity of floods and the resultant socio-economic and environmental loss. The risk of flooding needs to be lowered so that precious lives and environment can be saved from destruction and deluge. All sectors of society need to be developed so that all the people lie in the low-risk scenario and maintain equal coping mechanisms.

REFERENCES

Alexander, D. (1993). *Natural Disasters*. London: UCL Press.

Birkland, T. A. (1996). Natural disasters as focusing events: Policy communities and political response. *International Journal of Mass Emergencies and Disasters*, 14(2), 221–243.

Britton, N. R. (1998). *Safeguarding New Zealand's Future: Emergency Management's Role in Shaping the Nation*. Ministry of Civil Defence & Emergency Management.

Coppola, D. P. (2011). The management of disasters. DP Coppola, Introduction to international disaster management, 1–35.

Lawrence, W. R. (1895). The valley of Kashmir. H. Frowde.

Malik, I. H. (2021). Spatial dimension of impact, relief, and rescue of the 2014 flood in Kashmir Valley. *Natural Hazards*, 1–19.

Malik, I. H., & Hashmi, S. N. I. (2020). Ethnographic account of flooding in North-Western Himalayas: a study of Kashmir Valley. *GeoJournal*, 1–19.

Malik, I. H., & Hashmi, S. N. I. (2021). The great flood and its aftermath in Kashmir Valley: Impact, consequences and vulnerability assessment. *Journal of the Geological Society of India*, 97(6):661–669.

Meraj G, Romshoo SA, Yousuf AR, Altaf S, Altaf F. (2015). Assessing the influence of watershed characteristics on the flood vulnerability of Jhelum basin in Kashmir Himalaya. *Nat Hazards*, 77:153–175.

Meraj G, Yousuf AR, Romshoo SA. (2013). Impacts of the geo-environmental setting on the flood vulnerability at watershed scale in the Jhelum basin [MPhil dissertation]. Srinagar: University of Kashmir.

Mohammed AAA, Naqvi HR, Firdouse Z. (2015). An assessment and identification of avalanche hazard sites in Uri sector and its surroundings on Himalayan mountain. *Journal of Mountain Sciences* 12:1499–1510.

Oliver-Smith, A. (1999). What is a disaster? Anthropological perspectives on a persistent question. In The angry earth: Disaster in anthropological perspective (pp. 18–34).

Rashid, H., & Naseem, G. (2007). Quantification of loss in spatial extent of lakes and wetlands in the suburbs of Srinagar city during last century using geospatial approach. In Proc. of Taal: The 12th World Lake Conference (pp. 653–658).

Ray, P., Parvaiz, I., Jayangondaperumal, R., Thakur, V., Dadhwal, V., & Bhat, F. (2009). Analysis of seismicity-induced landslides due to the 8 October 2005 earthquake in Kashmir Himalaya. *Current Science*, 97(12), 1742–1751.

Romshoo SA, Bhat SA, Rashid I. (2012). Geoinformatics for assessing the morphometric control on hydrological response at watershed scale in the Upper Indus basin. *Journal of Earth System Science* 121:659–686.

Tali P. A. S. (2011) Land use/land cover change and its impact on flood occurrence: A case study of upper Jhelum floodplain. Dissertation, University of Kashmir.

United Nations International Strategy for Disaster Risk Reduction (UNISDR). (2009). A Guide to community based DRR in Central Asia, UNISDR.

United Nations International Strategy for Disaster Reduction (UNISDR). (2002). *Disaster reduction for sustainable mountain development: 2002 United Nations world disaster reduction campaign*. Geneva: United Nations International Strategy for Disaster Reduction.

Von Kotze, A. (1999). A new concept of risk. In: Holloway, A. *Risk, sustainable development and disasters: Southern perspectives*. Cape Town: Periperi Publications.

8 Remote Sensing Applications in Landslide Susceptibility Index Mapping of Rangamati District, Bangladesh

Nur Hussain and Muhammad Rizwan
McMaster University, Hamilton, Canada

Mohd. Shamsul Alam
Jahangirnagar University, Dhaka, Bangladesh

Li Suju
National Disaster Reduction Center of China, Beijing, China

CONTENTS

DOI: 10.1201/9781003224624-11

8.1 INTRODUCTION

Landslide is a common natural hazard in hilly and mountainous areas that causes considerable economic, social and ecological damage (Jian & Xiang-Guo, 2009; Chen et al., 2017). In mountainous and hilly territories, landslide often occurs during or after heavy rainfall, creating loss of life with natural, economic and social damage (Dai & Lee, 2002). In some cases, landslides can have a huge volume of debris, and almost 9 per cent of worldwide natural disaster events are mentioned as landslides (Gokceoglu et al., 2005). A mass rock movement, debris fall, or down-slope movement of the earth's surface subsequent to geomorphic alteration of the earth's surface which contributes to landscape evolution is debris referred to as a landslide (Avinash & Ashamanjari, 2010). In 2008, the Landslide Catalog notified 4 people's deaths by landslide per event among 67 countries (Kirschbaum et al., 2010). The vulnerability and fatality of landslides are higher in developing countries (Petley et al., 2007). The annual landslide occurrence maintenance cost is more than $2 billion for the United States of America (Spiker & Gori, 2003). In addition, Bangladesh is more vulnerable to landslides where more than 398 people were killed from 1999 to 2017 (Ahmed et al., 2014; BBC, 2013; GoB, 2010; Taj, 2013). Many countries and regions of the world are susceptible to landslide as noted in many studies; for example, the active tectonic region of Himalaya (Patwary, Ray & Parvaiz, 2009), the Lesser Himalaya of Nepal (Kumar et al., 2008), India (Devi et al., 2017), China (Shi-Biao et al., 2009; Chen et al., 2017; Jian & Xiang-Guo, 2009), Korea (Lee, 2001; Yeon, Han & Ho, 2010), Japan (Ayalew, Yamagishi & Ugawa, 2004), Vietnam (Tien et al., 2012), Thailand (Pradhan & Lee, 2010), Greece (Chalkias, Ferentinou & Polykretis, 2014), Italy (Ciampalini et al., 2015; Vanmaercke et al., 2017), European region (Mantovani, Soeters & Van Westen, 1996) and Colorado, USA (Regmi, Giardino & Vitek, 2010).

Landsliding occurs almost consistently in the southeastern mountain areas of Bangladesh during the heavy rainfall of the monsoon period. According to the Asian Disaster Preparedness Center (ADPC) at least 260 people were killed by landslides in Chittagong from 1999 to 2012 (Ahmed, 2013). Recently, landslide incidents have been increasing and occurring frequently. For instance, on 11 June 2007 more than 120 people were killed due to a landslide in Chittagong (GoB, 2010). The BBC (2012) and Rubel and Ahmed (2013) notified that, in June 2012, about 90 people were killed in different areas of the hilly region of Bangladesh due to the three-day-long torrential rainfall (Ahmed et al., 2014). On 13 June 2017, at least 107 people died in the Rangamati district of Bangladesh (BBC, 2013) though the number was more than 147 according to a leading national newspaper (*The Daily Star*, 19 June 2017). There is a substantial outcome of landslide observed in the exploration area, though no research work has been conducted yet. In this circumstance, Rangamati district is the most vulnerable area for landslide occurrence which passes on to the plan to prepare a landslide susceptibility map using a GIS-based weighted model.

Among geological conditions, rainfall, land use, elevation, vegetation, slope and stream were considered as the primary influencing variables of landslides occurring in the research area (Hussain et al., 2021). Ahmed (2015) conducted research using the landslide susceptibility index (LSI) which focused on landslide susceptibility and loss assessment in Chittagong metropolitan city. Similarly, Taj (2013) accomplished a study on a landslide disaster in Chittagong University Campus which represents only the Chittagong University area. According to the landslide records, a more hazardous landslide occurred in Rangamati district but there are no available research works about that landslide. From this perspective, this research emphasized LSI mapping on the Rangamati district of Bangladesh, employing remotely sensed data through a geographic information system (GIS)-based weighted overlay model (WOM). This research can assist in finding out the relative susceptible areas. The main purpose of this research is to indicate landslide susceptibility zones with geospatial location. The following objectives have been adopted to meet the key focus of this research: a) to explore the influencing factors of landslide; b) to compute LSI using a GIS-based WOM; c) to prepare a landslide susceptibility map. Ayalew et al. (2004) produced a landslide susceptibility map using a GIS-based weighted model (Ayalew, Yamagishi & Ugawa, 2004), while Roslee (2017) used a GIS-based WOM to prepare LSI with the consideration of landslide influencing factors (Roslee et al., 2017).

8.2 DATA AND METHODOLOGY

8.2.1 Study Area

The Rangamati district is in the southeast part of Bangladesh (Figure 8.1) including Chittagong tertiary hill tract which is also connected with the Indian and Myanmar border. In addition, the landform of the study area was developed during the tertiary period of the geological timescale (Brammer, 2014; Ahmed, 2015). The Rangamati district is located 91° 57′ 32″ east longitude to 92° 37′ 47″ east longitude and 21° 54′ 54″ north latitude to 23° 45′ 55″ north latitude. The total area of the Rangamati district is 6,116.11 km^2 and the population is 595,979 with a density of 97.44 people/km^2 (BBS, 2015). The elevation is 932 m above mean sea level. Figure 8.1 (a) represents the location of Bangladesh, (b) represents the location of the Rangamati district, and (c) shows the Google image of the study area.

Sultana and Casadevall (2020) compiled a list of 204 landslides in Bangladesh that details 727 fatalities and 1,017 injuries during the years 2000–2018. The normal numbers of yearly landslides are 19, with a 4 per cent yearly growth rate, which brings about 38 fatalities and 54 injuries on a yearly average. According to geological structure, this area is a landslide-hazardous region. In 2017 at least 132 people died and 5 were injured in the Rangamati landslide of Bangladesh (NIRAPAD, 2017). In this aspect, this research had selected Rangamati as a study area. The spatial dispersion of landslide-related casualties and injuries are shown in Table 8.1.

In June 2017, 147 people were killed due to severe landsliding in the Rangamati area. According to the table, landslide processes caused a large number of fatalities in the Rangamati area during June 2017. The monsoon season has a greater number of casualties than other seasons. In Bangladesh, the monsoon is the most ruinous season for landslides. The second most noteworthy fatalities occurred pre-monsoon.

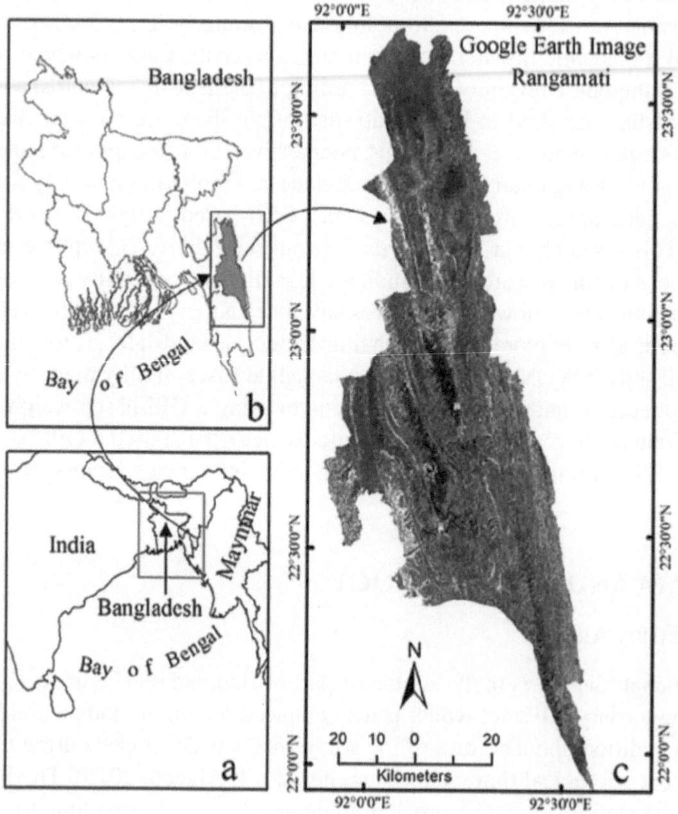

FIGURE 8.1 (a) Location of Bangladesh, (b) location of Rangamati district and (c) Google image of the study area

TABLE 8.1
Landslide Records of Study Area

Date	Consequence	Source
30 May 1990	Records not available	GoB, 2010
18 August 2007	74 people killed, 5 injured	*The Daily Star*, 2017
13 June 2017	147 people killed, 5 injured	*The Daily Star*, 2017
12 June 2018	12 people killed, 4 injured	*The Independent*, 2018

8.2.2 Data

Remote sensing has wide applications, and disasters, particularly landslides are not an exception to it (Singh, 2021). Exclusive data were selected according to influencing factors of landslides. The main factors of landslides are geological, meteorological and anthropogenic phenomena (Regmi, Giardino & Vitek, 2010). In this study,

TABLE 8.2
References of Satellite Data

Data	Reference	Date	Spatial Resolution
SRTM	SRTM1N20E092V3 SRTM1N21E091V3 SRTM1N21E092V3 SRTM1N22E091V3 SRTM1N22E092V3 SRTM1N23E091V3 SRTM1N23E092V3	23 September 23 2014	30 metres
TRMM	3B43.20140901.7.HDF	September 2014	0.25° × 0.25°
Landsat-5	LT05_L1TP_135045 LT05_L1TP_136044 LT05_L1TP_136045	3 November 2011 25 October 2011 25 October 2011	30 metres

elevation level, slope intensity, stream order, land use, rainfall intensity distribution and vegetation density are the main influencing factors of landslides in this study area. Shuttle Radar Topography Mission (SRTM) data were utilized to retrieve elevation, slope and stream order. Google Earth data have been used as land use and anthropogenic phenomena. The Tropical Rainfall Measuring Mission (TRMM), a joint mission of the National Aeronautics and Space Administration (NASA) and the Japan Aerospace Exploration Agency, provided monthly average rainfall data. Remote sensing-based TRMM data were used to retrieve rainfall intensity distribution. Landsat-5 data were used to determine the vegetation density of the study area, conducting an NDVI model. Table 8.2 represents the satellite data references.

8.2.3 Methodology

The collected data were analysed by various GIS tools to identify the LSI in this research. SRTM data have been used for stream ordering, slope measuring and exploring elevation by using geospatial analysis. A digital elevation model (DEM) has been used for elevation conferring a triangulated irregular network (TIN) method. Slope has been measured using hydrological analysis. Stream order has been retrieved by conducting hydrological fill, flow detection and flow accumulation of hydrological analysis with consideration of the Strahler (1952) stream order method. Google Earth-based land cover data have been analysed using the Euclidean distance algorithm to explore interaction between land use and anthropogenic characteristics. TRMM data have been used to retrieve rainfall intensity distribution simulating the Kriging interpolation model. Landsat 5 data have been used for NDVI to detect the vegetation density. All influencing aspects have been input on WOM. Figure 8.2 shows the flowchart of this research.

8.2.3.1 Land-use Exposure and Anthropogenic Phenomena

The Euclidean distance algorithm has been used for land-use linear data and stream ordering in this research. Capecchi et al. (2015) use Euclidean distance to evaluate

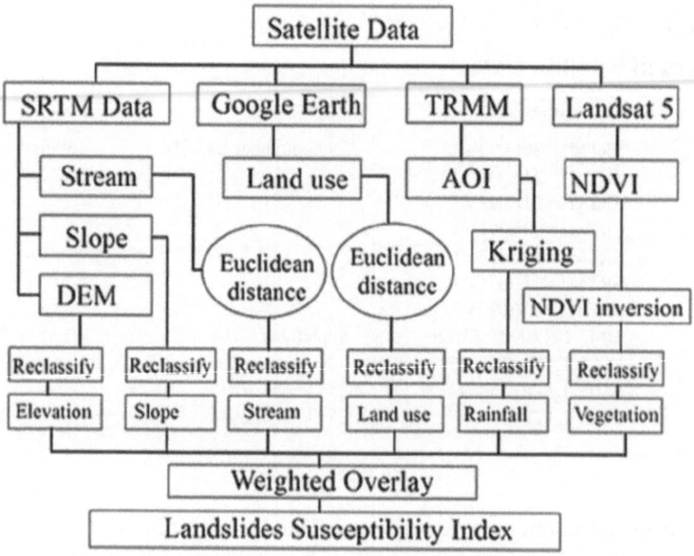

FIGURE 8.2 Data-processing approaches and methodological flowchart

the minimum distance between computing cells and the nearest river network in landsliding purpose (Capecchi, Perna & Crisci, 2015). This technique has likewise been used in similar works as a prompting factor, because it provides a possible account of activating functions related to erosion along the slope foot (Mossa et al., 2005; Mancini, Ceppi & Ritrovato, 2010). In the same way, Kalantar et al. (2018) conduct Euclidean distance analysis on topographic maps to produce distance of stream factor (Kalantar et al., 2018). The Euclidean algorithm works as follows for each cell. For this purpose, the distance to each source cell is controlled by working out the hypotenuse with x_max and y_max as the other two legs of the triangle (Figure 8.2a).

At the point when the information cell is at an equivalent separation to at least two sources of information, the data cell is allocated to the input that is first

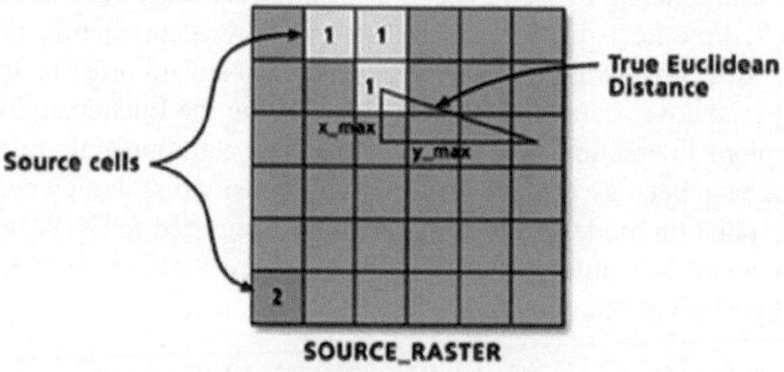

FIGURE 8.2A Euclidean distance algorithm (Esri, 2012)

experienced in the filtering procedure and provides the output raster (Esri, 2012). In addition, the Euclidean distance algorithm has been used to detect the linear distance to stream factor of this study. Similarly, Google Earth linear road and settlement data have been simulated in Euclidean distance analysis to explore the anthropogenic phenomena.

8.2.3.2 Rainfall Distribution and Kriging Interpolation

There is only one rainfall measuring rain gauge station in the Rangamati district; on account of this limitation, TRMM data have been used to retrieve rainfall distribution. TRMM data have 0.25° × 0.25° spatial pixel-size determination in terms of the study area that is very low resolution. In this circumstance, geostatistical Kriging interpolation has been used to explore rainfall distribution. Kriging is the geostatistical method for spatial data investigation that is typically utilized for modelling common phenomena (Zuvala, Fišerová & Marek, 2016; Roslee, 2012). The Kriging interpolation technique, a deterministic model, has been used to compute the factor of safety (FOS) and failure probabilities for the area (Roslee, 2012). Kriging considers the distance and direction among input points with spatial relationships that can be utilized to clarify disparity in the land surface (Esri, 2012). Catani et al. (2016) and Zuvala et al. (2016) conduct spatial patterns of landslide dimensions using Kriging interpolation (Catani, Tofani & Lagomarsino, 2016; Zuvala, Fišerová & Marek, 2016). The Kriging tool fits a mathematical function to a predefined number of focuses, or all points inside a predetermined range, to determine the output value for each location in the following equation:

$$\hat{Z}(s_0) = \sum_{i=1}^{N} \lambda_i Z(s_i) \tag{8.1}$$

where, $\boldsymbol{Z}\ (s_i)$ is the measured value at the ith location, $\boldsymbol{\lambda}_i$ is an unknown weight for the measured value at the ith location, s_0 is the prediction location and $\boldsymbol{N}$ is the number of measured values.

8.2.3.3 NDVI Inversion

Vegetation density is a key influencing factor of landslide because a high density vegetated area is freer from landslide while a low density vegetated area is more landslide-susceptible (Plank, Twele & Martinis, 2016). As a result, the NDVI has been inversed using raster reclassification due to influencing characteristics (Jian & Xiang-Guo, 2009). Three particular images of Landsat-5 were mosaicked and clipped to determine the vegetation cover of the study area following an NDVI model. Vegetation density was detected using the visible red band (0.63 μm– 0.69 μm) and near-infrared band (0.76 μm – 0.90 μm) channels of Landsat-5 in the following equation:

$$NDVI = \frac{b_4 - b_3}{b_4 + b_3} \tag{8.2}$$

where NDVI is normalized difference vegetation index, b_4 is near-infrared band (0.76 μm – 0.90 μm) and b_3 is visible red band (0.63 μm– 0.69 μm) of Landsat-5.

8.2.3.4 Landslide Susceptibility Mapping with Weighted Overlay Model (WOM)

The multi-criteria WOM is one of the best techniques for landslide susceptibility mapping (Ahmed, 2015). The LSI has been calibrated using WOM. WOM works with the consideration of ratio, interval, ordinal and nominal aspects of input parameters (Roslee et al., 2017; Ayalew, Yamagishi & Ugawa, 2004). Land-use and NDVI data have been assumed as the input parameter of anthropogenic phenomena, while rainfall data have been used for the meteorological parameter to simulate WOM. Similarly slope, aspect and elevation data have been used as geological input parameters of WOM. Weighted overlay has used a common measurement scale and weighted each according to its importance of input parameter. However, the WOM provides the LST according to importance of input parameter (Regmi, Giardino & Vitek, 2010; Roslee et al., 2017). In addition, the LSI has been conducted in this research using the geostatistical WOM of ArcGIS (Weighted Overlay; Spatial Analyst Tools, ArcGIS-10.7) following the formula:

$$LSI = \sum_{i=1}^{N} w_i.u_i \tag{8.3}$$

where LSI is the landslide susceptibility index, n is the total number of influencing factors, $w_i$wi represents the weight of factor, and $u_i$ui represents the contribution index of factor.

8.3 RESULTS

8.3.1 Influencing Factors of Landslides

The influencing factors of landslides were indicated according to the geological, meteorological and anthropogenic conditions of the study area. Rainfall intensity distribution, land use, elevation level, vegetation density, slope intensity and stream order distribution are the main influencing factors of landslides in the study area (Ahmed et al., 2014; Taj, 2013). All the factors have influencing characteristics that are complexly related to each other and to landslide occurrence. All factors have been categorized in the same scale from 1 to 8 in consideration of influencing aspect for weighted overlay simulation, because in applying this model common or same scale values conduct to different and divergent contributions to make an integrated analysis. In addition, WOM worked with the rational, ordinal, nominal and interval scale factors of input parameters. As a result, the same scale of input parameters was included to conduct WOM-based LSI (Figures 8.3 and 8.4).

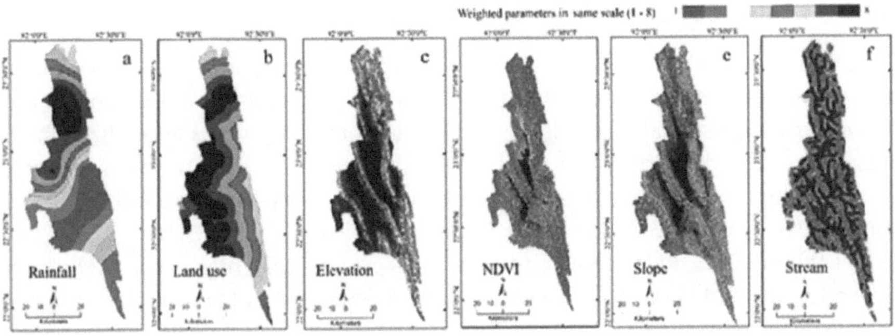

FIGURE 8.3 Influencing factors of landslide in same scale. We included the influencing factors of landslide events for the study area according to geological, meteorological and anthropogenic aspects; map 3(a) represents rainfall intensity distribution, map 3(b) represents land use, map 3(c) represents elevation level, map 3(d) indicates vegetation density, map 3(e) shows slope intensity and map 3(f) shows stream order distribution

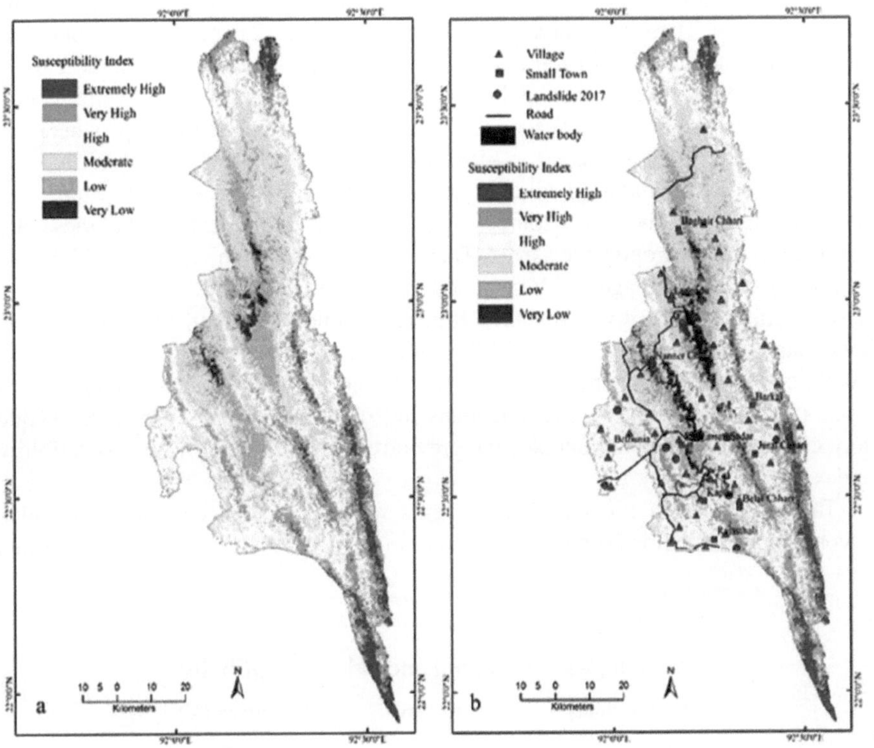

FIGURE 8.4 a) Landslide susceptibility index map and b) Landslide susceptibility index map with base map

8.3.2 Landslide Susceptibility Index (LSI)

The LSI delivered a defencelessness guide of the study zone, and the likelihood level of a landslide event was arranged in six categories. The landslide susceptibility was explored with geospatial distribution. The susceptibility index was normalized in six categories. According to Ayalew et al. (2004) the susceptibility categories assumed extremely high, very high, high, moderate, low and very low susceptibility. As a result, areas of about 3.76 km^2, 183.21km^2, 923.52 km^2, 3,442.51km^2, 629.97km^2 and 71.90km^2 were respectively included in the extremely high, very high, high, moderate, low and very low categories. Table 8.3 represents the spatial description of landslide susceptibility.

In this study area, the extremely high landslide susceptibility zone was 0.7 per cent; these areas are located in a fairly high piedmont landscape. These areas are ethnic tribal settlements; the inhabitants live in these susceptible hilly areas due to their religious beliefs and cultural customs and as a result they allow this vulnerable condition. Also, in the very high category was 3.48 per cent of the study area that also has piedmont land area. This category includes more inhabitants because this area is available for agricultural cultivation. The high category was 17.57 per cent of the total study area. These three high categories included 21.12 per cent of the total area of the Rangamati district. In addition, the previous landslides occurred in these three highly susceptible areas. Figure 8.4 shows the LSI map in section (a), while section (b) represents the landslide susceptibility category within a base map, where the previous events are denoted.

This research includes 77 villages according to Google image observation and national administrative data sources where human settlement exists; these are assumed as the anthropogenic aspect. There are 2 villages in the extremely high susceptibility zone, 5 villages with very high susceptibility and 12 villages with high susceptibility. These 19 villages have vulnerable conditions, while 40 villages have moderate susceptibility. A total of 18 villages are in low and very low susceptible condition. To ensure an emergency response is very difficult when a landslide event occurs because these are very remote areas, and communication is also damaged due to the landslide. Table 8.4 denotes the previous landslide events with susceptibility index.

The location of a previous seven events was detected; according to previous records, three landslide events were in a very high susceptibility zone and four events

TABLE 8.3
Spatial Description of Landslide Susceptibility

Susceptibility	Area (km^2)	Percentage (%)
Extremely High	3.76	0.07
Very High	183.21	3.48
High	923.52	17.57
Moderate	3,442.51	65.52
Low	629.97	12.11
Very Low	71.90	1.36

TABLE 8.4
LSI Level in Historical Landslide Events Eecords

Date	Relative Location	Geographic Location	Impact	LSI Level
30 May 1990	Jhagarbeel-Ghagra	92.014 E, 22.726 N	Not available	High
13 June 2017	Rangamati	92.258 E, 22.628 N	53 deaths	High
13 June 2017	Kaptai	92.343 E, 22.359 N	18 deaths	Very High
13 June 2017	Kawkhali	91.979 E, 22.518 N	23 deaths	High
13 June 2017	Jarachhari	92.329 E, 22.369 N	2 deaths	High
13 June 2017	Balaichhari	92.313 E, 22.487 N	3 deaths	Very High
12 June 2018	Naniarchar	92.111 E, 22.860 N	12 deaths	Very High

were in a high index zone. Furthermore, the landslide event locations were compared with GPS coordinate references, newspapers, Google images and the national Mouza (Village) database to justify the proper geolocation of events. These landslides occurred in very high and high zones of the susceptibility index map. The high LSI zone is actually a down-hill homestead area; as a result the events of this landslide were recorded by news agencies and the mass media. In addition, people vertically cut down the hilly apparent and established house; in this case, when the heavy rainfall started the landslide occurred in this area. As a result the lives of inhabitants were lost.

8.4 DISCUSSION

Land cover, slope, aspect, elevation, NDVI, distance to road, distance to waterbody, drainage density, geomorphological structure, urban growth and geology are the factors of landslides in the Chittagong urban area near the Rangamati district (Das & Raja, 2015; Plank, Twele & Martinis, 2016). The main factors of landslides are geological, meteorological and anthropogenic phenomena (Binaghi et al., 1998; Regmi, Giardino & Vitek, 2010). According to this concentration, rainfall intensity distribution, land use, elevation level, vegetation density, slope intensity and stream order distribution proved to be the main influencing factors of landslides in this study area. However, the fact that only six included influencing factors of landslide, due to national data-sharing restrictions, is the limitation of this study. Soil permeability data, topographical data and other ethnographic data are strictly confidential in the aspect of national security. In addition, this area is fully dominated by the military administration, due to previous conflict records between hilly ethnic tribes and the Bengali people. Furthermore, Ahmed et al. (2014) and Taj (2013) conducted landslide susceptibility research for Chittagong urban area using rainfall distribution, land use, elevation level, vegetation density, slope intensity, soil permeability and stream order data (Ahmed et al., 2014; Taj, 2013). In this circumstance, these six factors include the geological, meteorological and anthropogenic influencing phenomena of landslide events. All of the influencing factors are simulated in the same scale because WOM works with the rational, ordinal, nominal and interval scale factors with a common scale input parameter – for example, 1 to 9 or 1 to 10 scale, with the higher value being more significant, or the scale could be 0 to 1, defining the possibility of belonging to a particular set. According to this condition, all input parameters were normalized at 1 to 8 scale in this research to explore LSI. The LSI was normalized in six categories due to distinct susceptibility levels that would make more impression for awareness and preparedness.

8.5 CONCLUSION

Landslides have mostly occurred in the Rangamati area during pre-monsoon and full monsoon periods. The spatial distribution susceptibility of landslides is a result of the blend of many affecting parameters. A reliable and acceptable susceptibility map relies upon the inclusion and suitable determination of the purpose of those influencing parameters. In this research study, six landslide-controlling parameters – that is, rainfall, land use, elevation, vegetation, slope gradient and stream order distribution – were considered to satisfy geological, meteorological and anthropogenic contributions. In this study a WOM method was used wherein specific classes of each parameter are simulated using the geostatistical Euclidean distance algorithm, NDVI inversion and geospatial Kriging interpolation to produce a weighted input parameter. A weighted parameter is assigned in order to produce a landslide susceptibility index map. The outcome shows the susceptibility index analyses, and assessment allowed the research area to be divided into six zones of susceptibility: extremely high (0.07%), very high (3.48%), high (17.57%), moderate (65.52%), low (12.11%) and very low (1.36%). The findings of this research can be applied to track the way for further research by using this methodological framework to examine another region of study. Further analysis and discussion of these results will provide the base for further long-term studies on landslide-vulnerable areas and would have an implication for disaster management and disaster risk reduction strategies.

ACKNOWLEDGEMENTS

We would like to thank Ebadullah Khan, Associate Professor, Department of Geography and Environment, Jahangirnagar University, Bangladesh for his gracious support in revising the manuscript.

AUTHOR CONTRIBUTIONS

Nur Hussain conceived, designed, performed the data processing of the research and drafted the paper; Mohd. Shamsul Alam and Li Suju provided research suggestions and Muhammad Rizwan contributed to the manuscript revision.

REFERENCES

Ahmed, Bayes. 2013. "Understanding the Issues Involved in Urban Landslide Vulnerability in Chittagong Metropolitan Area, Bangladesh". Dhaka.

Ahmed, Bayes. 2015. "Landslide Susceptibility Mapping Using Multi-Criteria Evaluation Techniques in Chittagong Metropolitan Area, Bangladesh". *Landslides* 12 (September 2014): 1077–95. https://doi.org/10.1007/s10346-014-0521-x.

Ahmed, Bayes, Shahinoor Rahman, Sonia Rahman, Ferdous Farhana Huq and Sharmin Ara. 2014. "Landslide Inventory Report of Chittagong Metropolitan". Dhaka.

Avinash, K G, and K G Ashamanjari. 2010. "A GIS and Frequency Ratio Based Landslide Susceptibility Mapping: Aghnashini River Catchment, Uttara Kannada, India". *International Journal of Geomatics and Geosciences* 1 (3): 343–54.

Ayalew, Lulseged, Hiromitsu Yamagishi, and Norimitsu Ugawa. 2004. "Landslide Susceptibility Mapping Using GIS-Based Weighted Linear Combination, the Case in Tsugawa Area of Agano River, Niigata Prefecture, Japan". *Landslides* 1 (November 2003): 73–81. https://doi.org/10.1007/s10346-003-0006-9.

BBC. 2013. "Bangladesh: Rain Triggers Landslides and Leaves 107 Dead". *The BBC News*, June 2013.

BBS. 2015. "Population & Housing Census 2011".

Binaghi, E., L. Luzi, P. Madella, F. Pergalani, and A. Rampini. 1998. "Slope Instability Zonation: A Comparison between Certainty Factor and Fuzzy Dempster-Shafer Approaches". *Natural Hazards* 17 (1): 77–97. https://doi.org/10.1023/A:1008001724538.

Brammer, Hugh. 2014. "Climate Risk Management Bangladesh' s Dynamic Coastal Regions and Sea-Level Rise". *Climate Risk Management* 1: 51–62. https://doi.org/10.1016/j.crm.2013.10.001.

Capecchi, V., M. Perna, and A. Crisci. 2015. "Statistical Modelling of Rainfall-Induced Shallow Landsliding Using Static Predictors and Numerical Weather Predictions: Preliminary Results". *Natural Hazards and Earth System Science* 15 (2): 75–95. https://doi.org/10.5194/nhess-15-75-2015.

Catani, Filippo, Veronica Tofani, and Daniela Lagomarsino. 2016. "Spatial Patterns of Landslide Dimension: A Tool for Magnitude Mapping". *Geomorphology* 273: 361–73. https://doi.org/10.1016/j.geomorph.2016.08.032.

Chalkias, Christos, Maria Ferentinou, and Christos Polykretis. 2014. "GIS-Based Landslide Susceptibility Mapping on the Peloponnese Peninsula, Greece". *Geosciences* 4: 176–90. https://doi.org/10.3390/geosciences4030176.

Chen, Wei, Hongxing Han, Bin Huang, Qile Huang, and Xudong Fu. 2017. "Variable-Weighted Linear Combination Model for Landslide Susceptibility Mapping: Case Study in the Shennongjia Forestry District, China". *Internationa Journal OfGeo-Information* 6 (4): 133. https://doi.org/10.3390/ijgi6110347.

Ciampalini, Andrea, Federico Raspini, Silvia Bianchini, William Frodella, Federica Bardi, Daniela Lagomarsino, Federico Di, et al. 2015. "Geomorphology Remote Sensing as Tool for Development of Landslide Databases: The Case of the Messina Province (Italy) Geodatabase". *Geomorphology* 249: 103–18. https://doi.org/10.1016/j.geomorph.2015.01.029.

Dai, F C, and C F Lee. 2002. "Landslide Characteristics and Slope Instability Modeling Using GIS, Lantau Island, Hong Kong". *Geomorphology* 42: 213–28.

Das, Sourav, and Debasish Roy Raja. 2015. "Susceptibility Analysis of Landslide in Chittagong City Corporation Area", Conference Proceedings Paper: 1st International Electronics Conference on Remote Sensing, 22 June to 5 July 2015 (Open Access). 1–31.

Devi, Anita, M Israil, R Anbalagan, and Pravin K Gupta. 2017. "Subsurface Soil Characterization Using Geoelectrical and Geotechnical Investigations at a Bridge Site in Uttarakhand Himalayan Region". *Journal of Applied Geophysics* 144: 78–85. https://doi.org/10.1016/j.jappgeo.2017.07.005.

Esri. 2012. *ArcGIS Online Quick Start Guide*. California.

GoB. 2010. "National Plan for Disaster Management 2010–2015". Dhaka.

Gokceoglu, Candan, Harun Sonmez, Hakan A. Nefeslioglu, Tamer Y. Duman, and Tolga Can. 2005. "The 17 March 2005 Kuzulu Landslide (Sivas, Turkey) and Landslide-Susceptibility Map of Its near Vicinity". *Engineering Geology* 81 (March): 65–83. https://doi.org/10.1016/j.enggeo.2005.07.011.

Hussain, N., Firdaur, F., Rizwan, M. (2021). Remote Sensing of Photosynthesis, Vegetation Productivity and Climate Variability in Bangladesh. In Singh, R. (ed.) *Re-envisioning Remote Sensing Applications: Perspectives from Developing Countries*. First Edition. CRC Press. https://doi.org/10.1201/9781003049210

Jian, Wang, and Peng Xiang-Guo. 2009. "GIS-Based Landslide Hazard Zonation Model and Its Application". *Procedia Earth and Planetary Science* 1 (1): 1198–1204. https://doi.org/10.1016/j.proeps.2009.09.184.

Kalantar, Bahareh, Biswajeet Pradhan, Seyed Amir Naghibi, Alireza Motevalli, and Shattri Mansor. 2018. "Assessment of the Effects of Training Data Selection on the Landslide Susceptibility Mapping: A Comparison between Support Vector Machine (SVM), Logistic Regression (LR) and Artificial Neural Networks (ANN)." *Geomatics, Natural Hazards and Risk* 9 (1): 49–69. https://doi.org/10.1080/19475705.2017.1407368.

Kirschbaum, Dalia Bach, Robert Adler, Yang Hong, Stephanie Hill, and Arthur Lerner-Lam. 2010. "A Global Landslide Catalog for Hazard Applications: Method, Results, and Limitations". *Natural Hazards* 52 (3): 561–75. https://doi.org/10.1007/s11069-009-9401-4.

Kumar, Ranjan, Shuichi Hasegawa, Atsuko Nonomura, and Minoru Yamanaka. 2008. "Geomorphology Predictive Modelling of Rainfall-Induced Landslide Hazard in the Lesser Himalaya of Nepal Based on Weights-of-Evidence". *Geomorphology* 102 (3–4): 496–510. https://doi.org/10.1016/j.geomorph.2008.05.041.

Lee, Saro. 2001. "Statistical Analysis of Landslide Susceptibility at Yongin, Korea". *Environmental Geology* 40: 1095–1113. https://doi.org/10.1007/s002540100310.

Mancini, F., C. Ceppi, and G. Ritrovato. 2010. "GIS and Statistical Analysis for Landslide Susceptibility Mapping in the Daunia Area, Italy". *Natural Hazards and Earth System Science* 10 (9): 1851–64. https://doi.org/10.5194/nhess-10-1851-2010.

Mantovani, Franco, Robert Soeters, and C.J. Van Westen. 1996. "Remote Sensing Techniques for Landslide Studies and Hazard Zonation in Europe". *Geomorphology* 15 (3–4): 213–25. https://doi.org/10.1016/0169-555X(95)00071-C.

Mossa, Stefania, Domenico Capolongo, Luigi Pennetta, and Janusz Wasowski. 2005. "A GIS-Based Assessment of Landsliding in the Daunia Apennines, Southern Italy". *Polish Geological Institute Special Papers* 20 (January): 86–91.

NIRAPAD. 2017. "Landslides Situation in Southeast Bangladesh, June 2017".

Patwary, M.A.A., P.K. Champati Ray, and Irshad Parvaiz. 2009. "IRS-LISS-III and PAN Data Analysis for Landslide Susceptibility Mapping Using Heuristic Approach in Active Tectonic Region of Himalaya". *Journal of Indian Socciety of Remote Sensing* 37 (September): 493–509.

Petley, David N., Gareth J. Hearn, Andrew Hart, Nicholas J. Rosser, Stuart A. Dunning, Katie Oven, and Wishart A. Mitchell. 2007. "Trends in Landslide Occurrence in Nepal". *Natural Hazards* 43 (1): 23–44. https://doi.org/10.1007/s11069-006-9100-3.

Plank, Simon, André Twele, and Sandro Martinis. 2016. "Landslide Mapping in Vegetated Areas Using Change Detection Based on Optical and Polarimetric SAR Data". *Remote Sensing* 8 (4). https://doi.org/10.3390/rs8040307.

Pradhan, Biswajeet, and Saro Lee. 2010. "Environmental Modelling & Software Landslide Susceptibility Assessment and Factor Effect Analysis: Backpropagation Artificial Neural Networks and Their Comparison with Frequency Ratio and Bivariate Logistic Regression Modelling". *Environmental Modelling and Software* 25 (6): 747–59. https://doi.org/10.1016/j.envsoft.2009.10.016.

Regmi, Netra R, John R Giardino, and John D Vitek. 2010. "Geomorphology Modeling Susceptibility to Landslides Using the Weight of Evidence Approach: Western Colorado, USA". *Geomorphology* 115 (1–2): 172–87. https://doi.org/10.1016/j.geomorph.2009.10.002.

Roslee, Rodeano. 2012. "Intergration of GIS Using GEOSTAtistical INterpolation Techniques (Kriging) (GEOSTAINT-K) in Deterministic Models for Landslide Susceptibility Analysis (LSA) at Kota Kinabalu, Sabah, Malaysia". *Journal of Geography and Geology* 4 (1): 18–32. https://doi.org/10.5539/jgg.v4n1p18.

Roslee, Rodeano, Alvyn Clancey Mickey, Norbert Simon, and Mohd Norazman Norhisham. 2017. "Landslide Susceptibility Analysis (LSA) Using Weighted Overlay Method (WOM) Along The Genting Sempah To Bentong Highway, Pahang". *Malaysian Journal of Geosciences* 1 (2): 13–19.

Shi-Biao, BAI, Wang Jian, LUGuo-Nian, Zhou Ping-Gen, Hou Sheng-Shan, and Xu Su-Ning. 2009. "GIS-Based and Data-Driven Bivariate Landslide-Susceptibility Mapping in the Three Gorges Area, China". *Pedosphere* 19: 14–20. https://doi.org/10.1016/S1002-0160(08)60079-X.

Singh, R. (ed.) 2021 *Re-envisioning Remote Sensing Applications: Perspectives from Developing Countries*. First Edition. CRC Press. https://doi.org/10.1201/9781003049210

Spiker, C., and Paula L. Gori. 2003. "National Landslide Hazards Mitigation Strategy – A Framework for Loss Reduction". *U.S. Geological Survey*. https://doi.org/PSW-GTR-130.

Sultana, N. and Casadevall, S. R. 2020. "Analysis of landslide-induced fatalities and injuries in Bangladesh: 2000–2018". *Cogent Social Sciences*, 6 (1). DOI: 10.1080/23311886.2020.1737402

Taj, Susanta. 2013. "Landslide Disaster in Bangladesh: A Case Study of Chittagong University Campus". *International Journal of Research in Applied* 1 (6): 35–42.

Tien, Dieu, Biswajeet Pradhan, Owe Lofman, Inge Revhaug, and Oystein B Dick. 2012. "Catena Spatial Prediction of Landslide Hazards in Hoa Binh Province (Vietnam): A Comparative Assessment of the Ef Fi Cacy of Evidential Belief Functions and Fuzzy Logic Models". *Catena* 96: 28–40. https://doi.org/10.1016/j.ecatena.2012.04.001.

Vanmaercke, Matthias, Francesca Ardizzone, Mauro Rossi, and Fausto Guzzetti. 2017. "Geomorphology Exploring the Effects of Seismicity on Landslides and Catchment Sediment Yield: An Italian Case Study". *Geomorphology* 278: 171–83. https://doi.org/10.1016/j.geomorph.2016.11.010.

Yeon, Young-kwang, Jong-gyu Han, and Keun Ho. 2010. "Landslide Susceptibility Mapping in Injae, Korea, Using a Decision Tree". *Engineering Geology* 116 (3–4): 274–83. https://doi.org/10.1016/j.enggeo.2010.09.009.

Zuvala, Robert, Eva Fišerová, and Lukáš Marek. 2016. "Mathematical Aspects of the Kriging Applied on Landslide in Halenkovice (Czech Republic)." *Open Geosciences* 8 (1): 275–88. https://doi.org/10.1515/geo-2016-0023.

9 Remote Sensing Application in Landslide Assessments

Case of Kotropi in Himachal Himalayas

Kapil Batra and Neeru Kaushal
Lovely Professional University, Phagwara, India

CONTENTS

9.1 INTRODUCTION

A landslide is the movement of a mass of rock, particles, or earth down an incline, under the effect of gravity. Various factors are the reasons for landslides, including excessive or expanded precipitation, tremors, snow softening, and a repercussion of human games. An assortment of movements is related with landslide streaming, sliding, overturning or falling movements, and numerous landslides feature a blend of or additional assortments of movements (Varnes, 1978; Cruden & Varnes, 1996). In India, about 0.42 million km^2 or 12.6 per cent of land territory, barring snow-covered districts, is in danger of landslide hazard. Out of this, 0.18 million km^2 falls in northeast Himalaya, alongside Darjeeling and Sikkim Himalaya; 0.14 million km^2 falls in northwest Himalaya (Uttarakhand, Himachal Pradesh and Jammu and Kashmir); 0.09 million km^2 in Western Ghats and the Konkan slopes (of Tamil Nadu, Kerala, Karnataka, Goa and Maharashtra) and 0.01 million km^2 in the Eastern Ghats of Aruku area in Andhra Pradesh. The landslide-prone Himalayan territory falls inside the most quake-inclined zones (areas iv and v) (BIS, 2002) in which seismic tremors of

DOI: 10.1201/9781003224624-12

Modified Mercalli Intensity viii to ix can emerge, and consequently earthquakes occur, causing landslides. Combatting and alleviating the landslide hazard and danger is a considerable undertaking for the technocrats and determination producers worldwide as 80 per cent of the proposed fatalities due to landslide occur in developing countries.

In the state of Himachal Pradesh, landslide activities predominantly occur during the monsoon rainy season. On 13 August 2017, a gigantic landslide occurred at Kotropi on National Highway 154 (from Mandi to Pathankot) in the Himachal Himalayas. Media reports and data surveys recorded 46 fatalities from the occurrence (*The Tribune*, 2017, 14 August). Around 300 m of the highway was totally covered by debris from the landslide.

9.2 STUDY AREA

The area chosen for the study is along National Highway 154, passing through highly rugged terrain in the Mandi district of Himachal Pradesh, which is vulnerable to frequent landslides every year. The state of Himachal Pradesh is divided into three major divisions – namely, Shimla, Mandi and Kangra divisions. The tectonically active Himalayan mountains are known for their fragile landscape and frequent geological hazards, among which landslides are regular threats to this region.

The Kotropi region lies between 31° 91′ 21″ N latitude and 76° 88′ 79″ E longitude, which is geologically thrust between the Shivalik and Shali groups of rocks with dolomites, brick red shale, sandstones, clay and mudstones. These rocks, being weaker in strength, are highly prone to landslides (ISRO, 2015).

9.3 APPLICATION OF REMOTE SENSING IN LANDSLIDE STUDIES

The application of remote sensing in disaster monitoring and disaster management is a proven fact (Eguchi et al., 2008). Landslide hazards and the occurrence of such landslides can be very well analysed through an array of satellite images. Usually, landslides depend on many factors, such as heavy rainfall, earthquake, soil erosion, slope, construction activities, and land use of an area. Landslides can also be predicted and forecast, depending on the availability of detailed geospatial data (Chae et al., 2017). Various remote sensing applications in diverse fields have been elaborated through different case studies from developing countries (Singh, 2021).

Figure 9.1 depicts the Kotropi region in Sentinel images of two different times. Picture A portrays Kotropi on 11 August 2017, two days prior to the landslide, and shows the slight linear demarcation of the slip, already taking place, while picture B shows the area almost a week after the Kotropi landslide. One can easily visualize the area affected by the landslide on 13 August 2017.

9.4 RESULTS AND DISCUSSION

Experimental work was also performed earlier (Batra, 2018; Batra & Kaushal, 2018) after the occurrence of landslides to check the engineering properties of the samples. Engineering properties of the soil are elaborated here to analyse in detail the landslide causes from an engineering perspective. Sieve analysis of the soil samples from

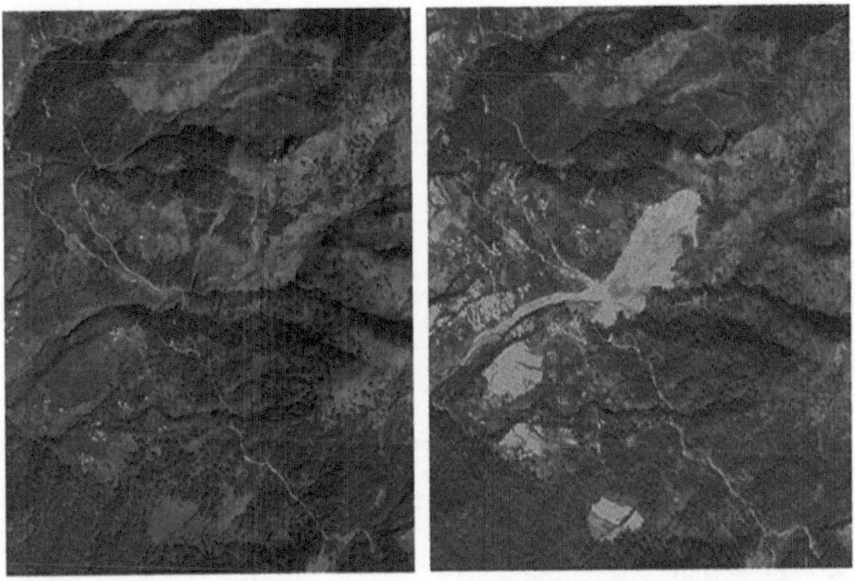

FIGURE 9.1 Staudy Area Before (11 Aug, 2017) and After (19 Aug, 2017) the Landslides (Sentinel Images)

TABLE 9.1
Sieve Analysis Results of Sampled Soil

Serial Number	Sieve Size (mm)	Retaining Weight of Soil (gram)	% of Retaining Soil
1	4.75	0	0
2	2.36	100	10
3	1	362	36.2
4	0.850	11.03	1.103
5	0.600	64	6.4
6	0.425	107.53	10.753
7	0.212	111.41	11.141
8	0.150	36.15	3.615
9	0.075	76	7.6
10	Pan	131.88	13.188

Source: Batra, 2018

the site are depicted in Table 9.1. Sieve analysis of the soil sample gets the details about MIT classification, which shows 10 per cent gravel, 69.212 per cent sand, and 20.788 per cent silt and clay particles found. Specific gravity of 2.677 reflects that percentage of chlorite was more in the soil sample.

The standard Proctor test helps to find the dry density and optimum moisture content (OMC) of the soil sample. Table 9.2 and Figure 9.2 show the result of the percentage of water content and dry density of the sampled soil. OMC of 16.3 per cent was observed, which was further used for the direct shear test and UCS test.

TABLE 9.2
Standard Proctor Test Results

Water Content in %	Dry Density
4.691	1.485
8.6	1.552
12.3	1.627
16.3	1.764
22.0	1.670
24.1	1.580
29.0	1.362

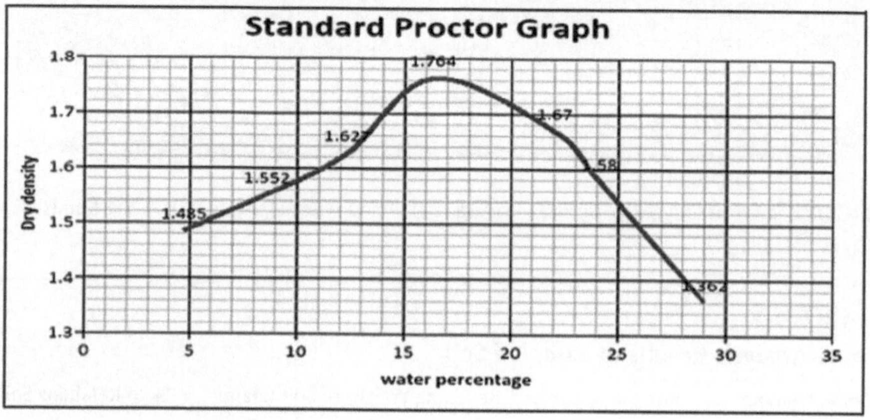

FIGURE 9.2 Standard Proctor Graph for Sampled Soil

Table 9.3 shows the shear strength of the soil with different shear stress. The direct shear test helps to find the shear strength, C and Φ of the given soil samples in Table 9.4. It shows the shear force failure results and calculates the value of the shear strength. It was observed from the experiment that the value of C found 0 and Φ of 33 degrees, as shown in Figure 9.2. Here that value of C – that is 0 – means the soil is cohesionless and also it was proved from the sieve analysis that this soil is sandy soil (Figure 9.3).

TABLE 9.3
Load vs Shear Strength

Normal Stress (Kg/cm²)	Shear Strength (Kg/cm²)
0.05	0.032
0.10	0.0649
0.15	0.0975

TABLE 9.4
Displacement vs Proving Ring, Samples 1, 2 and 3

Sample 1		Sample 2		Sample 3	
Normal Stress (Kg/Cm^2)	**0.05**	**Normal Stress (Kg/Cm^2)**	**0.10**	**Normal Stress (Kg/Cm^2)**	**0.15**
Displacement Ring (Dial Gauge)	Proving Ring	Displacement Ring (Dial Gauge)	Proving Ring	Displacement Ring (Dial Gauge)	Proving Ring
20	3.1	20	7.1	20	11
40	4.1	40	7.3	40	12.1
60	5.3	60	7.4	60	12.8
80	6.2	80	7.8	80	13
100	7.1	100	8.0	100	13.4
120	7.3	120	8.1	120	14
140	**7.3**	140	8.4	140	14.4
160	7.2	160	8.7	160	14.9
		180	9.1	180	15
		200	9.5	200	15.1
		220	9.8	220	15.2
		240	10	240	15.9
		260	10.2	260	16.1
		280	11	280	16.8
		300	11.3	300	17.2
		320	11.4	320	17.4
		340	12.9	340	18.3
		360	13	360	18.9
		380	**13**	380	19
		400	12.7	400	19.1
				420	19.4
				440	21.8
				460	21.9
				480	22
				500	**22**
				520	21.5

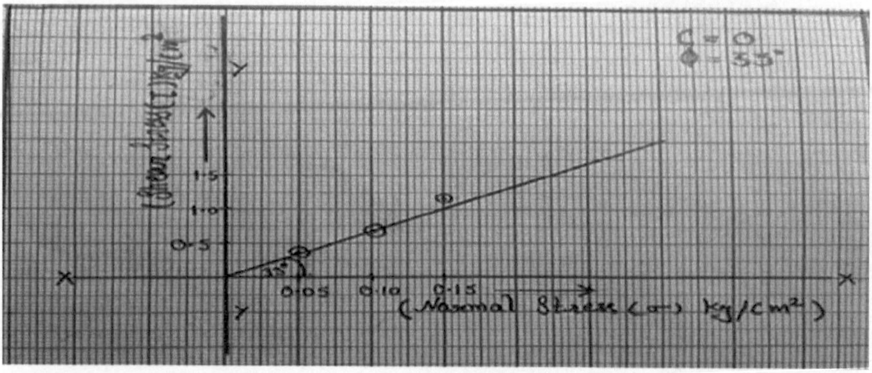

FIGURE 9.3 Association between the normal stress and shear strength

9.5 CONCLUSIONS AND MAIN FINDINGS

The massive slide of Kotropi is a third reactivation incidence of debris flow, which has occurred at 20-year intervals, in 1977, 1997 and 2017. The main causes of these slides have been the heavy rainfall and the instability of the slope. Due to old debris and post-slope failure, there is a possibility of a reoccurrence of slope failure from the upper slope of the summit area of the main slide as well as the debris that is already accumulated on the slope and valley side. Therefore, it is necessary to consider the temporal aspect of landslide hazard in order to carry out a complete quantitative assessment of the phenomenon. Sieve analysis results found that 69.212 per cent of the soil is sandy, due to which its plastic limit is impossible to find. OMC of 16.3 per cent has been observed. And in the direct shear test, Φ was 33°, C=0 and the mean shear strength was found to be 0.0648 kg/cm^2, which signifies a lower shear strength of soil which has been a cause of landslides.

To prevent the recurrence of landslides, adequate and properly designed culverts should be provided to channel the water coming from up-slope. Suitable vegetation should be planned to help slope stability by the removal of water and the reduction of surface soil erosion. The outcomes of this assessment will help in further research aimed at considering rainfall-induced landslides and their mitigation. Secondly, the relationship between landslide activity and rainfall can be established for the state and Western Himalayan region, by studying other variations in the rainfall pattern. A greater role can be played by remote sensing in preparing the landslide hazard zonation, based on various temporal and spatial resolution data in the future.

REFERENCES

Batra, K. (2018). *Engineering Analysis of the Kotropi Landslide*. M. Tech Dissertation submitted to NIT, Kurukshetra.

Batra, K. and Kaushal, N. (2018). A Case study of Kotropi Landslide in Mandi District. *Journal of Emerging Technologies and Innovative Research*, 5 (12): 1147–1154.

BIS. (2002). *Criteria for Earthquake Resistant Design of Structures: General Provisions and Buildings*, Bureau of Indian Standards (BIS), IS 1893: 2002, Part-1.

Chae, B. G., Park, H. J., Catani, F. et al., (2017). Landslide Prediction, Monitoring and Early Warning: A Concise Review of the State-of-the-Art. *Geosciences Journal*, 21: 1033–1070. doi:10.1007/s12303-017-0034-4

Cruden, D. M. and Varnes, D. J. (1996). Landslide types and processes. In: Turner, A. K. and Schuster, R. L. (eds.) *Landslides Investigations and Mitigation*, Transportation Research Board, US National Research Council. Special Report, 247. Washington, DC, Chapter 3, pp. 36–75.

Eguchi, R. T., Huyck, C. K., Ghosh, S. and Adams, B. J. (2008). The Applications of Remote Sensing Technologies for Disaster Management, *14th World Conference on Earthquake Engineering*, Oct. 12–17, 2008, Beijing, China.

ISRO (2015). *Version-2 Kotropi landslide, Mandi district, Himachal Pradesh: A preliminary report*, National Remote Sensing Centre (NRSC), ISRO, Hyderabad.

Singh, R. (ed.) (2021) *Re-envisioning Remote Sensing Applications: Perspectives from Developing Countries*. First Edition. CRC Press. doi:10.1201/9781003049210

The Tribune (14 Aug, 2017). 46 dead as landslide buries two buses in Mandi, rescue ops to resume tomorrow. Retrieved from: https://www.tribuneindia.com/news/himachal/46-dead-as-landslide-buries-two-buses-in-mandirescue-ops-to-resume-tomorrow/451552.html.

Varnes, D. J. (1978). Slope movement types and processes. In: Schulster, R. L. and Krizek, R. J. (eds). *Landslides, Analysis and Control, Special Report 176*: Transport Research Board, National Academy of Sciences, Washington, DC., pp. 11–33.

10 Geospatial Techniques to Quantify Urban Change

The Case of Harare, Zimbabwe

James Magidi

Tshwane University of Technology, Pretoria, South Africa

Luxon Nhamo

Water Research Commission of South Africa (WRC), Pretoria, South Africa

Webster Gumindoga

University of Zimbabwe, Mt Pleasant, Harare, Zimbabwe

Paidamwoyo Mhangara

University of Witwatersrand, Johannesburg, South Africa

Walter Musakwa

University of Johannesburg, Johannesburg, South Africa

Sylvester Mpandeli

Water Research Commission of South Africa (WRC), Pretoria, South Africa

University of Venda, Thohoyandou, South Africa

Tafadzwanashe Mabhaudhi

University of KwaZulu-Natal, Pietermaritzburg, South Africa

International Water Management Institute (IWMI-GH), West Africa, Kumasi, Ghana

DOI: 10.1201/9781003224624-13

CONTENTS

10.1 INTRODUCTION

Most cities in developing countries are struggling to cope with rapid urbanization emanating from high migration rates and increasing population (Beard, Mahendra & Westphal, 2016). The formulation of coherent adaptation strategies to manage rapid urbanization and cope with the high influx of people is a crucial pathway to enhance urban areas into centres of human development and transform them into sites for climate action (McGranahan, Schensul & Singh 2016). Achieving sustainable urban areas will go a long way toward achieving the United Nation (UN)'s Sustainable Development Goals (SDGs), particularly Goals 8 and 11. A recent study of 25 cities in Africa showed that most of them are growing at an average rate of between 4 to 5 per cent per annum, but they are failing to cope (Xu et al., 2019). The failure to cope with rapid urbanization is causing a host of social, economic and environmental challenges, such as overcrowding, poor service delivery, urban sprawl, pollution, poor amenities, fire hazards due to high temperatures, flood risks and high crime rates (Cohen, 2006). The UN has recognized the urgency of providing solutions to these challenges and has developed several targets to meet the 2030 global agenda on sustainable cities (Giles-Corti, Lowe & Arundel, 2019).

In the case of Zimbabwe, a series of events that took place since 1990 triggered rapid urbanization, particularly in the capital city of Harare. These events include the Economic Structural Adjustment Programme (ESAP) in 1992 (Potts & Mutambirwa, 1998; Ersado, 2006; Chattopadhyay, 2000), and the Fast Track Land Reform Programme in 2000 (Chambati, 2011; Groves, 2012). These two policies in particular caused economic recessions in the country that resulted in an exodus of people from rural areas into the more affluent urban areas, due to loss of livelihoods and employment opportunities in the once vibrant agricultural sector (Potts, 2006; Dzimiri & Runhare, 2012). The rural-urban migration has seen significant transformations in urban landscapes, as, for example, in Harare alone there was an increase in built-up area from 12.6 to 36.3 per cent between 1994 and 2013 (Kamusoko, Gamba & Murakami, 2013); Olawole et al., 2011). The built environment increased from 279.5 km^2 in 1984 to 445 km^2 in 2018 in Harare, of which low-income residential areas surged from 51.2 km^2 to 218.4 km^2 over the same period (Marondedze & Schütt, 2019). Such rapid growth has witnessed the deterioration of services such as water supply and sanitation, with most suburbs going without water for weeks.

Accurate assessment of urban change provides decision makers with evidence to make informed management and policy decisions (Magidi & Ahmed, 2018; Abebe, 2013). Geospatial applications provide the most time- and cost-effective methods to assess and map urban land-use change (Magidi & Ahmed, 2018; Abebe, 2013). The availability of long-term satellite images and the storage of demographic and climatic data in raster image format have facilitated quantifying changes in urban landscapes more accurately (Hegazy & Kaloop, 2015). In addition to traditional methods of land-use classification (supervised, unsupervised, or object-oriented classifications), recently developed techniques like machine learning algorithms are yielding more accurate classifications (Rodriguez-Galiano et al., 2012). For example, the random forest method (a tree-based classifier, and a non-parametric algorithm) yields highly accurate maps of complex land uses. The derived maps are combined with other geospatial metrics to assess changes in urban landscapes for sustainable land-use management (Abebe, 2013; McGarigal et al., 2002).

Shannon's diversity index (SHEI), and patch richness density (PRD) were the main landscape metrics that were used (Nhamo & Pius, 2013; Aithal & Ramachandra, 2013). Shannon's entropy is the spatial concentration or dispersion of land uses and is important for quantifying and differentiating urban sprawl (Aithal & Ramachandra, 2013). The entropy value ranges between 0 and 1. When the spread is clustered, the value is 0, and in evenly scattered distribution, the value is 1 (Aduah & Baffoe, 2013). The values are dependent on spatial characteristics (Jat, Garg & Khare, 2008). Configuration metrics relate to the spatial order of patches (Aithal & Ramachandra, 2013), and the class area (CA) index represents both the total built-up area and non-built-up area (Araya & Cabral, 2010).

The concentration of built-up areas, including impervious pavements, is the major reason for the rise in temperatures in urban areas, which is exacerbated by high concentration of pollutants and microclimates (Gaitani et al., 2011). Urban areas have been described as urban heat islands because of the impervious surfaces which cause high albedo, thermal capacity and heat conductivity, resulting in high temperatures (Sheng et al., 2017; Peres et al., 2018). Previous studies have found that the correlation between urban sprawl and land surface temperature (LST) is positive. Vegetation cover and soil background influence emissivity and vegetation vigour; hence the use of the normalized difference vegetation index (NDVI) in calculating emissivity (Sheng et al., 2017). When calculating LST, ground emissivity and atmospheric corrections are therefore considered.

Given these huge transformations taking place in the urban landscape, we used geospatial techniques to provide policy and decision makers with information on spatial changes taking place in urban landscapes and future impacts on socio-ecological systems. A discussion is made on the role of transformative approaches such as nexus planning, circular economy, and scenario planning in transforming urban areas into sites of climate action and adaptation in achieving SDGs 8 and 11.

10.2 METHODS

10.2.1 Study Area

Harare Metropolitan Province (Figure 10.1) includes the capital city of Zimbabwe (Harare city) and is one of the ten provinces in the country, incorporating Chitungwiza and Epworth. The city has an area of 960.6 km², stands at an altitude of 1,490 m, and has an estimated population of 1,530,000. It is the financial, commercial and communications hub of Zimbabwe (Gumbo et al., 2018). Temperatures are high in summer and low in winter. Its annual rainfall ranges between 470 mm and 1,350 mm.

Harare is the economic hub of Zimbabwe, dominated by agricultural activities. Harare has experienced exponential growth since the attainment of the country's independence in 1980 (Gumbo et al., 2018). However, such rapid urbanization has come with the challenges of failing infrastructure, poor service delivery and high unemployment rate.

10.2.2 Methodological Framework

The flowchart (Figure 10.2.) illustrates the procedure used to draw conclusions on the environmental, climatic, geographic and physical changes that have been taking place in Harare over the period from 1986 to 2018. The study acquired Landsat Thematic Mapper (Landsat 5) and Landsat Operation Land Imager (OLI) from the United States Geological Surveys' Earth Explorer, which were used in land cover

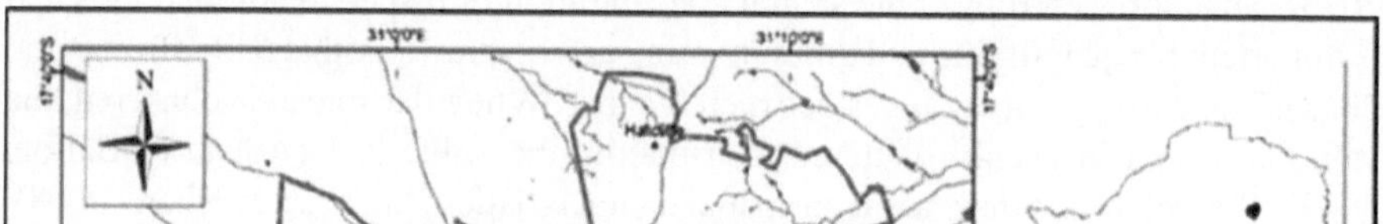

FIGURE 10.1 Locational map of Harare in Zimbabwe, showing the suburban areas

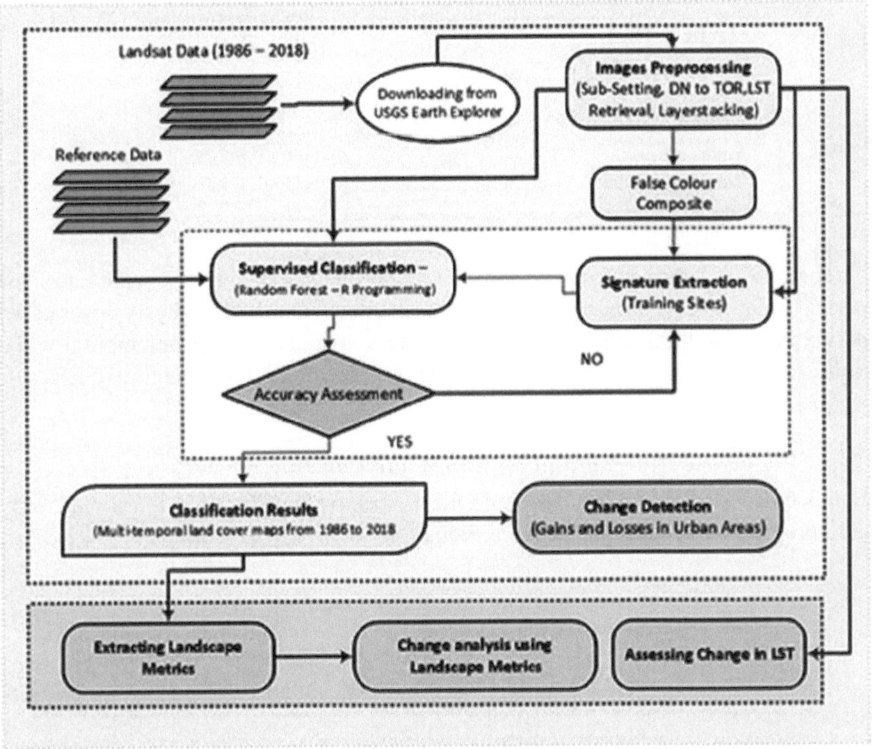

FIGURE 10.2 The flowchart showing the methodology used to extract urban areas

mapping. Data from the USGS data portal (Earth Explorer) were downloaded and verified geometrically and radiometrically using the QGIS semi-automatic classification tool (SCT). Georectified images were layer-stacked, resampled, and clipped to the boundary of the city of Harare. The R program was set on the QGIS software where it was used to classify the images with the random forest classifier. The R program managed to classify urban and non-urban areas (Table 10.1). The kappa accuracy model produced an acceptable accuracy of 75 per cent. Change detection was achieved through the landscape metrics to assess the gains in built-up area.

10.2.3 Data Sources and Land-use Classification

Landsat TM and OLI satellite images (1986, 1997, 2008 and 2018) were obtained from NASA's Earthdata portal and used to quantify changes in urban landscapes over the years. The random forest classifier separated the built environment from other land uses (Table 10.1) through the R programming language. The SCT converted digital numbers into top-of-atmosphere (TOA) reflectance. Classified images were sieved to eliminate small areas, and landscape metrics were used to classify the built environment. The kappa accuracy for 1986, 1997, 2008 and 2018 was 75.3%, 78.2%, 81.5% and 83.6%, respectively.

TABLE 10.1
Land-use/Cover Classes Used

Land-use/Cover Class	Description
Built-up areas	All built environment
Non-built-up areas	All ecological infrastructure

The LST was derived from the MOD11A2 products obtainable from the Earthdata portal (8-day interval and 1 km resolution dataset). The LST analysis covered the period between 2000 and 2019, and was used to evaluate the spatiotemporal variations. An algorithm to retrieve the LST was run on Google Earth Engine (GEE).

10.2.4 Retrieval of LST

Bands 6 and 10 for Landsat TM and OLI, respectively, were used to estimate temperature from spectral radiance using Equation 10.1 (Artis & Carnahan, 1982).

$$T_{Sat} = \frac{K_2}{\ln\left(\left(\frac{K_1}{L_\lambda}\right)+1\right)} \quad (10.1)$$

where: $T_{Sat} = At$ and represents satellite-derived temperature (K), L_λ = TOA and represents spectral radiance, K_1 and K_2 represent the band-specific thermal conversion constant (Table 10.2) (Kumar, Bhaskar & Padmakumari, 2012; Chander & Markham, 2003).

10.2.5 Calculating NDVI

The NDVI (Equation 10.2) was used as a standard algorithm that assesses vegetation greenness (Kumar & Shekhar, 2015). The NDVI evaluates the above-ground biomass quantity from red and infrared bands (Kumar & Shekhar, 2015). Green vegetation absorbs the red wavelengths because of chlorophyll while reflecting near-infrared wavelengths, and unhealthy leaves reflect the red band and absorb the near-infrared bands (Kumar & Shekhar, 2015). NDVI ranges between –1 and 1 and values between –1 and 0 where the NDVI above 0.1 is for vegetated areas (Laosuwan, Gomasathit & Rotjanakusol, 2017).

TABLE 10.2
Landsat Thermal Conversion Constants

Constant	Landsat 4	Landsat TM	Landsat ETM+	Landsat OLI
K1	671.62	607.76	666.09	In the metadata
K2	1,284.30	1,260.56	1,282.71	In the metadata

$$NDVI = \frac{(\text{NIR} - \text{RED})}{(\text{NIR} + \text{RED})} \tag{10.2}$$

where: NIR = near-infrared and RED = red band.

10.2.6 Calculating Emissivity

Emissivity is a wavelength function derived from environmental factors (water content, vegetation density, chemical composition, etc.). There is a correlation between emissivity and NDVI (Sobrino, Jiménez-Muñoz & Paolini, 2004). The conditional formulae used to compute emissivity are shown in Table 10.3.

10.2.7 Calculating LST

Satellite temperature (T_{Sat}) and emissivity were used to compute LST (LST/Tsurface) using Equation 10.3 (Chander & Markham, 2003; Artis & Carnahan, 1982; Weng, Lu & Schubring, 2004) and this was performed in QGIS.

$$T_{Surface} = \frac{T_{Sat}}{\left(1 + \left[\left(\frac{\lambda T_{Sat}}{\rho}\right)\ln(\varepsilon)\right]\right)} \tag{10.3}$$

where: T_{sat} is satellite temperature, λ = emitted radiance and average wavelengths and ε is the surface emissivity (Table 10.3) (Chander & Markham, 2003).

10.2.8 Population Data

Population data was obtained from the WorldPop population data (www.worldpop.org), which are recomputed every five years. Population datasets for the respective years were overlaid on the delineated built-up areas for the years under review to appreciate the changes in population over time.

TABLE 10.3
Land Surface Emissivity Estimation Equations

NDVI	Surface Emissivity (ε)
NDVI <-0.185	0.1
–0.185 ≤NDVI ≤0.157	0.97
0.157 ≤NDVI ≤0.727	1.0094+0.047*ln (NDVI)
NDVI >0.727	0.99

10.3 RESULTS AND DISCUSSION

10.3.1 Land-use/Cover Variations

The land-use/cover maps (Figure 10.3) were derived from supervised classification of Landsat 5 and 8. The accuracy ranged between 75.3 and 83.6 per cent. Transitions in the built environment over the review period are shown in Figure 10.3 representing a significant change in the impervious surface. The build environment increased from 18,237 ha in 1984 to 51,991 ha in 2018, representing an increase of 55.29 per cent (Table 10.4).

From 1986 to 1997, there were gains in built-up areas (Figure 10.4) of 10,497 ha, which is 84.08 per cent (Table 10.5) of the built-up areas in 1986. The rate of change between 1986 and 1997 was 954.27 ha per annum (7.64 per cent) (Table 10.5). For the period between 1997 and 2008 there was an increase in urbanized areas of 6,546 ha (23.72%) and the rate of increase was 595.08 ha (2.16%) between the two years (Table 10.5). Between 2008 and 2018 there was a 17,846 ha (52.26 per cent) increase in impervious surfaces. Overall, the change from 1986 to 2018 was 33,753 ha which

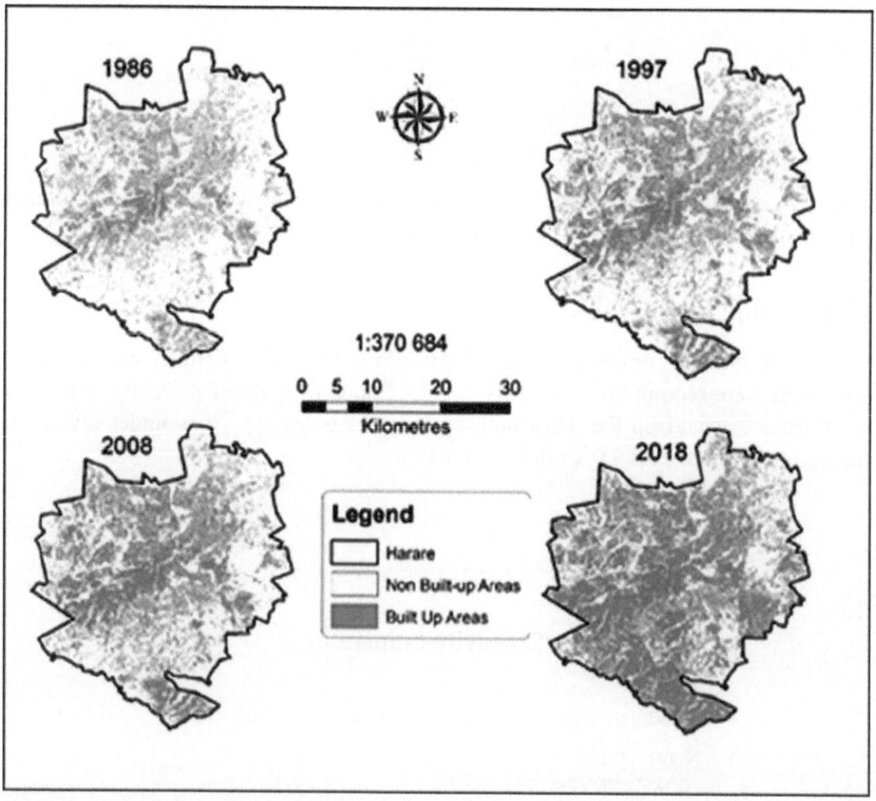

FIGURE 10.3 Land cover map of 1986, 1997, 2008 and 2018 extracted from the supervised classification of Landsat 5 and 8 images

TABLE 10.4
Changes in Built Environment Between 1986, 1997, 2008 and 2018

Year	Built Areas (ha)	Non-Built Areas (ha)	Proportion of Built Areas (%)
1986	18,237	75,786	19.40
1997	27,599	66,424	29.35
2008	34,145	59,878	36.31
2018	51,991	42,032	55.29

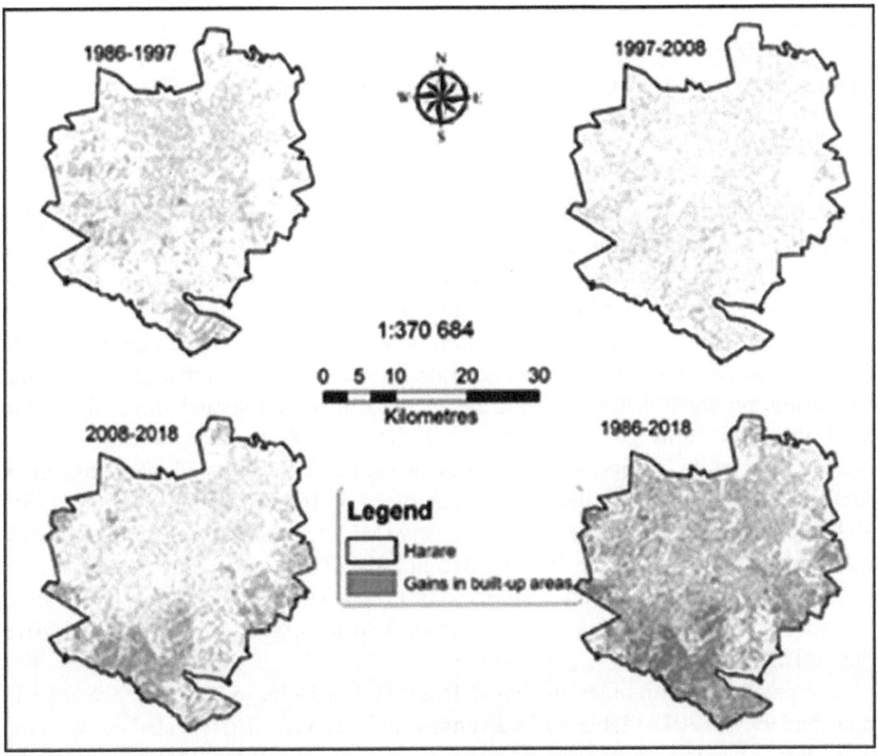

FIGURE 10.4 Gains in built-up area between 1986 and 1997

amounted to a 185.08 per cent increase, at a rate of 1,054.78 ha (5.28 per cent) per annum (Table 10.5).

Significant variations in built-up areas occurred between 1986 and 1997 which recorded the highest increase in the built environment, represented as follows: Harare Rural (89.47%), Chitungwiza (67.46%), Harare Urban (46.44%) and Epworth (42.56%) (Table 10.6). A similar trend followed between 1997 and 2008, but the most significant increases occurred between 2008 and 2018 with increases of 160.34%, 101.45%, 63.02 and 31.83% in Harare Rural, Chitungwiza, Epworth and Harare Urban, respectively (Table 10.6). Overall, the region that experienced a

TABLE 10.5
Land-use/Cover Changes from 1986 to 2018

From	1986	1997	2008	1986
To	1997	2008	2018	2018
Gains (Ha)	9 362	6 546	17 846	33 753
Time Span	11	11	10	32
Rate of Change per Year (ha)	851.09	595.09	1 784.6	1 054.78
Change in Urban Areas (%)	51.34	23.72	52.26	185.08
Rate of Change Per Year (%)	4.679	2.16	5,23	5.28

significant change from 1986 to 2018 was Harare Rural with 616.15% followed by Chitungwiza with 246.49%, then Epworth with 241.45% and lastly Harare Urban with 133.59%.

Based on landscape metrics there is a noticeable decrease in the number of patches from 20,877 in 1986, to 9,865 in 2018 (Table 10.7). The number of patches represents landscape aggregation or disaggregation, and these are noticeable in the study through several discontinuous patches. The number of patches is related to the aggregation index, which was 52.6 in 1986, 64.6 in 1997, 69.5 in 2008 and 80.4 in 2018 (Table 10.7). There is negative correlation between the patches and aggregation, as urban sprawl patches will amalgamate, thereby reducing the number of patches and increasing aggregation levels. With the increase in urban sprawl there, the mean patch increased from 0.874 in 1986 to 5.27 ha in 2018. Another indicator used to indicate changes in impervious surfaces is the standard deviation of patch area index (SDPAI), which increases with the values of 35.3, 104, 163 and 454 from 1986, 1997, 2008 and 2018. The mean patch density (MPD) revealed a negative trend with values changing from 22.2 in 1986 to 10.5 in 2019 (Table 10.7). The mean core area index revealed a positive trend from 0.536 in 1986 to 0.781 in 2018, and the standard deviation of core area index increased from 2.94 to 3.80 between 1986 and 2018 (Table 10.7).

The patch cohesion index increased from 98.2 in 1986, to 99.5 in 1997, 99.7 in 2008 and 99.9 in 2018 (Table 10.7). Evenness indices were also revealed between the two years. Shannon's evenness index increased from 0.710 to 0.992 between 1986 and 2018 (Table 10.7). Simpson's evenness index increased from 0.625 in 1986 to 0.989 in 2018 (Table 10.7). Changes in urban sprawl can also be quantified using the diversity index and in this study both Shannon's diversity index and Simpson's diversity index were used, and they revealed an increasing trend as depicted in Table 10.7. An analysis of urban sprawl in Harare in the past 32 years has shown some spatio-temporal changes in land use.

10.3.2 Change in Temperature

A progressive increase in temperature from 2000 through to 2018 is noted, as shown in Figure 10.5. LST in Harare increased from 29°C to 31°C between 2000 and 2018. The increase is mainly attributed to the increase in built environment (including

TABLE 10.6
Spatial Variations of Urban Sprawl in Different Regions in Harare

Region	Years				Percentage Changes			
	1986 (ha)	1997 (ha)	2008(ha)	2018 (ha)	1986–1997	1997–2008	2008–2018	1986–2018
Harare Urban	14,807.16	21,684.33	26,236.62	34,587.72	46.44	20.99	31.83	133.59
Harare Rural	1,502.82	2,847.43	4,133.97	10,762.47	89.47	45.18	160.34	616.15
Epworth	749.34	1,067.49	1,270.08	2,558.61	42.56	18.98	101.45	241.45
Chitungwiza	1,178.10	1972.89	2,503.98	4,082.04	67.46	26.92	63.02	246.49

TABLE 10.7
Landscape Metrics Derived from the Classified Images of 1986, 1997, 2008 and 2018

Landscape Metrics	1986	1997	2008	2018
Number of Patches (NP)	20,877	17038	16057	9865
Mean Patch Area (MPA)	0.874	1.62	2.13	5.27
Standard Deviation of Patch Area Index (SDPAI)	35.3	104	163	454
Mean Patch Density (MPD)	22.2	18.3	17.3	10.5
Mean Core Area (MCA)	0.536	0.607	0.601	0.781
Coefficients of Variation of Core Area Index (CVCA)	549	552	556	486
Standard Deviation of Core Area Index (SDCA)	2.94	3.35	3.34	3.80
Aggregation Index (AI)	52.6	64.6	69.5	80.4
Patch Cohesion Index (PCI)	98.2	99.5	99.7	99.9
Shannon's Evenness Index (SHEI)	0.710	0.873	0.945	0.992
Shannon's Diversity Index (SHDI)	0.492	0.605	0.655	0.688
Simpson's Evenness Index (SIEI)	0.625	0.829	0.925	0.989
Simpson's Diversity Index (SIDI)	0.313	0.415	0.463	0.494
Coefficient of Variation Fractal Dimension Index (CVFDI)	6.23	6.00	5.95	5.83

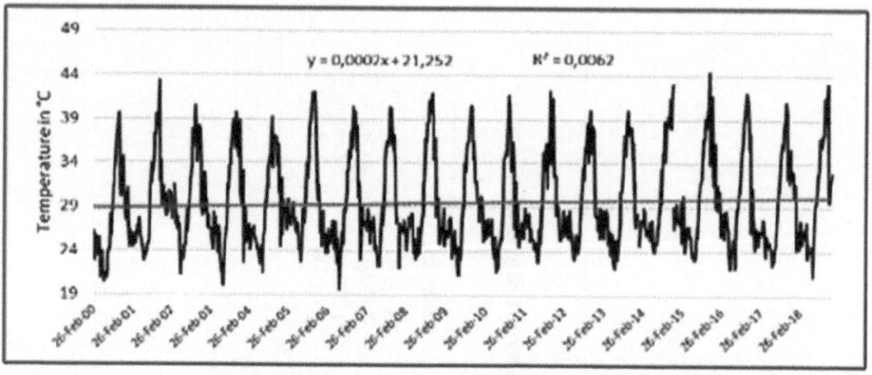

FIGURE 10.5 Variations and trends of LST in Harare between 2000 and 2019

impervious pavements). The results agree with the consensus that LST in urban areas is always higher than in the surrounding areas (Ranagalage et al., 2018; Li et al., 2018).

The image differencing change detection method was used to assess change between the different years. The normal distribution was used to identify negative or positive changes. Positive change implies an increase in LST between the two years while a negative change implies a decrease in LST. Between 1986 and 1997 the minimum change was –42.17°C, the mean is 3.23°C, and the maximum is 17.70°C, with a standard deviation of 2.11°C. Between 1997 and 2008 the minimum change was –47.21°C, the mean is –0.58°C, the maximum is 12.49°C, with a standard deviation of 1.86. Between 2008 and 2018 the minimum change was –14.47°C, the mean is

–2.40°C, the maximum is 12.53°C, with a standard deviation of 2.22. Overall, between 1986 and 2018 the minimum change was –6.58°C, the mean is 6.20°C, the maximum is 17.47°C, with a standard deviation of 1.98.

10.3.3 Spatial Profiles Along a Transect

A transect passing through different land-use classes was run over the study area. The spatial profiles show variations in LST between 1986 and 2018. The highest temperature recorded in 1986 was 28°C and the lowest was 23°C (Figure 10.6). In 2018, the highest temperature was 31.5°C and the lowest was 26°C (Figure 10.6). There were peaks and depressions in the spatial profiles in both 1986 and 2018. Peaks (high LST) were in areas with little or no vegetation such as roads and pavements (impervious surfaces) while depressions (low LST) were in vegetated areas. In vegetated areas there was low albedo as compared to bare and impervious surfaces.

Human activities caused land-use change because of urbanization, which accelerated urban land-use change, thereby affecting albedo, thermal capacity and heat conductivity. Many researchers support that there is a negative correlation between the amount of vegetation and urban heat islands (Kumar & Shekhar, 2015; Yu & Ng, 2007; Ma, Chen & Zhou, 2008; Adeyeri, Akinsanola & Ishola, 2017). There was an increase in temperature in many cities due to land-use change and in places where there is reduction in LST in vegetated areas and in water bodies (Kumar & Shekhar, 2015; Yu & Ng, 2007; Ma, Chen & Zhou, 2008; Adeyeri, Akinsanola & Ishola, 2017). Vegetation reduces incident radiation, thereby lowering LST of the

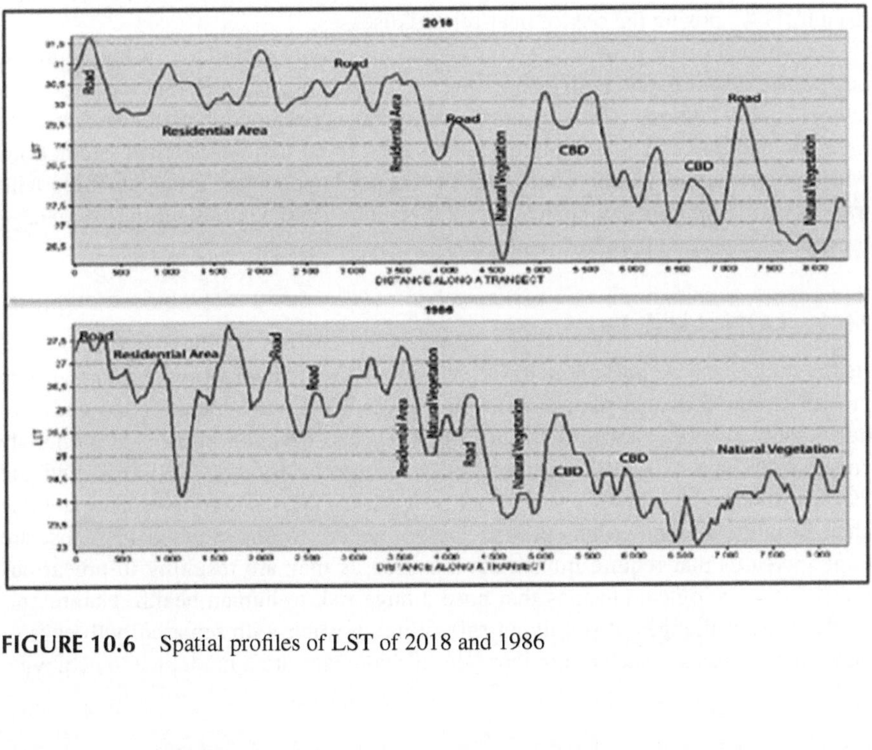

FIGURE 10.6 Spatial profiles of LST of 2018 and 1986

environment, and regulates carbon concentration, which in turn reduces LST (Li et al., 2013; Ali & Mohammed, 2016).

10.3.4 Population Increase and Implications

The population of Harare Metropolitan Province increased by 23 per cent between 2000 and 2018, an increase that has drastically transformed the urban landscape. The population increase in Harare is exerting pressure on available resources and infrastructure that was not initially designed to cater for the current large population. There are currently challenges of deteriorating roads, public health infrastructure and sanitation, which, if not solved soon, is a serious health bomb. There is now poor housing, and parking has become a nightmare for motorists. Congestion in the residential areas has seen the emergence of backyard industries and households, which poses a high health risk.

In most of the urbanized area, there are no more open spaces where air usually cleans up, posing a risk of the spread of disease. Most wetlands have been occupied by houses, increasing the risk of flooding. The encroachment into natural vegetation and other ecosystems has established novel human-wildlife interactions, posing the risk of novel pathogens from wildlife, as wildlife finds new habitat and food in the urban areas. There is a total change in the socio-ecological system, resulting in wildlife co-habiting with humans in the urban area, as wildlife finds readily available food due to poor waste management. As a result, there is an increase of water-borne and respiratory diseases, which is exacerbated by lack of clean water and increasing pollution, respectively. The increasing pollution and spread of impervious surface have witnessed an increase in average annual temperatures in Harare, which have risen to 31°C, posing the risk of heat-related diseases.

A combination of poor water, energy and air quality, unreliable water supplies, poor waste disposal, and high energy and food consumption in Harare has become a major cause of concern. This is exacerbated by the increasing population and changes in urban landscapes. This has affected and transformed socio-ecological interactions with a risk to human health. There is an urgent need to combine urban planning with transformative approaches, such as circular economy and nexus planning, to manage these challenges and build resilient cities.

10.4 CONCLUSION

In this study, we used geospatial techniques to monitor and quantify horizontal urban growth using landscape metrics. The process is cost- and time-effective and produces more accurate statistics to inform policy and decision making to formulate coherent strategies. The built environment increased by 186.08% between 1986 and 2018, with a yearly rate of 1,054.70 ha per year (5.28%). There have been notable changes in the built environment in the city of Harare from 1986 to 2019. These are huge increases that require immediate attention as they are resulting in precarious novel socio-ecological changes that pose a huge risk to human health. Harare currently faces challenges of supplying safe drinking water with reported outbreaks of water-borne diseases such as cholera. These challenges are a hindrance to achieving the 2030 global agenda on sustainable development. The procedure illustrated in this study offers quick assessments of the spatiotemporal patterns for urban areas. Particularly, landscape metrics performed quite well in evaluating urban structure more accurately. The results provide information for policy and decision making on formulating adaptation and response strategies on sustainable management of urban areas. This is of paramount importance as it provides pathways towards the 2030 global agenda on sustainable development.

REFERENCES

Abebe, Gerahegn Aweke. 2013. "Quantifying Urban growth pattern in developing Countries Using Remote Sensing and Spatial Metrics: A case study of Kampala, Uganda". *Msc Thesis*.

Adeyeri, O. E., A. A. Akinsanola, and K. A. Ishola. 2017. "Investigating surface urban heat island characteristics over Abuja, Nigeria: Relationship between land surface temperature and multiple vegetation indices". *Remote Sensing Applications: Society and Environment* 7:57–68. doi: 10.1016/j.rsase.2017.06.005.

Aduah, MS, and PE Baffoe. 2013. "Remote sensing for mapping land-use/cover changes and urban sprawl in Sekondi-Takoradi, Western Region of Ghana". *The International Journal of Engineering and Science (IJES)* 2 (10):66–72.

Aithal, Bharath, and TV Ramachandra. 2013. "Measuring urban sprawl in Tier II cities of Karnataka, India". *Global Humanitarian Technology Conference: South Asia Satellite (GHTC-SAS), 2013 IEEE*:321-329.

Ali, AR, and ES Mohammed. 2016. "Impact of industrial activities on land surface temperature using remote sensing and gis techniques-A case study in Jubail, Saudi Arabia". *J Geogr Nat Disast S* 6:2167-0587.

Araya, Yikalo H, and Pedro Cabral. 2010. "Analysis and modeling of urban land cover change in Setúbal and Sesimbra, Portugal". *Remote Sensing* 2 (6):1549–1563.

Artis, David A., and Walter H. Carnahan. 1982. "Survey of emissivity variability in thermography of urban areas". *Remote Sensing of Environment* 12 (4):313–329. doi: 10.1016/0034-4257(82)90043-8.

Beard, Victoria A, Anjali Mahendra, and Michael I Westphal. 2016. Towards a more equal city: framing the challenges and opportunities. In *GROWTH*. Washington D.C, USA: World Resources Institute (WRI).

Chambati, Walter. 2011. "Restructuring of agrarian labour relations after fast track land reform in Zimbabwe". *Journal of Peasant Studies* 38 (5):1047–1068.

Chander, Gyanesh, and Brian Markham. 2003. "Revised landsat-5 TM radiometric calibration procedures and postcalibration dynamic ranges". *IEEE Transactions on Geoscience and Remote Sensing* 41 (11):2674–2677.

Chattopadhyay, Rupak. 2000. "Zimbabwe: Structural adjustment, destitution & food insecurity". *Review of African Political Economy* 27 (84):307–316.

Cohen, Barney. 2006. "Urbanization in developing countries: Current trends, future projections, and key challenges for sustainability". *Technology in Society* 28 (1–2):63–80.

Dzimiri, Patrick, and Tawanda Runhare. 2012. "The human security implications of operation restore order on urban habitation in Zimbabwe". *Journal of Human Ecology* 38 (3):191–205. doi: 10.1080/09709274.2012.11906488.

Ersado, Lire. 2006. *Income diversification in Zimbabwe: Welfare implications from urban and rural areas*: The World Bank.

Gaitani, N, A Spanou, M Saliari, A Synnefa, K Vassilakopoulou, K Papadopoulou, K Pavlou, M Santamouris, M Papaioannou, and A Lagoudaki. 2011. "Improving the microclimate in urban areas: A case study in the centre of Athens". *Building Services Engineering Research and Technology* 32 (1):53–71.

Giles-Corti, Billie, Melanie Lowe, and Jonathan Arundel. 2019. "Achieving the SDGs: Evaluating indicators to be used to benchmark and monitor progress towards creating healthy and sustainable cities". *Health Policy* doi: 10.1016/j.healthpol.2019.03.001.

Groves, Zoe. 2012. "People and places: Land, migration and political culture in Zimbabwe". *The Journal of Modern African Studies* 50 (2):339–356.

Gumbo, Trynos, Manie Geyer, Inocent Moyo, and Thembani Moyo. 2018. "Mapping spatial locational trends of informal economic enterprises using mobile geographic information data in the city of in Harare, Zimbabwe". *Data in Brief* 20:1692–1699.

Hegazy, Ibrahim Rizk, and Mosbeh Rashed Kaloop. 2015. "Monitoring urban growth and land use change detection with GIS and remote sensing techniques in Daqahlia governorate Egypt". *International Journal of Sustainable Built Environment* 4 (1):117–124.

Jat, Mahesh Kumar, P. K. Garg, and Deepak Khare. 2008. "Monitoring and modelling of urban sprawl using remote sensing and GIS techniques". *International Journal of Applied Earth Observation and Geoinformation* 10 (1):26–43.

Kamusoko, Courage, Jonah Gamba, and Hitomi Murakami. 2013. "Monitoring urban spatial growth in Harare metropolitan province, Zimbabwe". *Advances in Remote Sensing* 2 (4):322–331.

Kumar, Deepak, and Sulochana Shekhar. 2015. "Statistical analysis of land surface temperature–vegetation indexes relationship through thermal remote sensing". *Ecotoxicology and Environmental Safety* 121:39–44. doi: 10.1016/j.ecoenv.2015.07.004.

Kumar, K Sundara, P Udaya Bhaskar, and K Padmakumari. 2012. "Estimation of land surface temperature to study urban heat island effect using LANDSAT ETM+ image". *International Journal of Engineering, Science and Technology* 4 (2):771–778.

Laosuwan, Teerawong, Torsak Gomasathit, and Tanutdech Rotjanakusol. 2017. "Application of remote sensing for temperature monitoring: The technique for land surface temperature analyis". *Journal of Ecological Engineering* 18 (3):53–60.

Li, Hui, John T Harvey, TJ Holland, and M Kayhanian. 2013. "The use of reflective and permeable pavements as a potential practice for heat island mitigation and stormwater management". *Environmental Research Letters* 8 (1):015023.

Li, Huidong, Yuyu Zhou, Xiaoma Li, Lin Meng, Xun Wang, Sha Wu, and Sahar Sodoudi. 2018. "A new method to quantify surface urban heat island intensity". *Science of the Total Environment* 624:262–272.

Ma, Wei, Yun-hao Chen, and Ji Zhou. 2008. "Quantitative analysis of land surface temperature-vegetation indexes relationship based on remote sensing". *Proc. 21st ISPRS Congress, Youth Forum*:261–264.

Magidi, James, and Fethi Ahmed. 2018. "Assessing urban sprawl using remote sensing and landscape metrics: A case study of City of Tshwane, South Africa (1984–2015)". *The Egyptian Journal of Remote Sensing and Space Science.* doi: 10.1016/j.ejrs.2018.07.003.

Marondedze, Andrew K, and Brigitta Schütt. 2019. "Dynamics of land use and land cover changes in Harare, Zimbabwe: A case study on the linkage between drivers and the axis of urban expansion". *Land* 8 (10):155.

McGarigal, Kevin, SA Cushman, MC Neel, and E Ene. 2002. "FRAGSTATS: spatial pattern analysis program for categorical maps". *Department of Agriculture, Forest Service, Pacifica Northwest Research Station.*

McGranahan, Gordon, Daniel Schensul, and Gayatri Singh. 2016. "Inclusive urbanization: Can the 2030 Agenda be delivered without it?" *Environment and Urbanization* 28 (1):13–34.

Nhamo, Luxon, and Chilonda Pius. 2013. "Validation of the rainfall runoff SCS-CN model in a catchment with limited measured data in Zimbabwe". *Journal of Water Resources and Environmental Engineering* 5 (6):295–303.

Olawole, MO, L Msimanga, SA Adegboyega, and FA Adesina. 2011. "Monitoring and assessing urban encroachment into agricultural land-A remote sensing and GIS based study of Harare, Zimbabwe". *IFE Journal of Science* 13 (1):149–160.

Peres, Leonardo de Faria, Andrews José de Lucena, Otto Corrêa Rotunno Filho, and José Ricardo de Almeida França. 2018. "The urban heat island in Rio de Janeiro, Brazil, in the last 30 years using remote sensing data". *International Journal of Applied Earth Observation and Geoinformation* 64:104–116. doi: 10.1016/j.jag.2017.08.012.

Potts, Deborah. 2006. "'Restoring order'? Operation Murambatsvina and the urban crisis in Zimbabwe". *Journal of Southern African Studies* 32 (2):273–291.

Potts, Deborah, and Chris Mutambirwa. 1998. ""Basics are now a luxury": perceptions of structural adjustment's impact on rural and urban areas in Zimbabwe". *Environment and Urbanization* 10 (1):55–76.

Ranagalage, Manjula, Ronald Estoque, Hepi Handayani, Xinmin Zhang, Takehiro Morimoto, Takeo Tadono, and Yuji Murayama. 2018. "Relation between urban volume and land surface temperature: A comparative study of planned and traditional cities in Japan." *Sustainability* 10 (7):2366.

Rodriguez-Galiano, V. F., B. Ghimire, J. Rogan, M. Chica-Olmo, and J. P. Rigol-Sanchez. 2012. "An assessment of the effectiveness of a random forest classifier for land-cover classification". *ISPRS Journal of Photogrammetry and Remote Sensing* 67:93–104. doi: 10.1016/j.isprsjprs.2011.11.002.

Sheng, Li, Xiaolu Tang, Heyuan You, Qing Gu, and Hao Hu. 2017. "Comparison of the urban heat island intensity quantified by using air temperature and Landsat land surface temperature in Hangzhou, China". *Ecological Indicators* 72 (Supplement C):738–746. doi: 10.1016/j.ecolind.2016.09.009.

Sobrino, José A, Juan C Jiménez-Muñoz, and Leonardo Paolini. 2004. "Land surface temperature retrieval from LANDSAT TM 5". *Remote Sensing of Environment* 90 (4):434–440.

Weng, Qihao, Dengsheng Lu, and Jacquelyn Schubring. 2004. "Estimation of land surface temperature–vegetation abundance relationship for urban heat island studies". *Remote Sensing of Environment* 89 (4):467–483.

Xu, Gang, Ting Dong, Patrick Brandful Cobbinah, Limin Jiao, Neema S. Sumari, Baohui Chai, and Yaolin Liu. 2019. "Urban expansion and form changes across African cities with a global outlook: Spatiotemporal analysis of urban land densities". *Journal of Cleaner Production* 224:802–810. doi: 10.1016/j.jclepro.2019.03.276.

Yu, Xi Jun, and Cho Nam Ng. 2007. "Spatial and temporal dynamics of urban sprawl along two urban–rural transects: A case study of Guangzhou, China". *Landscape and Urban Planning* 79 (1):96–109. doi: 10.1016/j.landurbplan.2006.03.008.

Part III

Remote Sensing Applications in Models and Planning

11 Remote Sensing & GIS-based Site Identification for Solid Waste Management of Amritsar Municipal Corporation, Punjab, India

Kanwar Deepak
DAV College, Jalandhar, India

CONTENTS

11.1 INTRODUCTION

Environmental pollution is becoming a part of human society, especially in Third World countries. All humans, through their actions, are contributing towards the poor health of the environment. Within human society there is a long list of reasons

DOI: 10.1201/9781003224624-15

responsible for environmental pollution. Municipal solid waste is one of the major reasons that contaminate the environment (Kaur et al., 2010). Numerous efforts are being made the world over to mitigate the environmental pollution caused by municipal solid waste. One of these efforts is to identify suitable sites for handling municipal solid waste. The present study focuses on the site identification for municipal solid waste of Amritsar city.

11.1.1 Meaning of Solid Waste

Several terms are used to define solid waste, such as garbage, trash, refuse, rubbish, litter and junk. In general, solid waste refers to material disposed of by people. 'Garbology', a new discipline, has arisen that studies waste (Baguchinsky, 1999). Waste is increasingly getting the attention of environmentalists, especially in the twenty-first century and as a result more and more studies are coming out throughout the world. The present study attempts to identify a suitable site for the municipal solid waste management of Amritsar city.

11.1.2 Role of Solid Waste in Environmental Pollution

It is common to witness very unruly and disorderly dumping of solid waste in open areas all across the city of Amritsar, as in many other parts of our country. This untreated solid waste not only spoils the landscape but also pollutes the environment. After being dumped, the biodegradable waste starts decomposing. In this process microbial and chemical reactions take place, resulting in the release of methane and carbon dioxide and other volatile organic compounds. This process starts one year after the dumping of solid waste in a landfill and continues for about 25 years (Clarke Energy, 2016). Both methane and carbon dioxide are greenhouse gases and their release into the atmosphere needs to be checked. Such sites witness the growth of insects, rodents and foul smells, besides landscape spoilage. The possibility of leaching of pollutants into the underlying soil layer and even groundwater increases when landfills are not lined. Increasing environmental pollution could be causing the effects of an urban heat island (Singh & Kalota, 2019).

11.2 STUDY AREA

The present study is focused on the Amritsar municipal corporation and its outgrowths (Figure 11.1). It is pertinent to mention here that the Ministry of Environment, Forests and Climate Change of the Government of India, via its notification dated 8 April 2016, laid down the revised rules of solid waste management, and it has been clearly mentioned that these rules will also apply to urban agglomerations. That is why the boundary of the study area extends beyond the administrative boundary of the municipal corporation to include the outgrowths. Moreover, waste generation is not limited to the administrative boundary but outgrowths also contribute their share to the solid waste. To find a suitable site for solid waste management a 10 km buffer is made around the Amritsar city boundary which basically marks the study area for the present work. This buffer is clipped with the district boundary; in the south the district boundary limits the study area (Figure 11.2).

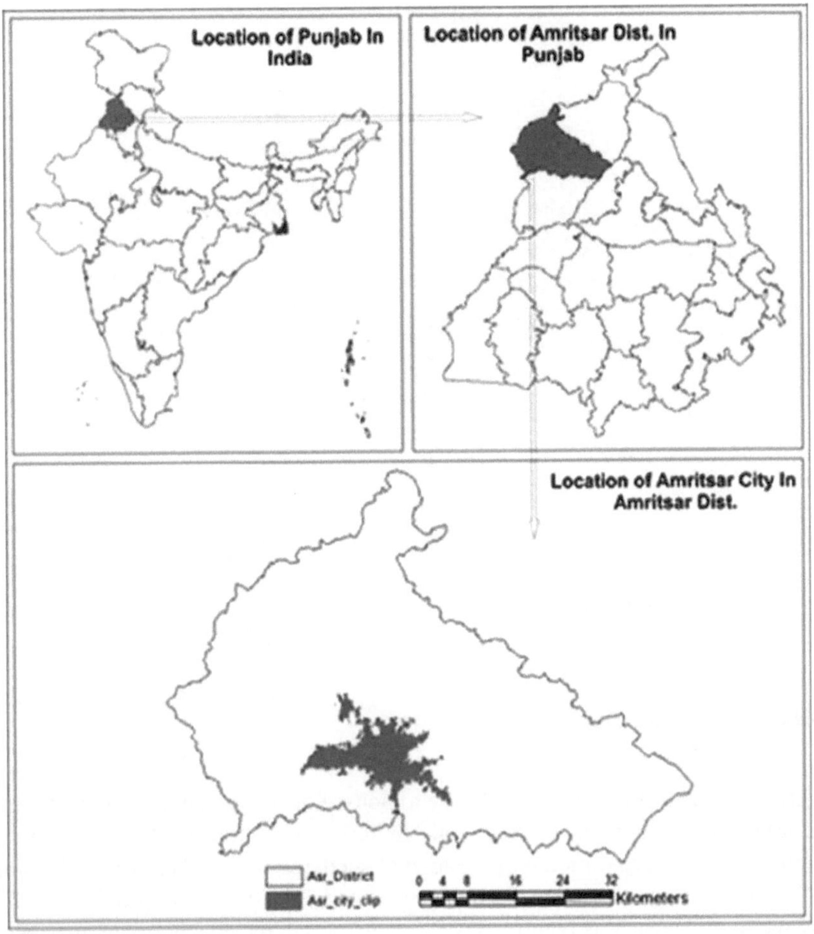

FIGURE 11.1 Location map

11.3 CURRENT SCENARIO

Around 600 tonnes of solid waste are generated each day in the city of Amritsar which is disposed of in the landfill site in Bhagtanwala. Solid waste is collected from house to house by sanitary workers. They are paid on a shared basis by the municipal corporation and the mohalla sudhar committees. In total there are 2,406 sanitary workers involved in solid waste collection, of which 1,443 are deployed by the municipal corporation and 963 by the mohalla sudhar committees (PUDA, Master Plan Report, Amritsar, 2010). In addition, there are 51 drivers, 24 sanitary inspectors, 20 heads of sanitary workers, 7 supervisors and 4 chief sanitary inspectors. There are around 500 waste collection trolleys and 120 rickshaw/rehris in operation for solid waste collection. In order to transport the solid waste from collection centres to the disposal site, the municipal corporation of Amritsar has deployed 10 trucks and 76 tractor trolleys in addition to 7 front-end loaders to load waste into

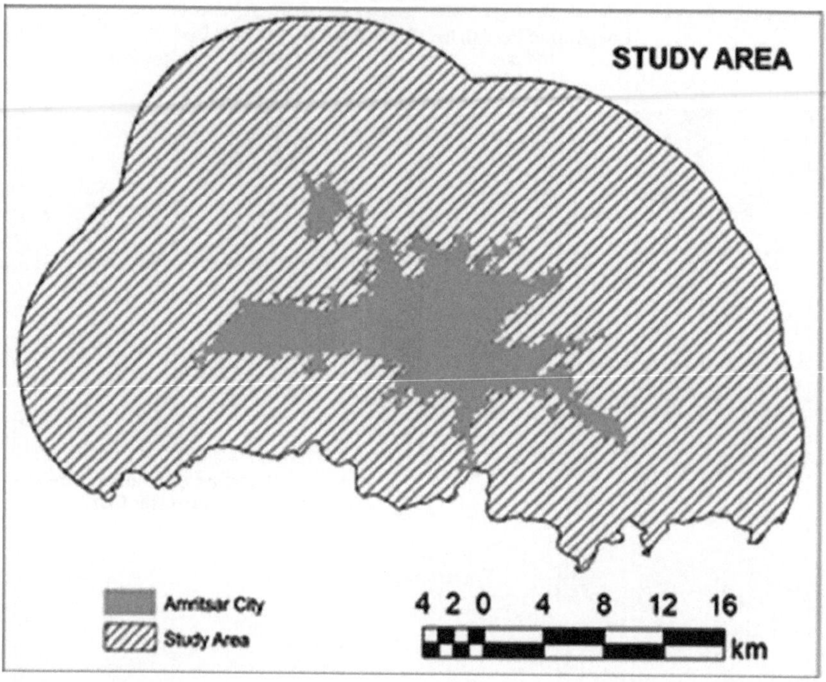

FIGURE 11.2 Study area

trucks and trolleys. In areas where house to house collection is not done, community bins are placed at strategic locations, but the inadequate number of community bins cannot handle the large amount of waste generated every day. Therefore, solid waste is often dumped in open areas, vacant plots, road sides and water bodies, etc. The solid waste of Amritsar is comprised of organic waste, recyclable waste, inert waste and industrial waste. In addition, medical waste and e-waste also contribute and their proportion is increasing day by day. The medical waste disposal is carried out by private companies but regulated by the municipal corporation.

11.3.1 Current Site

The municipal corporation of Amritsar has identified three sites for the disposal of solid waste: Bhagtanwala which is within the municipal limits (Punjab Pollution Control Board, 2017), (Figure 11.3), Fatehpur village and Bharariwal village. At present the waste is dumped at the 8.1 hectare Bhagtanwala without any pretreatment of the waste being undertaken. The odour control treatment is also only occasional. This site of Bhagtanwala has been in operation for 25 years and can be used for another 15 years. The other sites of Fatehpur village and Bhrariwal village are under development. The Fatehpur site is 6 km from the city with an area of 5.8 hectares, and the Bharariwal site is 5 km from the city with an area of 2.65 hectares. The waste of Rajasansi is disposed of at Dera Baba Nanak Road by the Ajnala municipal council

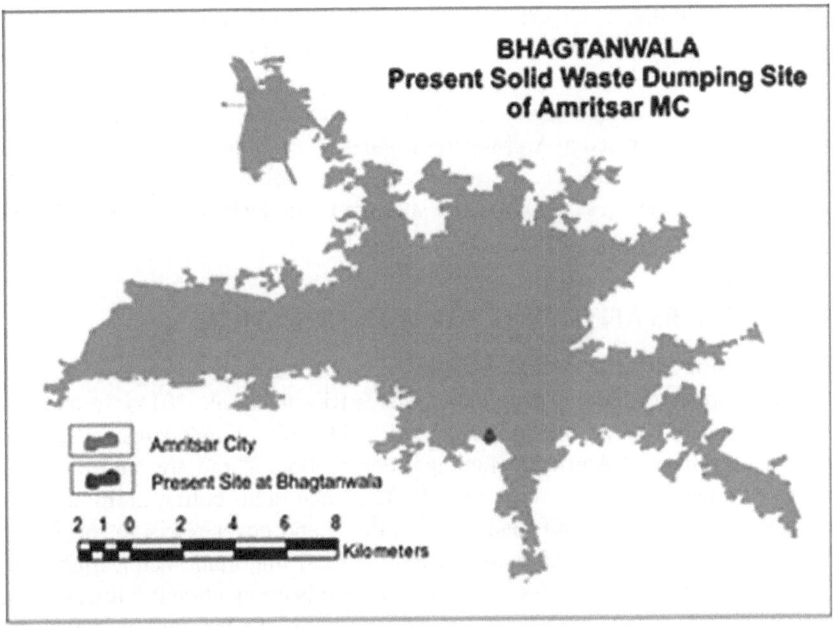

FIGURE 11.3 Bhagtanwala dumping site

because, according to the rules, there cannot be any solid waste dumping site within a radius of 10 km of Rajasansi Airport.

11.3.2 Ongoing Project

Amritsar municipal corporation has undertaken a project 'Integrated Solid Waste Management' to make the city of Amritsar clean, beautiful and healthy. This project has been divided into two phases. Phase 1 involves collection, segregation, storage and transportation of solid waste and this task has been assigned to the M/s Antony Waste Handling Company of Mumbai. The second phase involves the processing and disposal of solid waste and this task has been assigned to M/s AKC Developer of Noida. The first phase is already in operation but the second phase is yet to begin.

11.4 MATERIAL AND METHODOLOGY

Data for this study has been tapped from several remote sensing sources (Singh, 2021). Satellite data from the LISS-III sensor was obtained from the Bhuvan official site; satellite data from Landsat from the Global Landcover Facility (GLCF) official site; satellite data from Cartosat-I for slope from the Bhuvan official site; groundwater data from the Central Groundwater Board official site; soil data from the ENVIS official site. In addition to this, Google Earth data is also used. All data is processed in ArcMap from digitization and editing to analysis. Features that are digitized in ArcMap include landcover, soil, groundwater, slope, canals, drains and roads. After digitization multi buffers are created around city, villages/settlements, canals, drains and roads. Later, all the data of features and buffers saved in shape files are converted into raster datasets. This is followed by assigning importance to each raster as a percentage and assigning weights to subclasses on the scale of 5. Finally, weighted overlay analysis is applied in ArcMap to get the most suitable areas. Then all features including the most suitable areas are converted into kml files and viewed in Google Earth to find suitable sites. In total, three sites are located which are verified by overlaying on each raster in ArcMap.

11.5 FACTORS AFFECTING SITE IDENTIFICATION

Before selection of factors, similar studies were perused most of which were done on cities of southern India (Dharanikota, 2013; Rinsitha et al., 2014). In this study a total of nine factors are being considered to identify suitable sites for the solid waste management of Amritsar municipal corporation. These are land cover, soil drainage, groundwater depth, city multi buffer, canal multi buffer, drain multi buffer, village buffer, slope, and road multi buffer. Land cover refers to the features that cover a piece of land and is considered the most important factor. In the study area of Amritsar five subclasses of landcover have been identified: cultivated land, plantation/forest, village/settlement, industry, and city of Amritsar (Figure 11.4). It

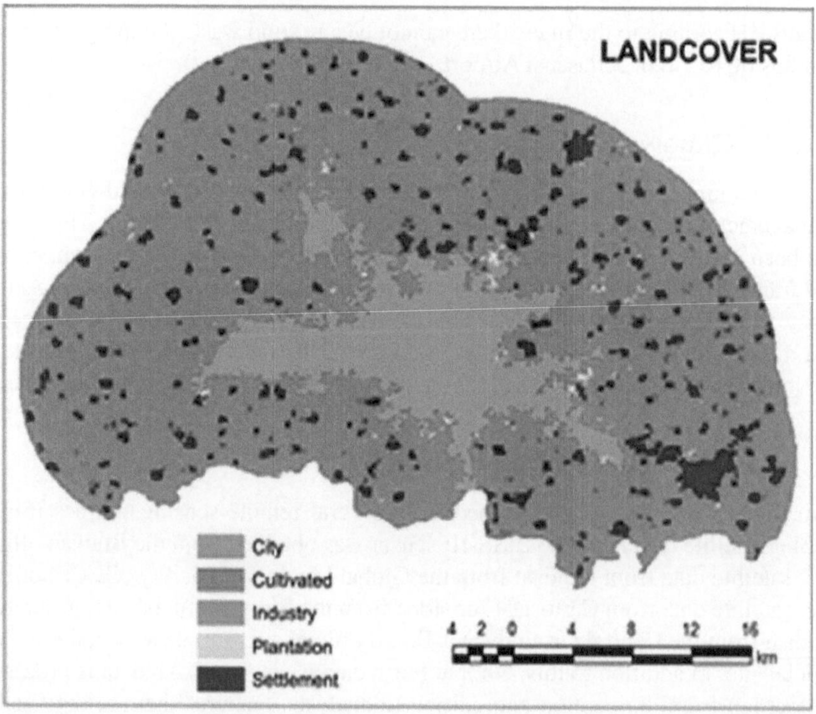

FIGURE 11.4 Land cover in study area

is important to mention here that 'Bhagtanwala' is the present site where municipal solid waste is being dumped. This site is situated in densely populated area of Amritsar city. In 2014 the municipal corporation of Amritsar decided to modernize this facility. Local residents strongly opposed this move and held a 42-day long protest (Tribune News Service, 2014, 19 December). The issue was discussed in both houses of Parliament also – i.e., on 13 February 2014 in Rajya Sabha and on 16 February 2014 in Lok Sabha. The parliamentary committee on urban development recommended closing all waste plants in residential areas because of serious health problems for the people living around such sites. In this study, efforts are made to avoid such populated areas for site identification. Plantation/forest areas are also avoided to protect the environment.

The next factor is that of soil drainage which refers to the movement of water in soil as a result of gravity. It is expected that some pollutants may leach down into the soil, especially after rainfall. This can happen when pollutants of solid waste mix with rainwater and enter into the soil. In the study area of Amritsar, three kinds of soil drainage exist: D4, D5 and D6 (Figure 11.5). Here D4 refers to moderately well-drained soil; D5 is well-drained soil; and D6 is somewhat excessively drained soil. Therefore, efforts are made to avoid the D6 type of soil drainage.

The next factor, that of groundwater depth, refers to the level of groundwater in metres below ground level (mbgl). As discussed above, the pollutants of solid waste may leach down into the soil by mixing with rainwater (Dharanikota, 2013). If this

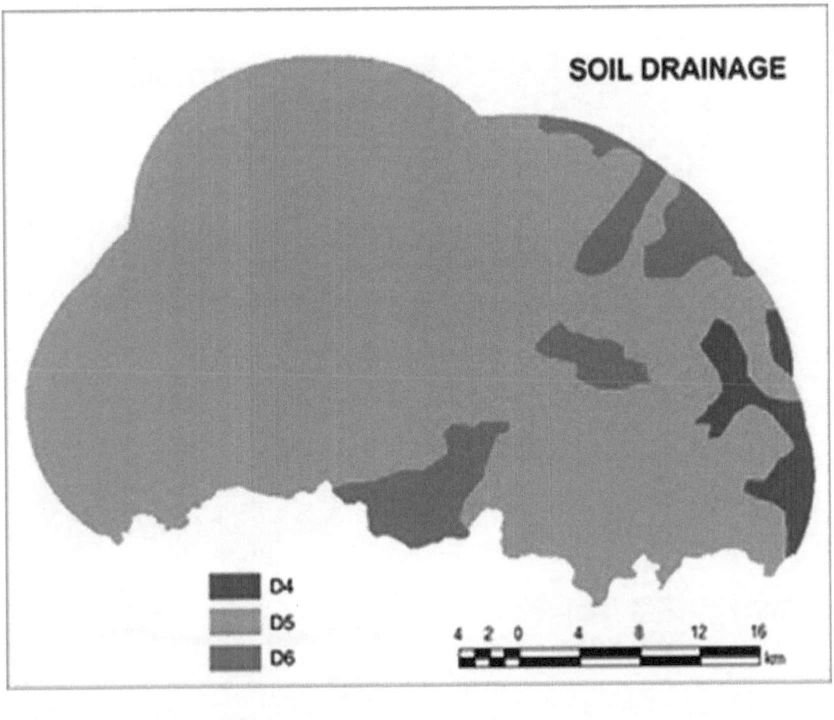

FIGURE 11.5 Soil drainage in study area

happens, the pollutants may reach groundwater level and pollute the groundwater. In such a scenario it will be impossible to clean the groundwater. Three levels of groundwater have been observed in this study area: <15 mbgl, 15–20 mbgl, and > 20 mbgl (Central Groundwater Board, 2013) (Figure 11.6). Efforts will be made to avoid groundwater pollution as a result of leaching down of solid waste pollutants. Therefore, shallow areas are avoided and areas with deeper groundwater level are preferred.

Distance from city or city multi buffers is the next factor to be considered for site identification. Cities often grow faster than villages because of higher growth rate. The decadal growth rate of Amritsar city from 2001 to 2011 is 15.5% (Census, 2011). If a site for solid waste management is located closer to the city boundaries, it may become part of the city in coming years, as happened in the case of the present site at Bhagtanwala. Therefore, a suitable site is preferred at a sufficient distance from the city. Five distance buffers have been made around the city of Amritsar. These are 0–2 km, 2–4 km, 4–6 km, 6–8 km and 8–10 km (Figure 11.7). Buffer zones that are closer to the city are avoided for site identification.

The next factor is that of canal buffers. The main canal of this study area is Upper Bari Doab canal which has several branches and distributaries. Fresh water flows in the canals which needs to be protected from the pollutants of the proposed solid waste management site. Five buffer zones have been marked on both sides of the canals, their branches and distributaries. These buffer zones are 0–250 m, 250–500 m,

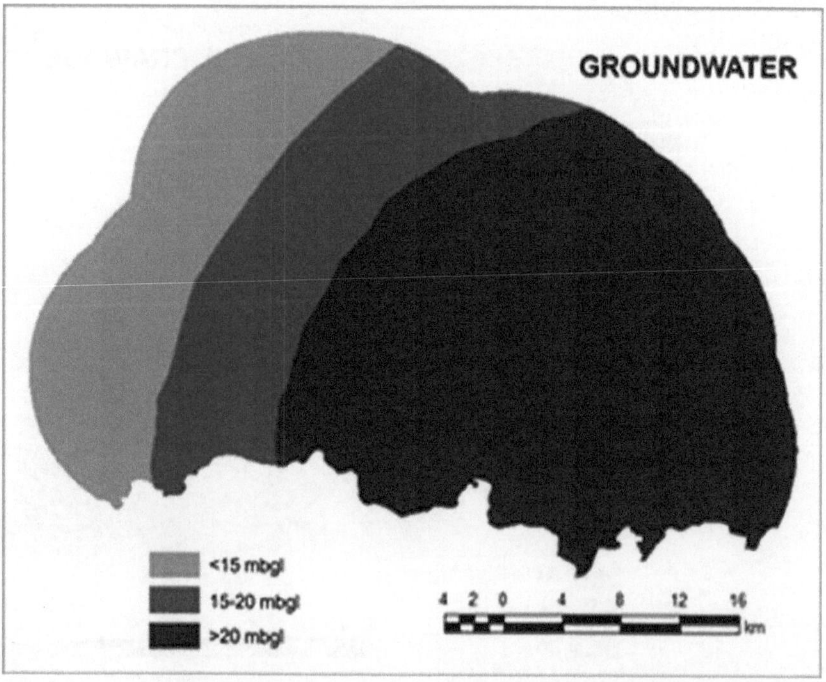

FIGURE 11.6 Groundwater in study area

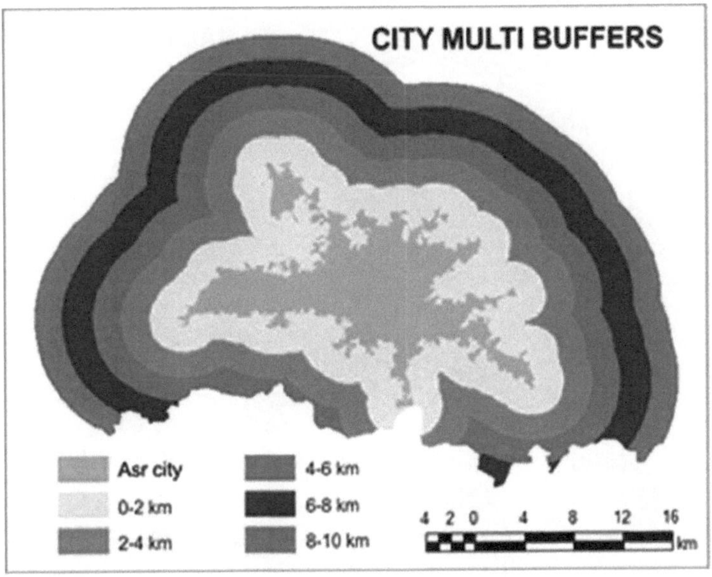

FIGURE 11.7 Distance buffers in study area

500–750 m, 750–1,000 m and > 1,000 m or 1 km (Figure 11.8). Efforts are made to keep the site at the maximum distance from the canals.

Drain buffers comprise the next factor that determines the site identification for solid waste management. There are four main drains in this area: Hudiara drain, Sakki nallah, Kasur nallah and Patti drain, in addition to several of their branches/distributaries. The purpose of these drains is to drain out the flood water, but nowadays these are becoming polluted because of sewage discharge. Five buffer zones have been created along both sides of all the drains as well as their branches/distributaries. These buffer zones are 0–250 m, 250–500 m, 500–750 m, 750–1,000 m and > 1,000 m or 1 km (Figure 11.9). Efforts are made to keep the site at the maximum distance from the drains.

The next factor is that of village buffer. A buffer of 500 m is made around all the villages/small towns (Figure 11.10). The purpose of making this buffer is to avoid areas that are very close to the settlements. Sites for solid waste management must not be in the vicinity of any populated areas.

Slope is the next factor that can be considered for site identification for solid waste management. Satellite data from Cartosat-I is used from the Bhuvan official website. The slope data is further reclassified into the following categories: 0–10 degrees, 10–20, 20–30, 30–40 and more than 40 degrees (Figure 11.11). Steep slope areas are generally avoided for site identification. Looking at the slope map, it is observed that most of the study area falls into the first two categories of 0–10 and 10–20 degrees of slope. That means the study area is generally plain in terms of relief and cannot be a dominant factor in site identification.

Lastly, the factor of road buffer is considered for site identification. The proposed site has to be well connected with the roads because every day the solid waste has to

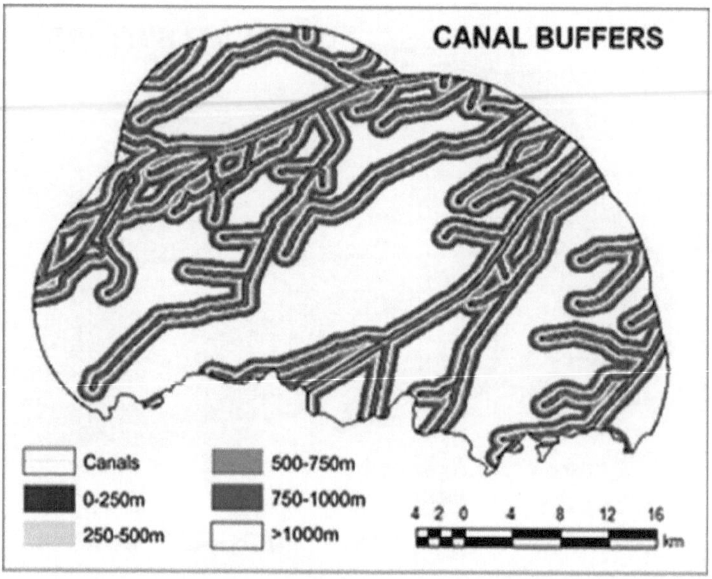

FIGURE 11.8 Canal buffers in study area

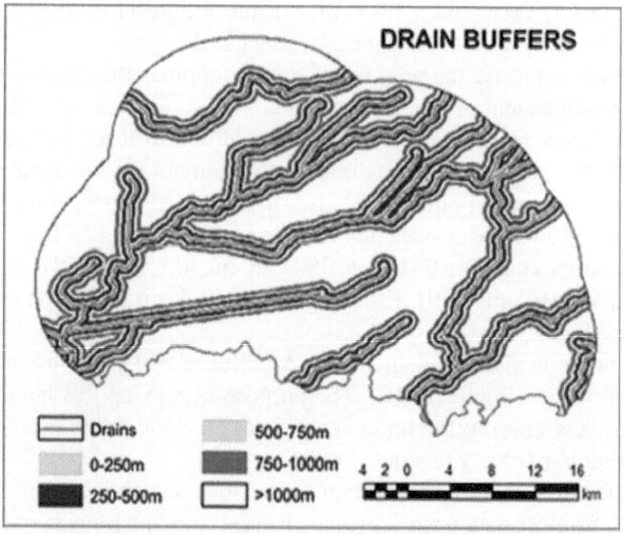

FIGURE 11.9 Drain buffers in study area

be brought from different locations of the city to this site. Looking at the map, it is observed that the study area has a very dense network of roads and almost all parts of the study area are well connected with the city. Further, five buffer zones are made on both sides of all the roads. These zones are 0–100 m, 100–200 m, 200–300 m, 300–400 m and more than 400 m (Figure 11.12). Efforts are made for the proposed site to be closer to the road for better connectivity.

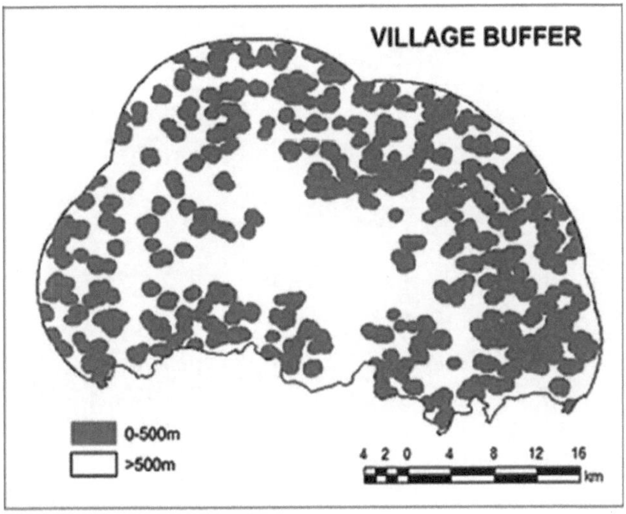

FIGURE 11.10 Village buffers in study area

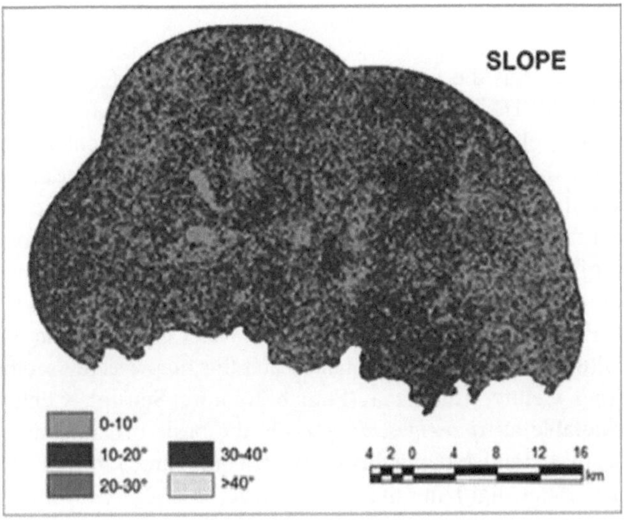

FIGURE 11.11 Slope in study area

11.6 CONVERTING DATA INTO RASTERS

In the present study a total of nine factors are being considered for site identification in the Amritsar study area to handle municipal solid waste. These are land cover, soil drainage, groundwater depth, city multi buffer, canal multi buffer, drain multi buffer, village buffer, slope and road multi buffer. In the beginning all these nine factors are digitized in ArcMap wherein each factor is drawn as a feature dataset in a separate layer. Each feature dataset is saved as a shape file. Land cover, being the

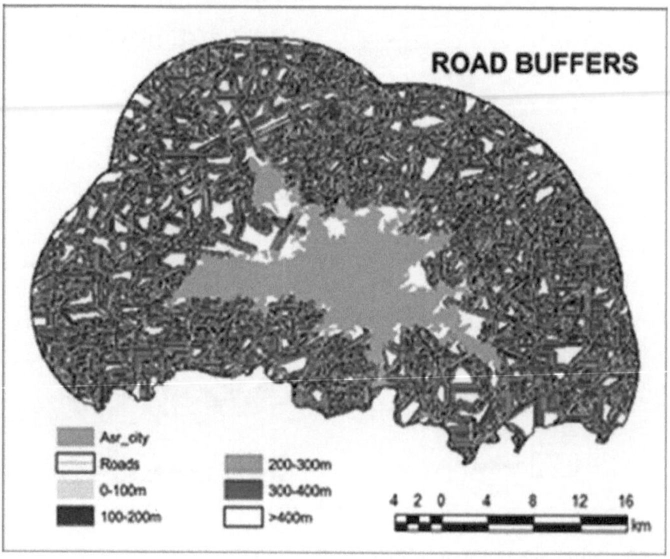

FIGURE 11.12 Road buffers in study area

most important factor, is the first to be made. Land cover refers to the features that cover a piece of land. The land cover layer is made by merging five feature datasets – i.e., cultivated land, plantations/forest, industry, village/settlement and Amritsar city. These five feature datasets are merged with the union tool to generate the land cover feature dataset. Then this shape file of land cover feature dataset is converted into a raster. The next factor is that of soil drainage which refers to water movement in the soil as a result of gravity. Soil drainage data was digitized by georeferencing the soils map of Punjab which was jointly prepared by the National Bureau of Soil Survey and Land Use Planning, Department of Soil and Water Conservation of Punjab, and Punjab Agriculture University of Ludhiana. Later this map was revised by the Punjab Remote Sensing Centre, Ludhiana (Punjab Remote Sensing Centre, Ludhiana, 2003). It is available at <*punenvis.nic.in*>. On the basis of soil drainage there are seven types of soil in Punjab but in this study area only three kinds of soils are found: D4, D5 and D6. After digitizing the soil drainage, it is also converted into a raster dataset. Next, groundwater data is also digitized by georeferencing the groundwater map obtained from the official website < http://cgwb.gov.in/District-GW-Brochures.html> of the Central Groundwater Board, Ministry of Water Resources, Government of India. Three levels of groundwater are found in this study area: <15 mbgl, 15–20 mbgl and > 20 mbgl. After digitizing the groundwater data, it is also converted into a raster dataset. Next, city multi buffer data is made by using the multi buffer tool on the city feature dataset. Five buffer zones are made around Amritsar city: 0–2 km, 2–4 km, 4–6 km, 6–8 km and 8–10 km. This feature dataset is also converted into a raster dataset for further analysis. Next, five multi buffers are made around the canals feature dataset: 0–250 m, 250–500 m, 500–750 m, 750–1,000 m and >1,000 m or 1 km. This is converted into raster datasets. Similarly, five multi buffers are made around drains and converted into raster datasets. Only one buffer of 500 m is made around villages and converted into a raster dataset. Next, slope data is not converted because that is already in raster format. Lastly, five multi buffers are made around roads: 0–100 m, 100–200 m, 200–300 m, 300–400 m and more than 400 m. This feature dataset is also converted into a raster dataset.

11.6.1 Assigning Importance and Weights

This is the most crucial stage for site identification. Although a variety of factors influence site identification for solid waste management, the influence of each factor varies. After comparing, we need to arrange all the factors according to their importance. The importance of each factor is assigned in terms of a percentage (Table 11.1). After assigning the percentage of importance to each factor, weights are assigned to subclasses of each factor. These weights to the subclasses are assigned on the evaluation scale of 1 to 5 by 1 (Table 11.2). That means if the scale value of 5 is assigned to a subclass, then that subclass will be most preferred for a site while the scale value of 1 will be least preferred. Some subclasses are assigned the restricted scale value and such areas are not considered for site identification. The importance and weights are assigned only after consulting scholars, engineers and planners associated with this field and also perusing several studies already undertaken on this topic.

11.7 APPLYING WEIGHTED OVERLAY ANALYSIS TOOL FOR SITE IDENTIFICATION

The tool of weighted overlay analysis is a methodology for suitability modelling and is used to find a suitable site for a phenomenon (Sharma, Pani & Mohapatra, 2015). Generally, weighted overlay analysis becomes a complex process; therefore, it is desirable to make submodels. Each factor may be a submodel or a part of it and must contribute to achieving the goal of site identification; therefore, selection of factors becomes crucial. Separate layers are created for all the factors. All factors may not be of equal importance, or the influence of each factor may vary. In such a scenario

TABLE 11.1
Percentage of Influence Assigned to Factors

Raster	% of Influence
Land cover	22
Soil	18
Groundwater	12
City multi buffers	12
Canal buffers	10
Drain buffers	08
Village Buffer	08
Slope	07
Road Buffers	03
TOTAL	100

TABLE 11.2
Percentage of Influence Assigned to Factors and Weights given to Subclasses

S. N.	Factor	% of Influence	Subclasses	Scale Value (0–5)
1	Land cover	22	Cultivated	5
			Industry	3
			Plantation	2
			Villages	0
			City	0
2	Soil Drainage	18	D4 (moderately well-drained)	5
			D5 (well-drained)	5
			D6 (somewhat excessively drained)	3
			D7 (excessively drained)	2
3	Groundwater	12	0–5 m	2
			5–10 m	3
			10–15 m	3
			15–20 m	4
			> 20 m	5
4	City Multi Buffers	12	0–2 km	1
			2–4 km	2
			4–6 km	3
			6–8 km	5
			8–10 km	4
5	Canal Buffers	10	0–250 m	1
			250–500 m	2
			500–750 m	3
			750–1,000 m	4
			> 1,000 m	5
6	Drain Buffers	8	0–250 m	1
			250–500 m	2
			500–750 m	3
			750–1,000 m	4
			> 1,000 m	5
7	Village Buffer	8	0–500 m	1
			> 500 m	5
8	Slope	7	0–10 degrees	5
			10–20 degrees	4
			20–30 degrees	3
			30–40 degrees	2
			40–90 degrees	1
9	Road Buffers	3	0–100 m	5
			100–200 m	4
			200–300 m	3
			300–400 m	2
			> 400 m	1

the influence of each factor must be assigned in terms of a percentage. For example, the factor of slope for housing a site in a plain area may not be important but for locating a skiing area in a hilly area it may be extremely important and must be given maximum importance. After assigning the percentage of influence to each factor, the next step is to assign weight to the subclasses of factors on an evaluation scale like 1 to 3 by 1 or 1 to 5 by 1 or 1 to 9 by 1. Before the tool of weighted overlay analysis is run it is important to give a path to the output raster, because a new raster dataset is generated after running the tool of weighted overlay analysis.

11.7.1 Major Steps in Weighted Overlay Analysis Tool

- The values of input raster are reclassified on a common evaluation scale.
- The cell values of each raster are multiplied by the percentage of influence of each raster.
- The resultant cell values are added together to generate the output raster.

In the present study, as mentioned earlier, a total of nine factors are being considered. All these nine factors are in raster format and assigned the percentage of influence and scale value to all the subclasses of factors on the scale value of 1 to 5 by 1 (Table 11.2). When the tool of weighted overlay analysis is run, the process of calculation begins for thousands of cells of each raster.

A new raster is generated with a suitability index (Figure 11.13). Areas that have a scale value of 5 are the most suitable areas for a solid waste management site, followed by areas with a value of 4 that are comparatively less suitable than 5, and then areas with a value of 3 that are least suitable, and finally areas with a value of 0 that are restricted areas and not considered for site identification. The most suitable areas with a value of 5 are saved in a separate kml file so that they can be viewed in Google Earth. Then an area of at least 50 acres (20 hectares) is sought for a suitable site within the

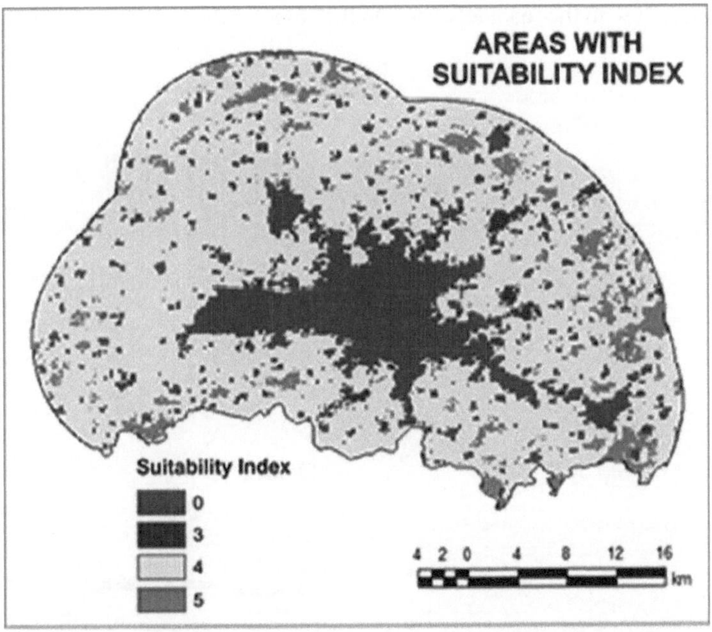

FIGURE 11.13 Areas with suitability index

most suitable areas with a scale value of 5 on the evaluation scale. Studies reveal that the present site and proposed sites range from 10 to 20 acres (4 to 8 hectares) in area. Given that the solid waste is likely to grow in coming years and the site should serve its purpose for at least 40 years, an area of 50 acres (20 hectares) is suggested here. Three sites of 50 acres (20 hectares) each are marked on a Google Earth image (Figure 11.14). Site 1 is extremely suitable, Site 2 is very suitable and Site 3 is suitable. For validation, these sites are brought back in ArcMap and overlaid one by one on all the nine raster datasets which are considered for site identification (Figure 11.15).

Here it is verified that all sites are located in the zones of high scale values of each raster. In the case of land use, all the sites are on cultivated land which has a scale value of 5. On the raster dataset of soil drainage, all sites are in the D5 category of soil drainage which has a scale value of 5. Next, on the groundwater raster, Sites 1 and 2 are in >20 mbgl category which has a scale value of 5 while Site 3 is in 15–20 mbgl which has a scale value of 4. Next, on city multi buffers raster, Sites 1 and 2 are in the 2–4 km category with a scale value of 2 while Site 3 is in 6–8 km category with a scale value of 5. It is worth mentioning here that only limited areas are available beyond 2–4 km category to the south of Amritsar city. This is so because the boundary of the study area ends and further south the district of Tarn Taran begins. Then, on the canal multi buffers raster, all three sites are in the category of >1,000 m or 1 km with a scale value of 5. That means all the sites are situated more than 1 km away from canals. Next, on the drains multi buffers raster, again all three sites are situated in the category of more than 1 km which has a scale value of 5. This means all three sites are more than 1 km away from drains. On the raster of road multi buffers, Sites 1 and 2 are situated in the 0–100 m buffer which has a scale value of 5 while Site 3 is situated in the 100–200 m buffer which has a scale value of 4. This means sites are well connected with the road network. On the slope raster, it is observed that all three

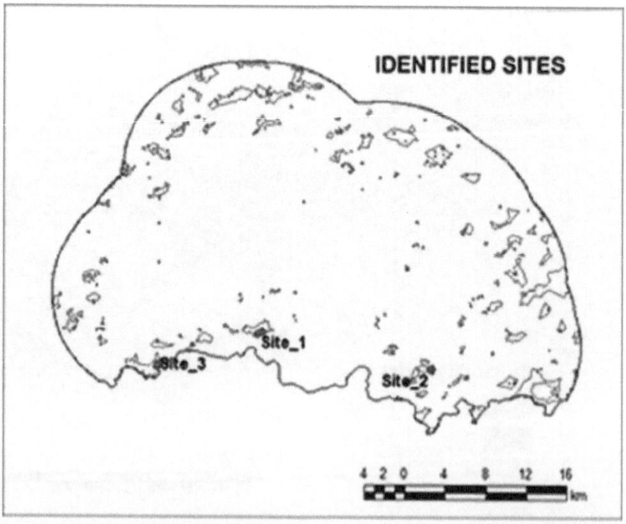

FIGURE 11.14 Identified dumping sites

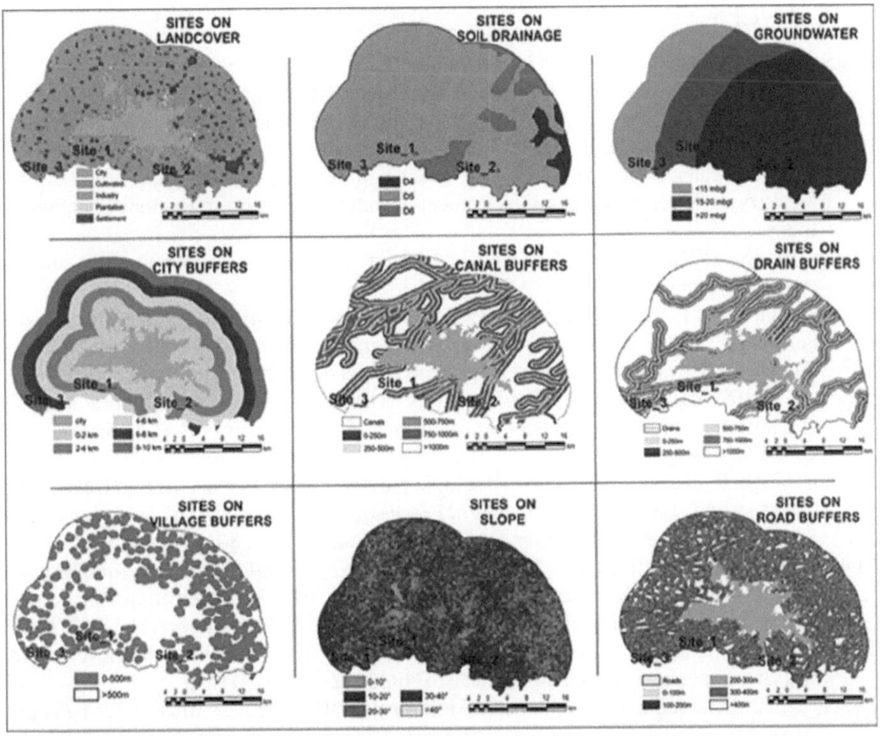

FIGURE 11.15 Dumping sites on selection criteria

sites have more than one category of slope but the categories of 0–10 and 10–20 degrees dominate which have scale values of 5 and 4 respectively. This means the sites are situated in plain areas. Lastly, on the raster of village buffer, it is found that all three sites are situated in the category of > 500 m buffer which has a scale value of 5. This means that all sites are more than 500 m away from the villages.

11.8 CONCLUSION

The present study has attempted to highlight the methodology of locating suitable sites for solid waste management for the municipal corporation of Amritsar. It has also demonstrated the potential of remote sensing and geographic information systems (GIS) in solving environmental problems. Three sites have been suggested: Site 1 is extremely suitable, Site 2 very suitable and Site 3 suitable for solid waste management. A clear understanding of all the nine factors is crucial for assigning the percentage of influence to each factor and also for assigning a scale value to the subclasses of each factor. Since all the factors influence differently towards site selection, weighted overlay analysis is the best solution to this problem. It allows more weight to be given to the dominant or extremely important factors and less weight to those factors that have limited influence on site selection. It is expected that such technologies will be used by city planners to solve urban and environmental problems.

REFERENCES

Baguchinsky, J. (1999). *Adventures in Garbology: What trash can tell us*. Florida Gulf Coast University. Retrieved from: http://itech.fgcu.edu/&/issues/vol2/issue2/garbology.htm

Census 2011. *Punjab Urban/Rural Population*. Retrieved from: https://www.census2011.co.in/census/state/punjab.html

Central Groundwater Board. (2013). *Ground Water Information Booklet, Amritsar District*. Ministry of Water Resources, Government of India, Northwestern Region, Chandigarh, India.

Clarke Energy. (2016). *Landfill Gas*. Retrieved from webpage: https://www.clarke-energy.com/landfill-gas/

Dharanikota, A. (2013). *Environment friendly site selection for municipal solid waste dumping in Kakinada city using remote sensing and geographical information system techniques* (Doctoral Dissertation). Acharya Nagarjuna University, Guntur, Andhra Pradesh.

Kalota D. (2015). Assessment of urban sprawl using landscape metrics: A temporal analysis of Ludhiana city in Punjab. *International Journal of Advances in Remote Sensing and GIS*, 4: 45–54.

Kaur, P., Singh, R. and Kamalpreet (2010). Municipal Solid Waste Management in Developing Countries: A Case Study of Chandigarh. *Proceedings of Venice 2010, Third International Symposium on energy from Biomass and Waste*, Venice, Italy, 8–11 November, 2010.

Punjab Pollution Control Board. (2017). *Annual Report submitted to Central Pollution Control Board for the period 01-01-2016 to 31-12-2016*. Retrieved from: www.ppcb.gov.in/Attachments/Municipal%20Solid%20Waste/MSWAnnualReport31.12.2016.pdf

Punjab Remote Sensing Centre. (2003). *Punjab: Soils*. PRSC, Ludhiana. Retrieved from: http://www.punenvis.nic.in/show_file.aspx?linkid=2-226951911.jpg

Punjab Urban Planning and Development Authority. (PUDA), (2010). *Master Plan Report, 2010–2031, Local Planning Area Amritsar*. Government of Punjab. Retrieved from: https://www.puda.gov.in/

Rinsitha, T., Manjubashini, A., & Satheesh, H.S.D. (2014). Solid waste dumping site suitability analysis for Chengalpettu town using GIS. *International Journal of Recent Development in Engineering and Technology*, 2: 82–85.

Sharma, M., Pani, P., & Mohapatra, S.N. (2015). Urban solid waste disposal site suitability analysis using geospatial technology: A case study of Gwalior city, Madhya Pradesh, India. *International Journal of Environmental Sciences*, 6: 352–367.

Singh, R., and Kalota, D. (2019). Urban sprawl and its impact on generation of urban heat island: A case study of Ludhiana City. *Journal of Indian Society of Remote Sensing*, 47(9): 1567–1576. Retrieved from: https://doi.org/10.1007/s12524-019-00994-8

Singh, R. (ed.) (2021) *Re-envisioning remote sensing applications: Perspectives from developing countries*. First Edition. CRC Press. https://doi.org/10.1201/9781003049210

Tribune News Service. (2014, December 19). Bhagtanwala Residents Protest MC's Decision to Dump Garbage. *The Tribune*. Amritsar Edition. Retrieved from: https://www.tribuneindia.com/news/amritsar/bhagtanwala-residents-protest-mc-s-decision-to-dump-garbage/19934.html

12 Use of Remote Sensing Techniques in Land Degradation Mapping

Nusrat Rafique, Durdanah Mattoo, and Sajad Ahmad Mir
University of Kashmir, Srinagar, India

Tahir Hussain Muntazari
National Institute of Technology, Srinagar, India

CONTENTS

DOI: 10.1201/9781003224624-16

12.1 INTRODUCTION

Land is characterized from multiple points of view. The United Nations Convention to Combat Desertification (UNCCD) (2013) defines land as 'the terrestrial bio productive system that comprises soil, vegetation, other biota and the ecological and hydrological processes that operate within the system'. It is one of the most open assets, creating items that are fundamental for human civilization to thrive and shapes the basis of a strong Earth. In any case, because of the wild and steady utilization of this asset, it is confronting a genuine danger of diminishing in its efficiency and quality (Tagore et al, 2012). This danger is in the shape of land deterioration or degradation. Degradation has a long history since the settling of human beings (Scherr, 1999). In the fall of extraordinary civilizations in earlier history, land degradation is accepted to be an incredible contributor (Daily, 1997). Over 2,000 years ago the incomparable Mayan civilization in Mexico was crushed by the consumption of soil fruitfulness and soil erosion as the population expanded which prompted extreme land corrosion and the exhaustion of land fertility until the 'tragedy of the commons' brought about an exorbitant decay of the soil (Olson, 1981).

Land deterioration is one of the genuine threats the cosmos is confronting these days. The United Nations Convention to Combat Desertification (UNCCD) (2004) and the Millennium Ecosystem Assessment (MEA) (2005) define it as *the persistent reduction or loss of land ecosystem services, notably the primary production service* (Vogt et al., 2011; Safriel, 2007). It is likewise characterized as the bringing down of the efficiency limit of land. Land degradation is happening in nearly all areas in every nation, whether developed or developing. It legitimately influences around 250 million individuals and furthermore is the explanation of the decrease of quality as well as quantity of fresh water and soil profitability (Aggarwal et al., 2010). About two billion hectares of agricultural land, pasture and woodlands are influenced by this very reason (Oldeman et al., 1990; Al-Dousari et al., 2000). However, the poor are influenced most who are completely subject to this asset for their livelihood. The seriousness and magnitude in numerous parts of the world can be seen by the fact that over 30 per cent of timberland, 20 per cent of all cultivated areas and 10 per cent of prairies are experiencing deterioration (Bai et al., 2017). Millions of hectares of land of every dimension are deteriorating in every single climatic region and more than 2.6 billion individuals are influenced in more than 100 nations, impacting over 33 per cent of the world's land territory (Adams & Eswaran, 2000). The whole environment has been nearly immersed by the degradation, yet individual aspects concerning land, backwoods (forests), water assets (surface, ground), grasslands (rangelands), biodiversity (animals, vegetative cover, soil) and croplands (rain fed, irrigated) are also concerning (Brinkman, 2005).

In the dry lands which spread practically over almost 40 per cent of the Earth's surface this wonder is more articulated there (Dobie, 2001). Likewise, 73 per cent of rangelands in various regions are as of now degraded, with 47 per cent of minor rain-fed yield lands and some portions of irrigated yield lands. Around 20 per cent of the world's rangelands and grasslands have been harmed by overgrazing (FAO, 1996).

Practically, all agro climatic regions of the world are influenced by land debasement (Brinkman, 2005). About 15 per cent of the world's land and 13 per cent of the

Pacific region is degraded as evaluated by the Global Assessment of Soil Degradation (GLASOD). The denuding in China is approximately 2.67 million km^2, that is about 28 per cent of China based on the watched outcomes is influenced (Brinkman, 2005). The reason for desertification in China is climate, yet anthropogenic elements are predominant (Brinkman, 2005). About 20–30 per cent land and about 250 million individuals in sub-Saharan Africa are influenced via land degradation which is across the board there (Snel & Bot, 2003). Moreover, Snel and Bot (2003) reported that land debasement is far-flung and harsh in Latin America and Asia as well as in almost all other regions of the cosmos; 16 per cent of the ground area in South America and the Caribbean is influenced, while in Middle America the impact more serious (arriving at 26 per cent of the aggregate, or around 63 million hectares) than in South America (where it influences 14 per cent of the aggregate or around 250 million hectares). In Europe, a significant impediment to sustainable agriculture is soil erosion which is one of the most considerable and far-reaching types of land deterioration (Gobin et al., 2004).

Land degradation as a term portrays how various land assets (soil, water, vegetation, rocks, air, climate, relief) have changed to a more regrettable condition.

12.2 DIRECT CAUSES OF LAND DEGRADATION

a. Moving development without sufficient dormant periods.
b. Overcutting of greenery.
c. Ill-advised soil protection.
d. Overgrazing.
e. Developing terrains of longer potential as well as high normal perils.
f. Cultivating areas of longer prospect and/or great instinctive jeopardy.
g. Erroneous crop rotations.
h. Uneven fertilizer use.
i. Improper direction and provision of canal irrigation.
j. Previous draining of groundwater.

12.3 UNDERLYING CAUSES OF DEGRADATION

a. Land shortage.
b. Land regime: holding and open access resources.
c. Monetary pressure and mentalities.
d. Poverty.
e. Population growth. (Young, 1994)

12.4 TYPES OF LAND DEGRADATION

Land degradation takes place due to natural hazards; there are some direct and underlying causes, which erase the fruitfulness of the soil (FAO, 1994). The primary driver of land degradation is soil erosion due to water, considering both immediate and direct causes. Various kinds of land degradation and their effects are shown in the following diagram (Figure 12.1.).

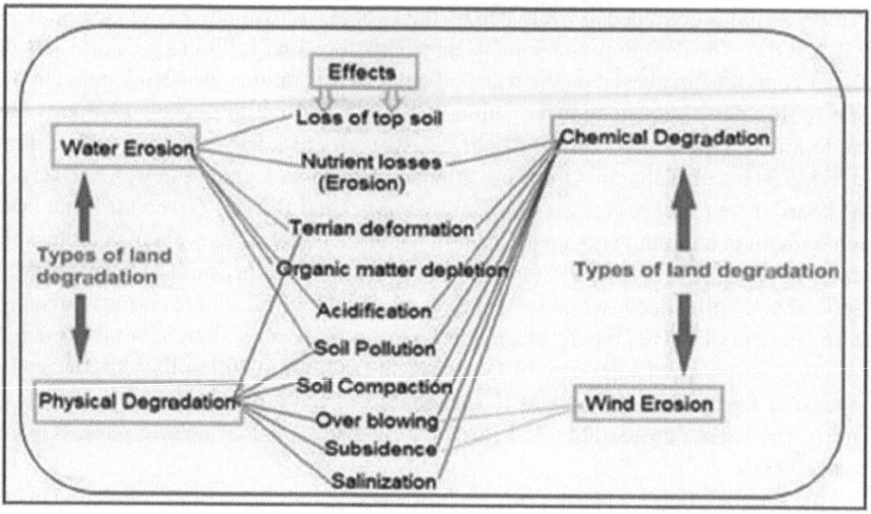

FIGURE 12.1 Cases of land degradation and their effects (IUSS, 2001)

Soil Erosion by Water: this is erosion due to water and precipitation. Practically all parts of the world, including both hot and cool deserts, are influenced by this reason. Sheet and rill erosion is arranged into the category of slight erosion, narrow and shallow as moderate, and ravines as severe. Landsliding heightened by people because of freeing from vegetation, street development and so forth is likewise remembered for it.

Soil Erosion by Wind*:* additionally, this is referred to as Aeolian disintegration, and is characterized as disintegration by the action of wind. It happens for the most part in dry territories where winds disintegrate sand, specifically in three categories for suspension, saltation and creep. It happens for the most part in territories without greenery or in meagrely vegetated regions. Wind expels for the most part free top soil which is wealthy in plant and creature supplements, decreasing the capacity of the soil and reducing its future manageability (SAC, 2016).

Water Logging: this is also called physical deterioration of earth, and is caused when an under-depleted bundle of land gathers water for a longer time, hence influencing the efficiency of the land. This circumstance can prompt salinization and a rise in the water table like flood, salt-rich soil dish, over-abundant water system, wrong waste arrangement, and so on.

12.4.1 Salinization/Alkalization

This is generally the issue of parched and semi-dry districts and can be both characteristic and synthetic. The fundamental drivers are abundance in evapo-transpiration, dry season, overabundant water system, increment in irrigation, and rise in groundwater table to between 2.5 and 3.5 metres from the surface of the soil. Narrow activity brings minerals from the groundwater to the outside of the soil, with pace water gets dissipated and the minerals remain. Overwatering system and overabundant utilization of composts and different chemicals are the primary drivers of salinization in inundated territories.

12.4.2 Frost Heaving

This is an upward growing of soil because of the development of ice during freezing conditions. As a rule it happens when the frigid temperature infiltrates the soil and transforms the current dampness into ice, in this way creating an upward development in the soil. This happens for the most part in snowbound territories.

12.4.3 Anthropogenic or Manmade

This incorporates every one of those land degradation forms which are actuated straightforwardly or in a roundabout way by people. It incorporates mechanical effluents, unearthing sand, rocks, soils, brick kilns, mining regions and dump regions.

12.4.4 Mass Movement

The descending development of soil and shakes, rock trash, mud, and so on, affected by gravity (yet without the active activity of moving liquids), is incorporated under the common term of mass movement. Downslanting development of soils, overload or bedrock under the immediate impact of gravity is remembered for it. It speaks to the unconstrained giving away of earth materials when gravitational power surpasses the interior quality of the stuff. It includes rolling, skidding and flowage of heaps of soil, overburden and bedrock.

12.4.5 Remote Sensing and Land Degradation

For land degradation appraisals, earlier ages did not have quantitative information accessible for mapping, and therefore to a great extent depended on specialists' suppositions (Oldeman, 1992). The development improvements in satellite remote sensing permitted later investigations to use satellite picture information – for example, from the advanced very-high-resolution radiometer (AVHRR).

Remote sensing gives quick and economic data about the ground evaluation when no different methods are accessible (Singh, 2021); likewise it gives spotless and away from physiographic limits. It is an extraordinary method for taking tedious regular or annual tests.

The act of remote sensing and geographic information is shown in Figure 12.2.

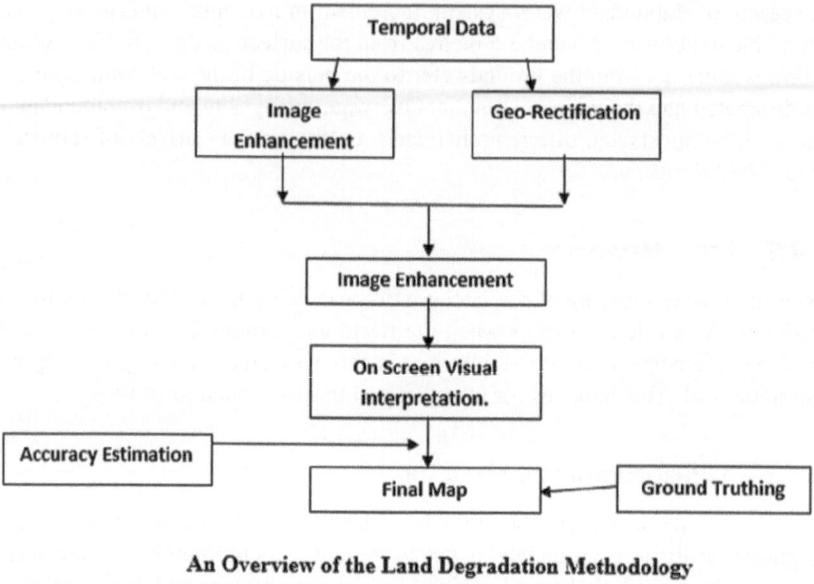

FIGURE 12.2 An overview of the land degradation methodology

Topic	Geographical Representation	Properties Required	Technique
Land deterioration	To display significant kinds of land degradation working.	Kinds of land deterioration. Seriousness level. Cause depiction. Geography. Land-use type.	Visual interpretation of satellite imagery with much ground information.
Land degradation hotspots	Shows the extent of territories, which are degraded, during period extent, ordered according to level of degradation.	Degradation group. Degradation class. Significant causes.	Arrangement of NDVI and land-cover maps. Identification of land degradation hotspots. Categorization of areas into different types and levels of degradation.
Land degradation prediction	Shows the degrees of areas most destined to degrade in not-so-distant future and their arrangements as indicated by the likelihood of its propensity to become degraded.	Evaluation of expected land degradation. Level of certainty. Anticipated period to arrive at the end.	Principal drivers to be recognized from land degradation change maps. Use of authentic historical and current land degradation data.

12.5 MAPPING LAND DEGRADATION

This section will deal with the onscreen visual interpretation of satellite information for mapping land degradation areas. It comprises remote sensing data, preliminary work and the technique of onscreen digitizing. The methodology for the most part generally consists of mapping at any scale, arrangement of base map, on-screen visual rendering/interpretation of the satellite information, formation of legend, ground proofing, soil sampling, characterization of degradation classes and conclusion of map. The procedure is shown in the following Figure 12.2.

12.5.1 Case Study

This is the shared work of different institutes of the Indian Council of Agricultural Research (ICAR) and the Department of Space.

12.5.2 Study Objectives

Spatial distribution and status of degraded and wastelands of India.

12.5.3 Methodology

The whole process was made in geographic information system (GIS) conditions; the accessible spatial layers of salt-affected soils, water erosion (soil loss), acid soils, thick-forested and open-forest layers, and lands influenced by wind erosion were taken into consideration. In the initial step, the soil loss classes of >10 tonnes/ha/yr (assembled as 10–15, 15–20, 20–40, 40–80 and >80 tonnes/ha/yr) were taken into consideration. They were assembled, and generated as layer one. Afterwards, the acid soil layer (<5.5 pH) and soil loss layer were superposed, and layer two was produced, and then the Aeolian erosion layer (intermediate to extremely severe classes) and layer two were overlaid upon each other and layer three was generated. The total area remained the same in the case of the wind erosion areas, since it falls under neither dense nor open-forest cover. The dense/impenetrable woodlands (>40% canopy), open woodlands (<40% canopy) (FSI, 1999) and layer three were superimposed, and the range of water erosion and acid soils under dense forest (>40% canopy) were deducted. The land debasement category amalgamated with open woodlands generated was taken as layer four. In step five, the salt-affected layers with two categories (saline and sodic) and land debasement classes were integrated and all feasible combinations of classes were analysed to generate the fifth layer. In the last move, the physical degradation layers, such as mining and industrial badlands, barren rock and stony badlands, snow-covered and ice caps and soaked, were merged with other land debasement layers, and the resulting land debasement map of India with all potential compound categories was generated. The whole breakup of the methodology is shown in the flowchart (Figure 12.3).

12.6 RESULTS AND DISCUSSIONS

The different types of debased and wasteland areas under the various categories of wind, physical and chemical degradations, and water have been carved out (Figure 12.4). The National Bureau of Soil Survey and Land Use Planning (2019) has classified the degradation classes which have been grouped into 19 classes, and the area under each class has been assessed and identified.

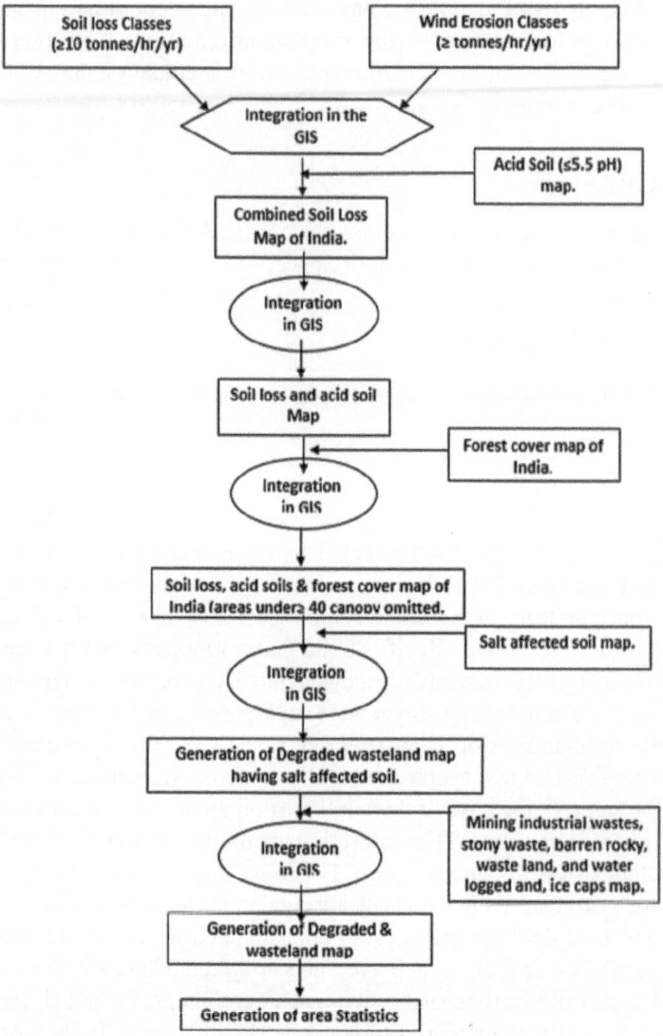

FIGURE 12.3 Methodology flowchart

It has been found that the highest degree of degraded areas by water erosion is in Uttar Pradesh while the least is found in the Andaman and Nicobar Islands with 54 per cent and 0 per cent respectively. A total of 91 per cent of the area of Nagaland is found to be affected by acid soil while the least acid soil (about 0 per cent) is found in Utter Pradesh, Bihar, Haryana, Jammu and Kashmir, Andhra Pradesh, Punjab, Delhi, Gujarat, Andaman and Nicobar Islands.

Nine per cent of the area of Andaman and Nicobar is found to be affected by saline soils, followed by Gujarat (8 per cent), Rajasthan (5 per cent) and Maharashtra, Orissa and Haryana with 1 per cent.

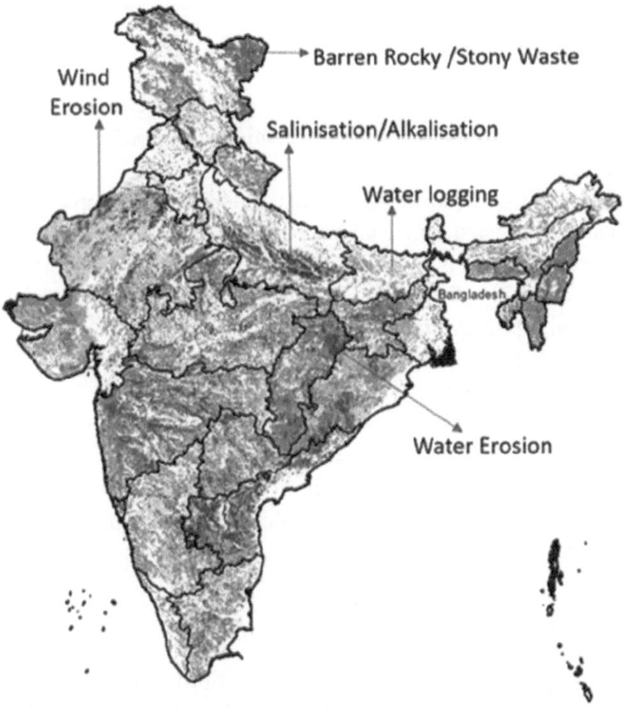

FIGURE 12.4 Land degradation map of India (generated using LISS-III data of 2015–16) (Source: https://www.isro.gov.in/earth-observation/land-degradation)

The methodologies adopted here for the compatibility of different datasets (Chen et al., 2003), with the use of progressive means like remote sensing and GIS, have proved to be the one-stop answer for providing the degraded and wastelands of the country. The whole work is done in the GIS environment. The remote sensing technique gives a summarized view of the inaccessible areas where humans cannot think of going.

12.6.1 Remote Sensing and Soil Mapping

The soil surveys studies mapping in India goes back to the early twentieth century when irregular endeavours were made to characterize soils, generally on edaphic contemplation. Dr F. F. Rickens, a soil specialist from the USA, cleared the way for precise soil surveys in the nation with the foundation of the All-India Soil and Land Use Survey Organization, while the improvement in space technology opened application plausibility in soil mapping (Karale et al., 1983).

a. **Aerial Photo Interpretation**:
   Over the last few decades panchromatic vertical aerial photographs have been utilized for soil mapping. The major ways followed in photo interpretation are (i) pattern analysis; (ii) physiognomic; (iii) element analysis;

and (IV) physiographic. The sample area technique and the adjusted photo-interpretation technique are the two most commonly followed techniques. The physiographic and photo element approach and the sample area technique supported by limited field check have been widely used because of their adaptability to different terrain conditions and various scales of mapping.

Different works done in such a manner by a few people and associations in this way assisted with making elevated photo understanding procedures operational in soil mapping at semi-nitty gritty and reconnaissance levels (Karale et al., 1970; Hilwig & Karale, 1970; Murthy & Hirekerur, 1972; Srinivasan, 1972; Mathur et al., 1980; Kolarkar et al., 1980; Joshi & Dhir 1980; Ahuja & Manchanda, 1980).

b. **Space-borne Studies:**

Another period began in the asset study with the beginning of the Landsat arrangement. They gave a one-of-a-kind arrangement of attributes, – for example, succinct view, availability of information on the phase of vegetation and state of soil, fleeting highlights, obtaining and handling of information in various groups.

12.6.2 Soil Moisture and Remote Sensing

In land surface hydrology, soil dampness is one of the significant factors in the act of commanding the measure of water that invades in the direction of the soil and renews the aquifer. In horticulture, nature, untamed life and general wellbeing, it shapes the most significant association between the water cycle and the life of creatures, plants and people.

The impact of geographical varieties of soil wetness and flora on the advancement and seriousness of extreme tempests has appeared in the investigations of Chang & Wetzel (1991) while the capacity of soil dampness to impact surface moisture angles and to parcel approaching radiative vitality into reasonable and idle warmth has appeared in the investigations of Engman (1997). In enormous scope, the soil wetness and surface conditions are crucial factors in choosing the profundity of the planetary boundary layer and flow and wind designs (Mahfouf et al., 1987; Zhang & Anthes, 1982).

Researchers the world over having perceived the requirement for soil moisture perceptions on an enormous scope considering the spatial degree on a mainland scale, in 1980 examiners utilized the exceptional sensor microwave imager (SSM/I) (Hollinger et al., 1990; Njoku et al., 1998) and checking multichannel microwave radiometer datasets. To contemplate the soil moisture recoveries, affectability and scaling on a mainland degree, the scanning multichannel microwave radiometer (SMMR) information has been utilized (Paloscia et al., 2001; Guha & Lakshmi, 2004). The SSM displaying has additionally been utilized in catchment contemplation (Lakshmi, 1998). Missions included the Tropical Rainfall Measuring Mission (TRMM)'s Microwave Imager (TMI) and WindSat (GAO et al., 2006). Analysts of NASA Goddard Space Flight focus and the Vrije University Amsterdam created soil dampness information (Owe, 2008).

Microwave remote sensing has come in useful in soil dampness estimation over the last 35 years (Njoku et al., 2002; Schmugge et al., 2002). For latent microwave remote sensing, the delegate models are GOM, POM and SPM (Ulaby et al., 1983), IEM (Fung, 1994), AIEM (Choudhury et al., 1979) and Q/H (O'Neill, 1996). A delegate soil moisture estimation model for greenery regions is the MIMICS model (Wang & Choudhury, 1981). Researchers created diverse soil moisture displaying endeavours from dynamic microwave sensors based on the connection between backscatter coefficients and soil moisture – for example, empirical models (Schaefer et al., 2007), semi-experimental models (Merlin et al., 2010) and truly based models (Merlin et al., (2009). At the territorial scale, soil dampness has been measured by utilizing remote detecting methods, which speak to another period in soil dampness estimations.

Classification	Techniques	Merits	Demerits
Optical	(a) Visual-based technique (b) Thermal infrared-based technique	(a) Estimable geographical resolution, multiple bands accessible (b) Great geographical resolution, numerous satellites accessible	Vegetation impedance, nocturnal impacts and poor. Vegetation impedance, overcast pollution, nocturnal effects, poor physical resolution and meteorological impacts temporal resolution.
Passive microwave	(Semi) experimental, substantial-based approaches	Huge certainty for exposed ground surfaces, boundless by clouds and/or daylight circumstances, high sensual resolution	Off colour geographical resolution, determined by floral cover and exterior bumpiness.
Active microwave	(Semi) experimental, substantial-based approaches	Exceptional geographical resolution, inexhaustible by clouds and/or daylight conditions	Sway by exterior bumpiness and flora, offset amount, off colour profane durations resolution.
Collective methods	(a) Active and passive MW (b) MW and optical	(a) Better temporal and spatial resolution (b) lessened flora and surface bumpiness effects	(a) SMC mounting and verification needs, non-identical SMC measurement depth. (b) SMC clamber and affirmation demand heed, non-identical SMC quantifying depth.

Remote Sensing Methods used in Soil Dampness Estimation (Petropoulos et al., 2015)

12.6.3 Soil Fertility Assessment and Remote Sensing

Natural carbon, soil nitrate levels, soil dirt and thickness are the fundamental segments of soil fruitfulness attended to by remote detecting. To the extent that supplements of plants are viewed, nitrogen is one of the principal supplements augmenting the plant yield and financial returns to ranchers.

SPOT and Landsat and a large group of other open and private sensors have demonstrated their value for crop distinguishing proof and wetland stock. The unpredictable idea of the connection between soil parameters and reflectance is the serious issue of getting data from remotely detected information. A remarkable effect on reflectance has been seen because of organic issue, dampness, surface, cation-exchange capacity (CEC), mineral oxides and surface conditions. These connections are frequency-dependent; accordingly, they influence soil parameters which may influence reflectance contrastingly in various regions of the range.

12.6.4 Organic Matter

Organic matter assumes a significant job in agri-business creation and ecological research. Lab-based reflectance spectroscopy related to high elevation remote detecting has risen as one of the significant stages during the most recent three decades.

The spectral response of soils due to colour has been associated with soil fertility as:

a. By and large soils darker in shading are higher in natural issue and are less intelligent than those soils with lower natural issue content.
b. The soils of glaciated regions, prairie grass vegetations, high organic soils, and the soils of forest vegetation can be promptly separated in high-resolution satellite images.
c. Phantom reactions in uncovered soil zones are commonly veiled with high build-up content from past yield deposits. In such circumstances, utilization of high spatial computerized landscape models (DEM) delivered by laser or remote microwave procedures can give a superior comprehension of soil arrangement and surface dampness development and subsequently help in deciphering surface soil's natural issue content (Krishan et al., 2014).

12.6.5 Nitrogen Deficiency

Nitrogen lack causes abatement in leaf chlorophyll focuses, prompting an expansion in leaf emultance in the noticeable area (400–700 nm). In any case, increment in plant reflectance because of decreased measure of chlorophyll may likewise happen because of a few different components like bugs and illnesses (Carter et al., 2001). In the red edge area almost between 680 and 780 nm of the electromagnetic spectrum a state of inflection occurs termed the red edge position (REP) which is influenced by biophysical and biochemical parameters. This deviation in the REP to shorter or longer frequencies is intended to gauge changes in foliar appendages' chlorophyll content or nitrogen content as the articulation point and furthermore as a marker of vegetation strain. A direct extrapolation strategy for removing REP utilized by Cho & Skidmore (2006) has demonstrated high relationships with a wide scope of foliar nitrogen focuses for both tight and wide bandwidth of hue cycle.

12.6.6 Soil Resource Studies

Soil is as significant an asset as water; it gives supplements as well as solid footings to the foundations of plants, therefore it is a basic fixing for their sound development and yield. It is a perplexing blend of natural and mineral substances which is continually being framed by the enduring of rocks.

The soil order frameworks utilized far and wide are the mixture of various assets, such as individuals' need, innovation, preparing utilization, and information accessibility. For mapping soils with the utilization of remote sensing, it positively affected the earth on the grounds that for doing present-day cultivating, remote sensing images having adequate goals to catch characteristic soil fluctuation in enough detail to help 'exactness cultivating' proposals was just conceivable with the advances, for example, in GIS, remote sensing and global positioning systems (GPS).

To take care of the different issues of characteristic assets, land assessment gives a level-headed premise to dissect soil, atmosphere and land parameters to show up at ideal arrangement. In this manner in assessing forms, GIS turns into a significant instrument to coordinate the unpredictable choices to be taken under multivariate circumstances of the asset.

The major GIS applications relevant to soils are:

a. Crop suitability studies
b. Land efficiency evaluation
c. Land irrigability estimation
d. Land productivity evaluation
e. Soil erosion modelling irrigation
f. Irrigation management directions in command areas
g. Prioritization of sub-watersheds/micro-watersheds in a given watershed
h. Soil suitability evaluation for various purposes like specific crops, industries, forestry, etc.
i. To identify critical areas in watershed/micro watershed
j. To generate excellent land-use plans, etc.
k. Quantification of soil loss
l. Planning inner-city expansion
m. Making reclamation plans of degraded lands.

12.7 CONCLUSION

The remote sensing application is almost three decades old. The potential of this technology needs to be employed at regular intervals for monitoring the soil. Advances in this technology are paving new methods to monitor soil details and agriculture. The advance in stereoscopic satellite information in soil mapping has not been fully utilized and greater ambition is seen in this field. For the categorization of soils and estimating the systematic anatomy of soil map units, ground-penetrating radar has been proven to be very beneficial. For scrutinizing the geographical extensions along with profane durations and fluctuations in soil dampness, synthetic aperture radar (SAR) data needs to be explored. Geographically accurate soil data assists in the use of fertilizers and pesticides, thereby paving the way for the originator to increase profits and to approach land in an environmentally friendly way for producing food.

REFERENCES

Adams, C.R., and H. Eswaran. "Global land resources in the context of food and environmental security". *Advances in land resources management for the 20th century. Soil Conservation Society of India*, New Delhi (2000): 35–50.

Aggarwal, Pramod K., W. E. Baethegan, P. Cooper, R. Gommes, B. Lee, H. Meinke, L. S. Rathore, and M. V. K. Sivakumar. "Managing climatic risks to combat land degradation and enhance food security: Key information needs". *Procedia Environmental Sciences* 1 (2010): 305–312.

Ahuja, R. L., and M. L. Manchanda. "Use of aerial photo-interpretation technique for soil survey of a part of Ganges alluvial plain in Muzzaffarnagar District, UP". In *Seminar on application of photo photointerpretation and remote sensing techniques for natural resources survey and environmental analysis*, IIRS, Dehradun. (1980).

Al-Dousari, A. M., Rafat Misak, and Shaad Shahid. "Soil compaction and sealing in Al-Salmi area, western Kuwait". *Land Degradation & Development* 11, no. 5 (2000): 401–418.

Bai, Z. G., D.L. Dent, Lennart Olsson, and Michael E. Schaepman. "Global assessment of land degradation and improvement 1: identification by remote sensing". Food and Agriculture Organization – International Soil Reference and Information Centre (2017).

Chang, Jy-Tai, and Peter J. Wetzel. "Effects of spatial variations of soil moisture and vegetation on the evolution of a prestorm environment: A numerical case study". *Monthly Weather Review* 119, no. 6 (1991): 1368–1390.

Chen, Kun-Shan, Tzong-Dar Wu, Leung Tsang, Qin Li, Jiancheng Shi, and Adrian K. Fung. "Emission of rough surfaces calculated by the integral equation method with comparison to three-dimensional moment method simulations". *IEEE Transactions on Geoscience and Remote Sensing* 41, no. 1 (2003): 90–101.

Cho, Moses Azong, and Andrew K. Skidmore. "A new technique for extracting the red edge position from hyper-spectral data: The linear extrapolation method". *Remote sensing of environment* 101, no. 2 (2006): 181–193.

Choudhury, B. J., Thomas J. Schmugge, A. Chang, and R. W. Newton. "Effect of surface roughness on the microwave emission from soils". *Journal of Geophysical Research: Oceans* 84, no. C9 (1979): 5699–5706.

Daily, G. C. *Nature's Services: Societal Dependance on Natural Ecosystems*. Washington DC: Island Press (1997).

Dobie, Philip. "Poverty and the Dry lands". *Challenge Paper Series, UNDP Dry lands Centre, Nairobi* (2001).

Engman, Edwin T. "Soil moisture, the hydrologic interface between surface and ground waters". *Iahs Publication* (1997): 129–140.

FAO. "Land degradation in South Asia, its severity, causes and effects upon the people". World Soil Resources Report No. 78. Food and Agriculture Organization (1994).

FAO. "Rome declaration on world food security". World Food Summit, Food and Agriculture Organization, Rome (1996).

FSI. "State of Forests: 1998". Forest Survey of India, Dehradun (1999).

Fung, Adrian K. *Microwave scattering and emission models and their applications*. Boston: Artech House. (1994).

Gao, Huilin, Eric F. Wood, T. J. Jackson, M. Drusch, and R. Bindlish. "Using TRMM/TMI to retrieve surface soil moisture over the southern United States from 1998 to 2002". *Journal of Hydrometeorology* 7, no. 1 (2006): 23–38.

Gobin, Anne, R. Jones, M. Kirkby, P. Campling, Gerard Govers, C. Kosmas, and A. R. Gentile. "Indicators for pan European assessment and monitoring of soil erosion by water". *Environmental Science & Policy* 7, no. 1 (2004): 25–38.

Guha, Aniruddha, and Venkataraman Lakshmi. "Use of the Scanning Multichannel Microwave Radiometer (SMMR) to retrieve soil moisture and surface temperature over the central United States". *IEEE transactions on geoscience and remote sensing* 42, no. 7 (2004): 1482–1494.

Hilwig, F. W. and R. L. Karale. "Aerial photo-interpretation for detailed soil survey of salt affected soils near Taleta, Meerut District, U.P."All India Symposium on Soil Salinity, Kanpur. (1970).

Hollinger, James P., James L. Peirce, and Gene A. Poe. "SSM/I instrument evaluation". *IEEE Transactions on Geoscience and Remote Sensing* 28, no. 5 (1990): 781–790.

IUSS. *The Land Degradation and Desertification Report*. The Working Group on Land Degradation and Desertification, International Union of Soil Science, Vienna, Austria. (2001).

Joshi, D. C. and S. P. Dhir. "Experience of small-scale soil mapping in Nagaur using aerial photo analysis technique", presented at the seminar on Application of photo-interpretation and RS techniques for natural resources survey and environmental analysis, Dehradun (1980).

Karale, R. L., Y. P. Bali, and KV Seshagiri Rao. "Soil mapping using remote sensing techniques". *Proceedings of the Indian Academy of Sciences Section C: Engineering Sciences* 6, no. 3 (1983): 197–208.

Karale, R. L., K. R. Venugopal, and F. W. Hilwig. "Reconnaissance soil survey in the Ganges alluvial plain in Meerut District". Unpublished report. (1970).

Kolarkar, A. S., R. P. Dhir, and N. Singh. "Salt affected lands in south-eastern arid Rajasthan-their distribution and genesis as studied from aerial photographs". *Proceedings of the Seminar on Application of Photointerpretation and Remote Sensing Techniques for natural resources survey and environmental analysis*. Indian Society of Photo Interpretation and Remote Sensing. (1980): 132–136.

Krishan, G. S. K., Saha, S. Kumar, and N. R. Patel. "Remote sensing in soil fertility evaluation and management". In R. Chandra and K. P. Raverkar, eds. *Bioresources for sustainable plant nutrient management*. New Delhi: Satish Serial Publishing House. (2014): 509–533.

Lakshmi, V. "Special sensor microwave imager data in field experiments: FIFE-1987". *International Journal of Remote Sensing* 19, no. 3 (1998): 481–505.

Mahfouf, Jean-François, Evelyne Richard, and Patrick Mascart. "The influence of soil and vegetation on the development of mesoscale circulations". *Journal of Climate and Applied Meteorology* 26, no. 11 (1987): 1483–1495.

Mathur, A., A. K. Sharma and T.R. Rathore. "Soil survey of a part of Tara and Bhabar Forest using aerial photographs as base maps", presented at the seminar on *Application of Photo-Interpretation Techniques for Natural Resources Survey and Environmental Analysis*, Dehradun (1980).

Merlin, Olivier, Ahmad Al Bitar, Jeffrey P. Walker, and Yann Kerr. "A sequential model for disaggregating near-surface soil moisture observations using multi-resolution thermal sensors". *Remote Sensing of Environment* 113, no. 10 (2009): 2275–2284.

Merlin, Olivier, Ahmad Al Bitar, Jeffrey P. Walker, and Yann Kerr. "An improved algorithm for disaggregating microwave-derived soil moisture based on red, near-infrared and thermal-infrared data". *Remote Sensing of Environment* 114, no. 10 (2010): 2305–2316.

Murthy, R. S. and L. R. Hirekerur. "Use of air photo interpretation techniques: Mapping in soil surveys", presented at appreciation *Seminar on Use of API in survey and mapping of natural resources*, Dehradun (1972).

Njoku, E., B. Rague, and K. Fleming. "The Nimbus-7 SMMR Pathfinder brightness data set". *Jet Propulsion Lab, Pasadena, Calif, USA* (1998).

Njoku, Eni G., William J. Wilson, Simon H. Yueh, Steve J. Dinardo, Fuk K. Li, Thomas J. Jackson, Venkat Lakshmi, and J. Bolten. "Observations of soil moisture using a passive and active low-frequency microwave airborne sensor during SGP99". *IEEE Transactions on Geoscience and Remote Sensing* 40, no. 12 (2002): 2659–2673.

Oldeman, L. R., R. U. Hakkeling, W. G. Sombroek. *World map of the status of human-induced soil degradation: An explanatory note*. Wageningen: International Soil Reference and Information Centre, (1990).

Oldeman, L. R. "Global extent of soil degradation". In *Bi-Annual Report 1991–1992/ISRIC*, pp. 19–36. ISRIC, (1992).

Oldeman, L. R. "Soil Degradation: A Threat to Food Security?". In *International Conference on Time Ecology: "Time for Soil Culture-Temporal Perspectives on Sustainable Use of Soil"*, April 6–9, 1997, *Tutzing*, pp. 15–25. (1997).

Olson, Gerald W. "Archaeology: Lessons on future soil use". *Journal of Soil and Water Conservation* 36, no. 5 (1981): 261–264.

O'Neill, P. E., N. S. Chauhan and T. J. Jackson. "Use of active and passive microwave remote sensing for soil moisture estimation through corn". *International Journal of Remote Sensing* 17, no. 10 (1996): 1851–1865. DOI: 10.1080/01431169608948743.

Owe, Manfred, Richard de Jeu, and Thomas Holmes. "Multisensor historical climatology of satellite-derived global land surface moisture". *Journal of Geophysical Research: Earth Surface* 113, no. F1 (2008).

Paloscia, Simonetta, Giovanni Macelloni, Emanuele Santi, and Toshio Koike. "A multifrequency algorithm for the retrieval of soil moisture on a large scale using microwave data from SMMR and SSM/I satellites." *IEEE Transactions on Geoscience and Remote Sensing* 39, no. 8 (2001): 1655–1661.

Petropoulos, George P., Gareth Ireland, and Brian Barrett. "Surface soil moisture retrievals from remote sensing: Current status, products & future trends". *Physics and Chemistry of the Earth, Parts A/B/C* 83 (2015): 36–56.

SAC, ISRO. "Desertification and land degradation Atlas of India (based on IRS AWiFS data of 2011–13 and 2003–05)". *Speace Applications Centre, ISRO, Ahmedabad, India, Ahmedabad* (2016).

Safriel, Uriel N. "The assessment of global trends in land degradation." In M. V. K. Sivakumar and N. Ndiang'ui, eds. *Climate and land degradation*, pp. 1–38. Springer: Berlin, Heidelberg (2007).

Schaefer, Garry L., Michael H. Cosh, and Thomas J. Jackson. "The USDA natural resources conservation service soil climate analysis network (SCAN)." *Journal of Atmospheric and Oceanic Technology* 24, no. 12 (2007): 2073–2077.

Scherr, Sara J. "Soil degradation: a threat to developing-country food security by 2020?". *International Food Policy Research Institute* 27, (1999).

Schmugge, Thomas J., William P. Kustas, Jerry C. Ritchie, Thomas J. Jackson, and Al Rango. "Remote sensing in hydrology." *Advances in Water Resources* 25, no. 8–12 (2002): 1367–1385.

Singh, R. (ed.) *Re-envisioning remote sensing applications: perspectives from developing countries*. First Edition. CRC Press. (2021). doi:10.1201/9781003049210

Snel, M., and A. Bot. "Draft paper: suggested indicators for land degradation assessment of drylands". *FAO, Rome* (2003).

Srinivasan, T. R. "Photo-interpretation for land and soil resource appraisal". In *Appreciation seminar on use of API in survey and mapping of natural resources, Forest Research Institute*, Dehradun, May. (1972).

Tagore, G. S., G. D. Bairagi, N. K. Sharma, R. Sharma, S. Bhelawe, and P. K. Verma. "Mapping of degraded lands using remote sensing and GIS techniques". *Journal of Agricultural Physics* 12, no. 1 (2012): 29–36.

Ulaby, Fawwaz T., Mohammad Razani, and Myron C. Dobson. "Effects of vegetation cover on the microwave radiometric sensitivity to soil moisture". *IEEE Transactions on Geoscience and Remote Sensing* 1 (1983): 51–61.

Vogt, J. V., Uriel Safriel, Graeme Von Maltitz, Youba Sokona, Robert Zougmore, Gary Bastin, and Joachim Hill. "Monitoring and assessment of land degradation and desertification: towards new conceptual and integrated approaches". *Land Degradation & Development* 22, no. 2 (2011): 150–165.

Wang, J. R., and B. J. Choudhury. "Remote sensing of soil moisture content, over bare field at 1.4 GHz frequency". *Journal of Geophysical Research: Oceans* 86, no. C6 (1981): 5277–5282.

Young, Anthony. *Land degradation in South Asia: its severity, causes and effects upon the people*. Food and Agriculture Organization, Rome. (1994).

Zhang, Dalin, and Richard A. Anthes. "A high-resolution model of the planetary boundary layer—Sensitivity tests and comparisons with SESAME-79 data". *Journal of Applied Meteorology* 21, no. 11 (1982): 1594–1609.

13 Feasibility Analysis of a Railway Route Using Remote Sensing/Geoinformatics and Analytical Hierarchy Process (AHP) Techniques

Rashmi Chandan and Chandan Kumar Boraiaha
Central University of Kerala, Kasaragod, India

CONTENTS

13.1 INTRODUCTION

Transportation networks are a critical part of the economy (Railway Ecology, 2017). Railway networks are a cost-effective and efficient means of travel and freight movement. When it comes to investigating the needs of a new railway line to be laid,

DOI: 10.1201/9781003224624-17

certain survey techniques need to be followed. Pre-investment decision surveys involve prioritizing and long-term planning. These surveys generally take the form of reconnaissance surveys (Railway Engineering Code, 2018) and are termed feasibility studies. When planning a new rail link, several parameters, such as the need for a railway link, economic viability, security, ground condition and environmental impact, are to be taken into consideration by the decision makers and transportation experts.

The significance of remote sensing lies in its ability to reach inaccessible places and generate a synoptic view. Any developmental projects need prior planning before implementation. Unfortunately, our current urban setup makes such pre-planning surveys time consuming as well as costly. However, the tremendous advances in remote sensing technology made in recent decades has made pre-planning surveys cost-effective by providing satellite imagery of the desired resolution, which can be further processed and manipulated in geographic information system (GIS) environs to get the desired results.

Geoinformatics (the combination of remote sensing and GIS technology) can aid quick decision-making process in a variety of ways. For instance, site selection for urban development and route selection for transportation planning have been made simpler by land suitability mapping techniques using geoinformatics technology (Tischler, 2017).

The integration of the analytical hierarchy process (AHP) (Saaty, 1980) with geoinformatics serves as a powerful tool in the multi-criteria decision method (MCDM). Multi-criteria decision analysis was developed as a part of operational research. It makes use of computational and mathematical tools in order to perform subjective evaluation of performance-based criteria (Mardani et al., 2015). Due to its precise evaluation, MCDM is extensively used in various sectors that need prior planning – for example, urban planning, energy, environment, business and economy.

AHP is proposed by Saaty (1986) as a decision analysis technique to evaluate complex multi-attribute alternatives among one or more decision makers. AHP determines the percentage importance of different parameters/criteria which are used to determine site suitability. Land/site suitability analysis is a geoinformatics-based process which involves targeting a suitable location that fulfils the selection criteria (such as environmental, social and economic) for a specific purpose (Bozdag et al., 2016).

The combination of geoinformatics-based MCDM with AHP serves as an important tool in land suitability studies. This is because the process of identifying the significance of various criteria used in land suitability studies and assigning them suitable ranks based on certain calculated weights that makes use of scale-of-importance and expert opinion is made simple by integrating AHP with geoinformatics (Aburas et al., 2015). Various workers have made an attempt to apply geoinformatics and MCDM-based land suitability studies in various domains. For example, land suitability for urban planning (Joerin et al., 2001; Bozdag et al., 2016), urban transportation facility (Farkas, 2009), sanitary land fill (Karakus, 2019), water storage (Ahmad & Verma, 2017), photovoltaic farms (Merrouni et al., 2017), and solar wind hybrid power station (Wu & Geng, 2014).

The current study is based on the integration of geoinformatics (remote sensing and GIS) and AHP for investigating the feasibility of our proposed Shivamogga-Harihar railway route. Multiple criteria for site suitability analysis are developed, based on literature review, field investigation, surveying and Indian railway engineering codes. An AHP-based pair-wise comparison matrix is obtained, by giving priorities to different parameters considered for the feasibility study. Weights of values are derived for each criterion which is utilized in GIS weighted overlay analysis to derive a suitability (feasibility) map of the study area.

13.2 MATERIAL AND METHODS

Based on a literature survey and engineering codes (Indian railway engineering code for engineering department, revised edition 1982, correction slip 08/01/2020; Farkas, 2009; Borda-de-Água et al., 2017; Railway Handbook, 2013; International Energy Agency, 2013, www.iea.org), the following criteria were identified which determines the suitability of land to lay new railway track in the study area:

Geotechnical parameters: Lithology, soil, slope, land use/land cover (LULC), geomorphology

Environmental parameters: Forest cover, water bodies/wet lands, minimal landscape modification (environmental parameters' criteria are included in LULC classification).

Raster data set on each of the above indicators are generated and processed. Each data layer is categorized into three classes – i.e., most suitable, moderately suitable and least suitable for laying new railway track. Weights for each of the selected eight criteria are calculated using AHP pair-wise comparison method using Excel spreadsheet (Goepel version 2018, http://bpmsg.com). The weights of each criterion are used to derive a suitability map by using the weighted overlay tool in ArcGIS 10.3. A flowchart depicting the methodology followed in this study is given in Figure 13.1.

13.2.1 Generation of Thematic Maps for the Criteria Set

- Land use/land cover (LULC) map: The LULC map of the study area is extracted from IRS LISS III data. Eight LULC classes were made by supervised classification techniques, with the built-up area again updated from Google Earth. Forest, eucalyptus plantation, stony waste, water body, settlement, double crop, kharif and plantation are the eight land-use classes derived.
- Lithology map: The lithological map of the study area is extracted from the Lithological Map of Karnataka (KSRSAC, 2004). Further, it is updated based on our field studies and is used for categorization of the area into three suitability rankings.

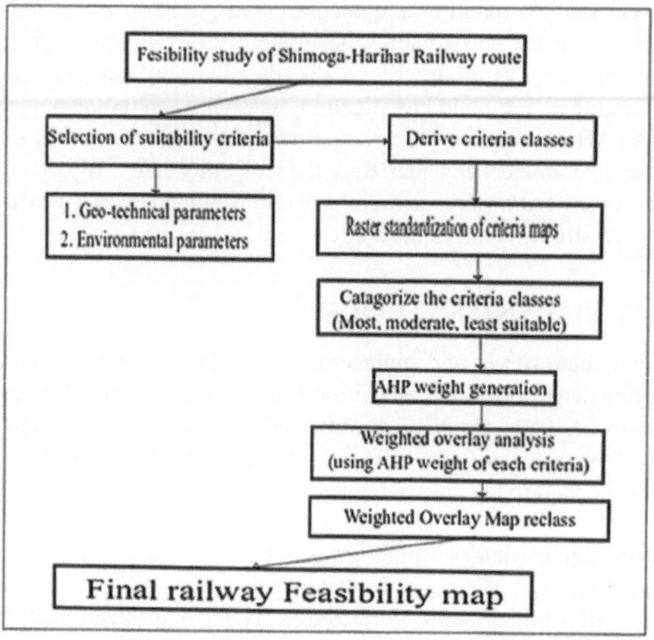

FIGURE 13.1 Flowchart of methodology adopted

- Soil map: The soil map is extracted from the National Atlas and Thematic Mapping Organization (NATMO) and the Geological Survey of India (GSI) map.
- Geomorphology map: A geomorphology map of the study area is prepared using satellite imagery and topo sheets, by applying supervised classification and manual digitization techniques.
- Map indicating proximity influence on railway track is prepared by buffer analysis technique, available in ArcGIS 10.3a – Distance from water bodies.

13.2.2 Study Area

The study area covers a part of the Shivamogga and Davanagere districts of Karnataka state in south India (Figure 13.2). A linear stretch to the left of the Tunga and Tungabhadra rivers is taken for the feasibility analysis. The study area lies between 75°30′ to 75°49′12″ longitudes and 13°54′ to 14°34′48″ latitudes. The total geographical area of the current study is 1,411 km^2.

13.2.3 Data Used

13.2.3.1 Primary Data

Survey of India toposheets (48 O/9, N/10, N/11, N/12, N/14, N/15)
Satellite data: IRS LISS III 2016

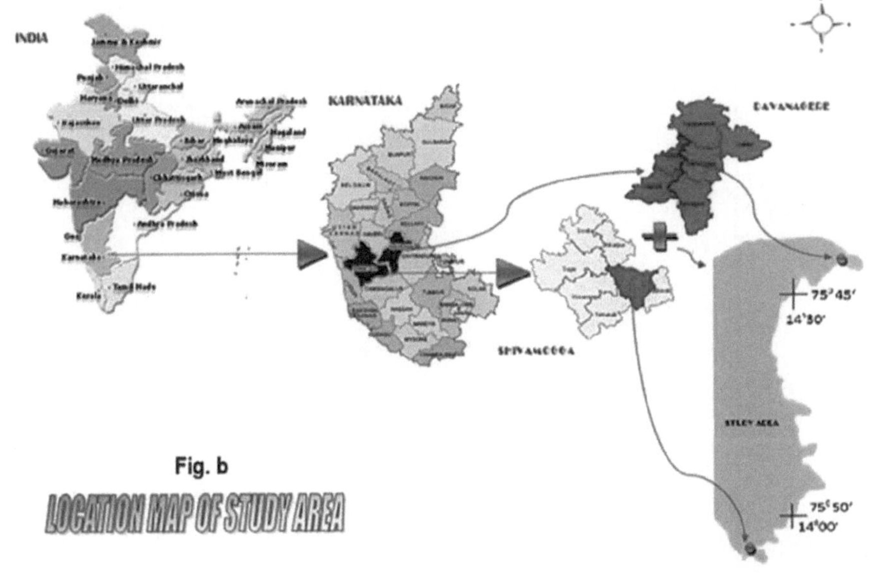

FIGURE 13.2 Location map of study area

13.2.3.2 Secondary Data

Literature review, Indian Railway Engineering Code 2020, Indian Railway Permanent Way Manual 2004, Socio-economic Data 2006.

13.2.3.3 Generation of Criteria Maps

The priorities given for each class of different criteria map are based on the reasons derived from the literature survey and engineering codes (Indian railway engineering code for engineering department, revised edition 1982, Correction slip 08/01/2020; Farkas, 2009; Borda-de-Água et al., 2017; Railway Handbook, 2013; International Energy Agency, 2013, www.iea.org) and expert opinion.

Thus, the following layers are derived:

1. Suitability ranking for land use/land cover in study area (Figure 13.3)
2. Suitability ranking for hydrogeomorphology in study area (Figure 13.4)
3. Suitability ranking for soil in study area (Figure 13.5)
4. Suitability ranking for lithology in study area (Figure 13.6)
5. Suitability ranking for slope class in the study area (Figure 13.7).

13.3 DRAINAGE BUFFERING

Water in the topsoil or topmost layer of the earth's surface poses a great threat for railway track construction. The surface and subsurface water should be well away from the path of the railway track. The source of moisture in the railway track is surface water, soaked water, seepage water and hydroscopic water. The presence

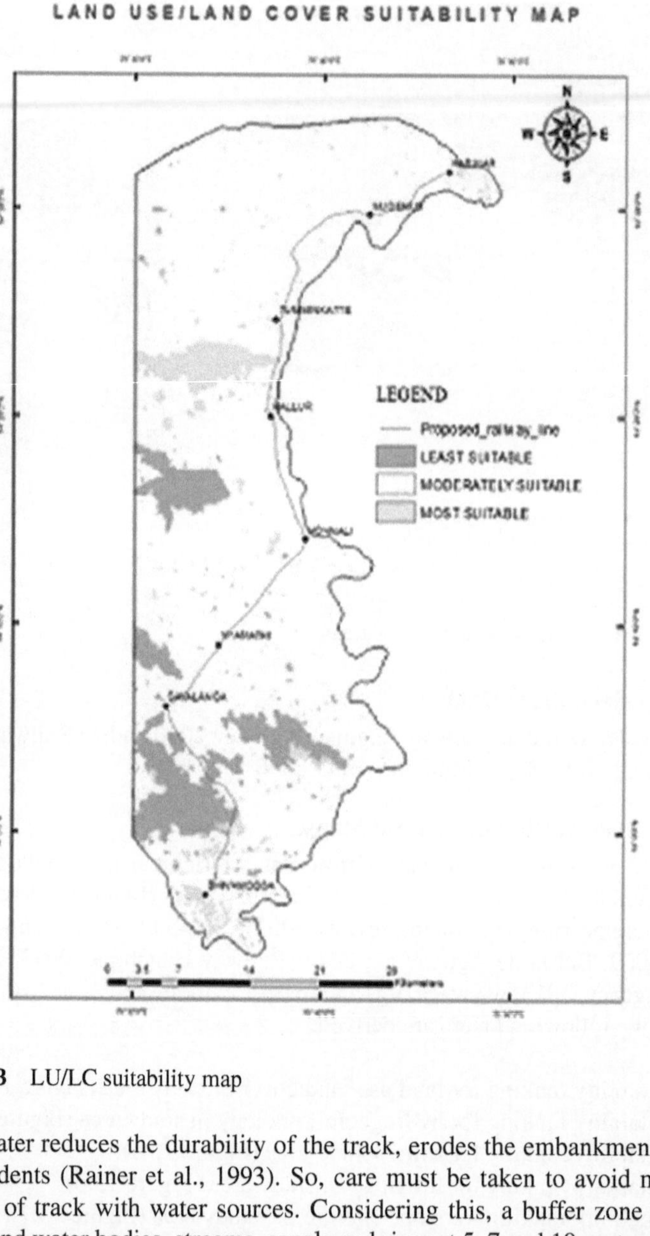

FIGURE 13.3 LU/LC suitability map

of excess water reduces the durability of the track, erodes the embankment and can lead to accidents (Rainer et al., 1993). So, care must be taken to avoid maximum intersection of track with water sources. Considering this, a buffer zone has been created around water bodies, streams, canals and river at 5, 7 and 10-metre distances (Figure 13.8), respectively. The route is aligned in such a way as not to intersect with these buffer zones.

13.4 STANDARDIZATION OF CRITERIA MAPS

Each criteria map was standardized in GIS environment to a uniform unit before weighted overlay analysis. Standardization technique employs conversion

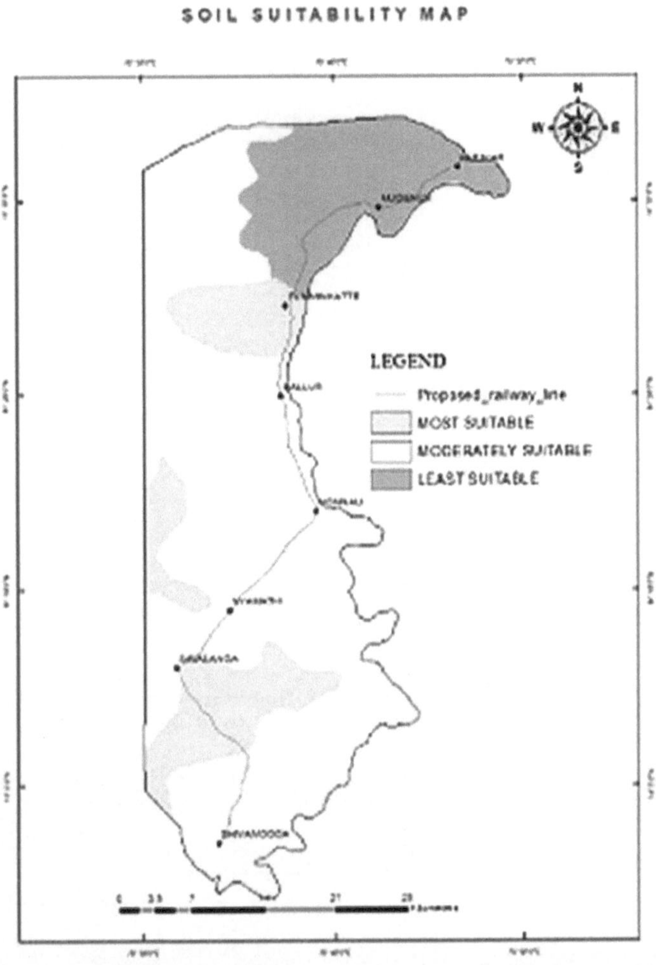

FIGURE 13.4 Geomorphology suitability map

of measurements in each thematic (criteria) map into a uniform scale (Yalew, 2016). Standardization requires conversion of the criteria maps from vector to raster format. Further, the raster map needs to be reclassified into suitable classes – most, moderately and least suitable. Weights of each criteria map are calculated using AHP. Finally, the suitability/feasibility map is produced by weighted overlay analysis.

13.5 WEIGHT CALCULATION FOR CRITERIA MAPS

Lithology, land use/landcover, soil, geomorphology and slope are the five important parameters considered for laying a new railway line (Railway Handbook, 2013). Weights of the same are obtained by AHP pair-wise comparison matrix.

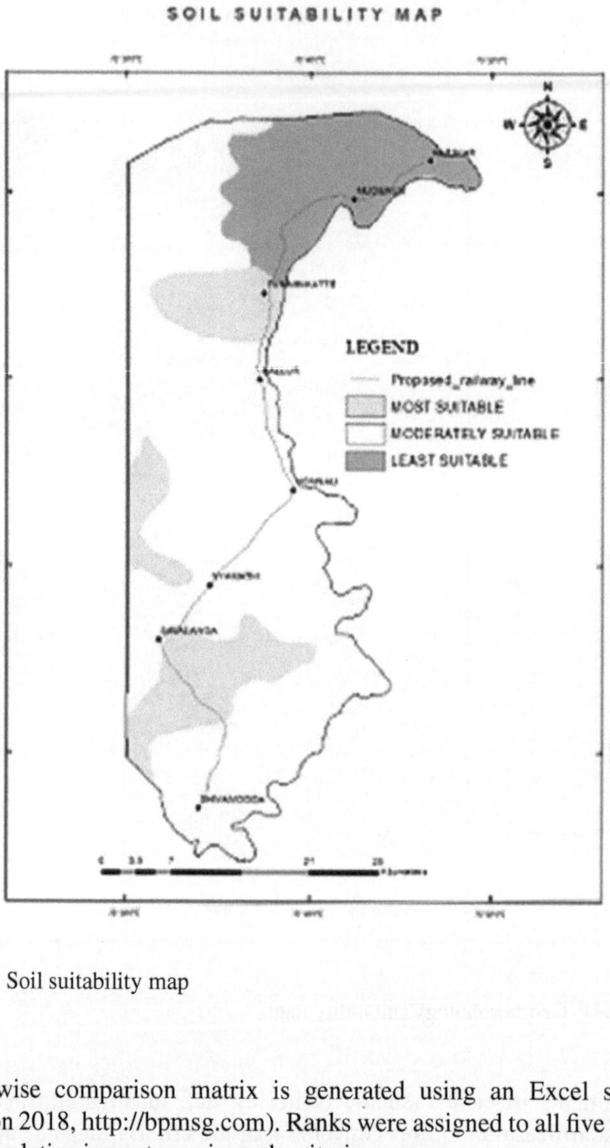

FIGURE 13.5 Soil suitability map

AHP pair-wise comparison matrix is generated using an Excel spreadsheet (Goepel version 2018, http://bpmsg.com). Ranks were assigned to all five parameters based on their relative importance in each criteria map.

Prioritization of parameters in AHP is based on the 9-point scale (Saaty, 2008) (Table 13.1) measurement which finally creates a matrix of pair-wise comparison (Saaty, 1980). This pair-wise comparison allows independent evaluation of each factor, which eases the multiple decision problem. 1 corresponds to equal weightage

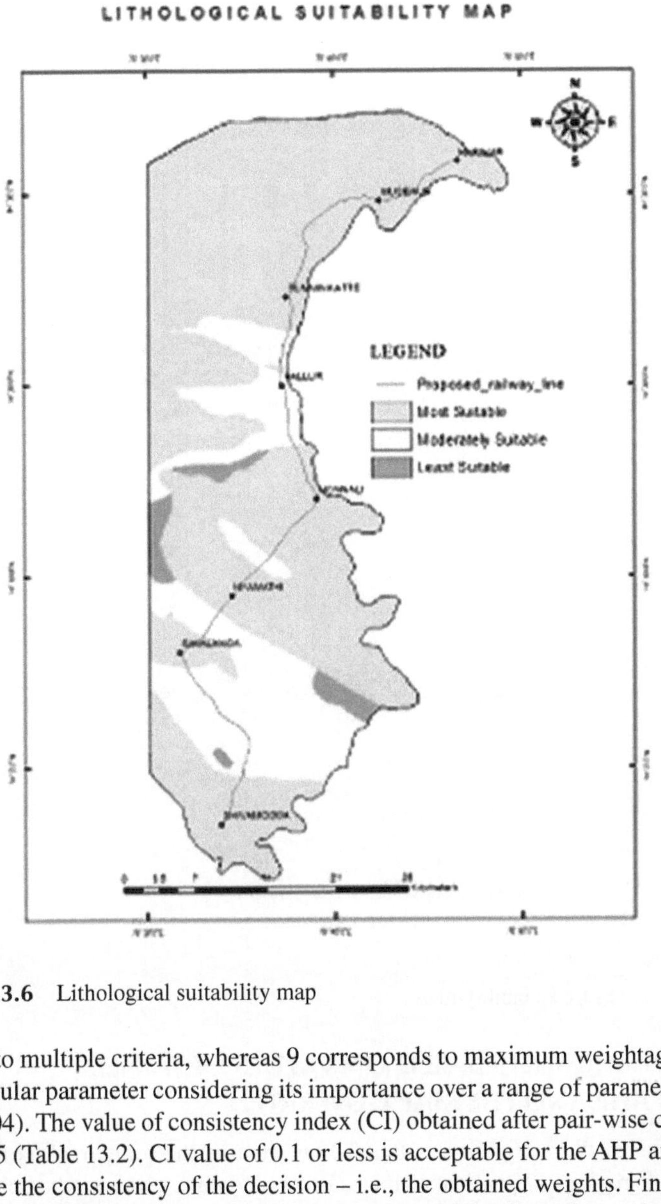

FIGURE 13.6 Lithological suitability map

assigned to multiple criteria, whereas 9 corresponds to maximum weightage assigned to a particular parameter considering its importance over a range of parameters (Saaty, 1980, 2004). The value of consistency index (CI) obtained after pair-wise comparison is 0.09025 (Table 13.2). CI value of 0.1 or less is acceptable for the AHP analysis and to validate the consistency of the decision – i.e., the obtained weights. Finally, all the map layers were integrated in a GIS environment (weighted overlay analysis) to generate the final suitability map (Figure 13.9) indicating three classes – i.e., most suitable, moderately suitable and least suitable for the proposed railway track.

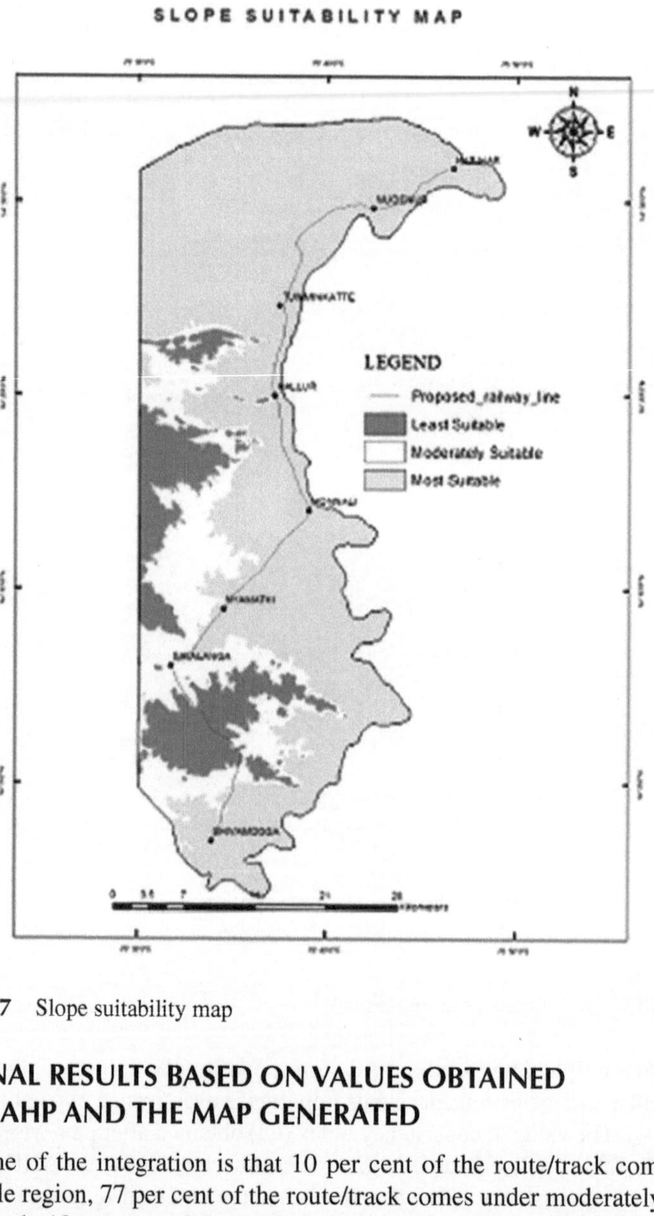

FIGURE 13.7 Slope suitability map

13.6 FINAL RESULTS BASED ON VALUES OBTAINED IN AHP AND THE MAP GENERATED

The outcome of the integration is that 10 per cent of the route/track comes under most suitable region, 77 per cent of the route/track comes under moderately suitable region and only 13 per cent of the track comes under least suitable region.

The gross outcome of the study is that the proposed route is moderately to most suitable for a railway track. Overall, the entire study reveals that the route is 87 per cent (that is, route length of about 73.73 km out of 84.33 km route, including both moderately to most suitable land) feasible and could be executed. One bridge is to be

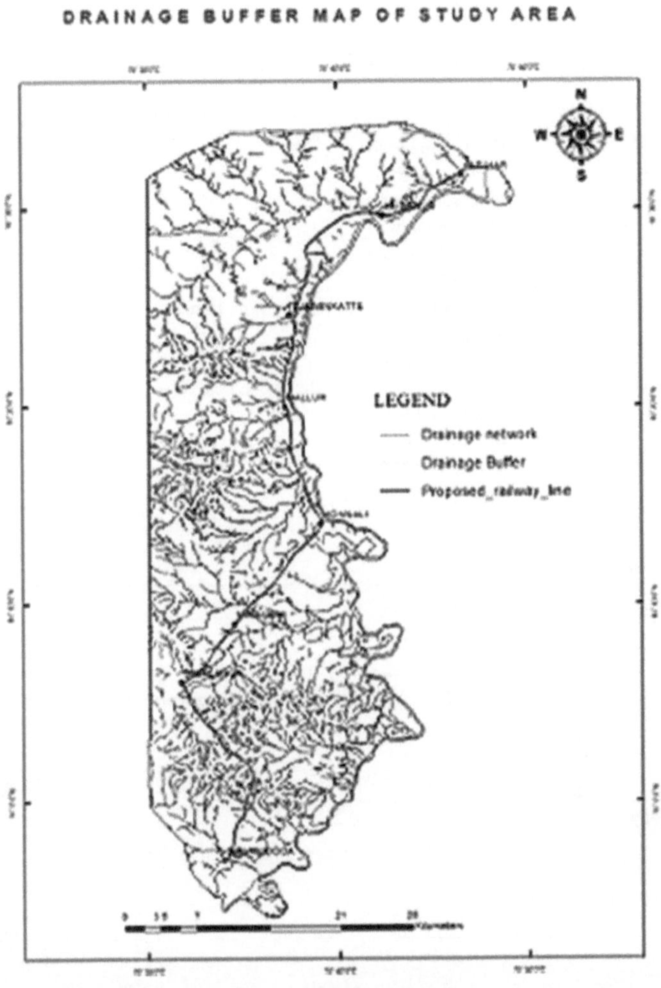

FIGURE 13.8 Drainage buffer zonation map of study area

constructed and 11 canal waterways, 47 stream waterways and 8 level crossings to be built along the route. The route was properly aligned following all necessary criteria and the results of the study are positive.

13.7 CONCLUSION

Adopting remote sensing and GIS as major geotechnical tools, the study has been carried out along the course of the proposed railway route and positive results have

TABLE 13.1
Fundamental Scale of Absolute Numbers (Saaty, 2004)

Intensity of Importance	Definition
1	Equal importance
2	Weak or slight
3	Moderate importance
4	Moderate plus
5	Strong importance
6	Strong plus
7	Very strong or demonstrated importance
8	Very, very strong
9	Extreme importance
Reciprocal of above	If activity *i* has one of the above non-zero numbers assigned to it when compared with activity *j*, then *j* has the reciprocal value when compared with *i*
1.1–1.9	If the activities are very close

TABLE 13.2
Pair-wise Comparison Matrix

Criteria	Soil	LULC	Lithology	Geomorphology	Slope	Weights
Soil	1	1	2	5	4	0.323
LULC	1	1	2	5	3	0.312
Lithology	1/2	0.5	1	5	4	0.222
Geomorphology	0.2	0.2	0.2	1	3	0.081
Slope	0.25	0.333333	0.25	0.333333	1	0.061

Consistency Index CI = 0.09025<0.1

been obtained. The remote sensing and GIS-based approach adopted in the study gives a synoptic, holistic and generalized view of the terrain and helps in better criteria development for route alignment.

The terrain under study has yielded positive results for railway track establishment. The area is also good from a socio-economic point of view for the establishment of a rail route.

Altogether, the multidisciplinary nature of both remote sensing and GIS has proved powerful in the establishment of feasibility criteria and in analysing the proposed route. It is now in the hands of the government to think over this endeavour and provide better connectivity to Shivamogga – the gateway of Malenadu.

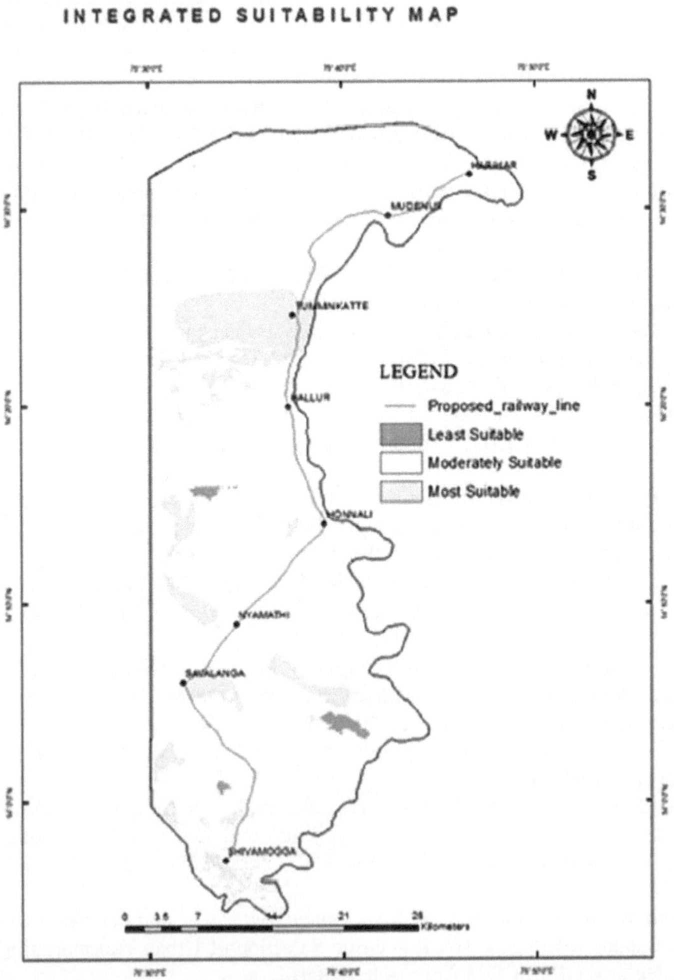

FIGURE 13.9 Integrated suitability map of study area

REFERENCES

Aburas, M.M., Abullah, S.H., Ramli, M.F. and Ash'aari, Z.H. A review of land suitability analysis for urban growth by using the GIS-based analytic hierarchy process. *Asian Journal of Applied Sciences*, 3(6). 2015.

Ahmad and Verma. Application of analytic hierarchy process in water resources planning: A GIS based approach in the identification of suitable site for water storage. *Water Resources Management*, 32, 5093–5114 (2018).

Ahmad, I. & Verma, M.K. GIS based analytic hierarchy process in determination of suitable site for water storage. *European Water* 60: 139–146, 2017. E.W. Publications.

Borda-de-Água, L., Barrientos, R., Beja, P. and Pereira, H.M., *Railway Ecology* (p. 320). Springer Nature (2017).

Bozdag, A. et al., AHP and GIS based land suitability analysis for Cihanbeyli (Turkey) County. *Environment and Earth Science* (2016) 75. doi: 10.1007/s12665-016-5558-9.

Chen, W. et al., GIS-based landslide susceptibility mapping using analytical hierarchy process (AHP) and certainty factor (CF) models for the Baozhong region of Baoji City, China *Environment and Earth Science*. doi: 10.1007/s12665-014-3749-9. Springer-Verlag (2014).

de FSM Russo, R., and Roberto C. Criteria in AHP: a systematic review of literature. *Procedia Computer Science* 55: 1123–1132. (2015)

Graymore et al. A GIS-based multiple criteria analysis decision support system for progressing sustainability. *Ecological Complexity* 6 (2009) 453–462. http://www.engineeringarticles.org/railway-survey-new-railway-line-survey/2015

Indian Railways Code for The Engineering Department. (Revised Edition 1982) (Fourth Reprint) 2012, Govt of INDIA (Embodying All Correction Slips issued up to 57, Dated 08.01.2020).

Jankowski, P. and Richard, L. Integration of GIS-based suitability analysis and multicriteria evaluation in a spatial decision support system for route selection *Environment and Planning. B, Planning & Design* 21(3):323–340 (1994).

Joerin F. et al. Using GIS and outranking multicriteria analysis for land-use suitability assessment. *International Journal of Geographical Information Science* 15:153–174 (2001). DOI: 10.1080/13658810051030487

Karakus, C. B. Evaluation of groundwater quality in Sivas province (Turkey) using water quality index and GIS-based analytic hierarchy process. *International Journal of Environmental Health Research*, 29(5), 500–519 (2019).

KSRSAC., https://kgis.ksrsac.in/kgisdocuments/KGIS2021/MapReports/State%20Maps/Lithology.jpg (2004)

Mardani, A., Jusoh, A., Nor, K., Khalifah, Z., Zakwan, N. and Valipour, A. Multiple criteria decision-making techniques and their applications–a review of the literature from 2000 to 2014. *Economic Research* 28(1), 516–571. doi: 10.1080/1331677X.2015.1075139. (2015)

MEB. *7th International Conference on Management, Enterprise and Benchmarking* June 5–6, 2009 Budapest, Hungary 169 Route/Site Selection of Urban Transportation Facilities: An Integrated GIS/MCDM Approach. (2009)

Merrouni, A.A., et al. A GIS-AHP combination for the sites assessment of large-scale CSP plants with dry and wet cooling systems. Case study: Eastern Morocco. *Solar Energy*, 166, 2–12. doi: 10.1016/j.solener.2018.03.038 (2017).

Mishra, D. Site suitability analysis of solar energy plants in stony wasteland area: A case study of trans-Yamuna Upland region, Allahabad District, India. *Journal of the Indian Society of Remote Sensing* 48(4):659–673 (2020)

Nishizawa, K. The improvement of pairwise comparison method of the alternatives in the AHP. doi: 10.1007/978-3-319-19857-6_41 (springer 2015).

Parry, J.A. et al. GIS based land suitability analysis using AHP model for urban services planning in Srinagar and Jammu urban centers of J&K, India. *Journal of Urban Management* 7 46–56 (2018).

Railway engineering code. Indian railway-Indian railway permanent way manual 2004, by Indian railway institute of civil engineering, chapter II pp 15–108, chapter IX pp. 214–241. (2018)

Report of social research project on "The role of youth in national development" with special reference to the linking of Shivamogga and Harihar through railway line (2007), pp. 6–9.

Rainer et al., Competition, regulation and nationalization: The Prussian railway system in the nineteenth century. *Scandinavian Economic History Review*, (1993):129–154. DOI: 10.1080/03585522.1993.10415864

Saaty, T.L., *The Analytic Hierarchy Process*, McGraw-Hill, New York, (1980).

Saaty, T.L., Axiomatic foundation of the analytic hierarchy process, *Management Science*, Vol. 32, No. 7 (Jul., 1986), pp. 841–855.

Saaty, T. L. Fundamentals of the analytic network process—Multiple networks with benefits, costs, opportunities, and risks. *Journal of Systems Science and Systems Engineering*, 13(3), 348–379. (2004).

Saaty, T. L. Decision making with AHP. *International Journal Services Sciences*, 1, 1, 83–98. (2008).

Singh, R. (ed.) *Re-envisioning remote sensing applications: Perspectives from developing countries*. First Edition. CRC Press. https://doi.org/10.1201/9781003049210.(2021)

Tischler, S. Finding the right way-a new approach for route selection procedures?. *Transportation Research Procedia* 25 (2017): 2809–2823.

Yalew, S. G. Land suitability analysis for agriculture in the Abbay basin using remote sensing, GIS and AHP techniques. *Model Earth System Environment*. doi: 10.1007/s40808-016-0167-x. (2016).

14 Comparing Performance of Inverse Distance Weighting and Modified Shepard Interpolation Models with Different Sampling Arrangements

A Case Study

Dheera Kalota

Lovely Professional University (LPU), Phagwara, Punjab, India

Mohammad Firoz Khan

Jamia Millia Islamia University, New Delhi, India

CONTENTS

14.1 INTRODUCTION

The accuracy of an interpolation model has been an issue of research in various disciplines for generation of raster or continuous data. The application of interpolation, being a significant tool for data generation, has increased in fields like

DOI: 10.1201/9781003224624-18

DEM/DTM generation (Kim et al., 1999; Ali, 2004; Chaplot et al., 2006; Aguilar et al., 2005; Li, 1992), climatic research (Kurtzman & Kadmon, 1999; Baigorria & Bowen, 2001; Vicente-Serrano et al., 2003; Li et al., 2005; Nikolova & Vassilev, 2006) ecological and environmental research (Abdi & Nandipati, 2009) and many other fields, including application of remote sensing (Singh, 2021). In the field of remote sensing, it has been found useful to estimate concealed or missing remotely sensed data (Rossi et al., 1994), for resampling of satellite images (Teoh et al., 2008); to enhance visibility in either remotely sensed images or other images (Shi et al., 2007) and a variety of other applications in remote sensing.

Inverse distance weighting (IDW) is an interpolation model closely related to nearest neighbour technique (Burrough & McDonnell, 1998). IDW works under the assumption that each point, considered for interpolation, influences the resulting surface up to a fixed distance (Mitas & Mitasova, 1999). In IDW, the effect of the surrounding values on the grid node is determined by the weights assigned to the distance in terms of the power assigned to it. The IDW interpolation has its basis on the assumption derived from Tobler's law that the value at any unknown point is directly proportional to the distance from known points (Zhang et al., 2015). On the other hand, modified Shepard's method can be termed as an advanced form of IDW interpolation algorithm. IDW was found to assign too much influence to data points even if they were far away from the point of approximation (Thacker et al., 2010). To this effect various modifications were proposed by different researchers. The original method of Shepard was modified by Franke and Neilson (1980) and further by Robert J. Renka (1988), which resulted in the present modified Shepard method.

Thus, both models run on a similar basic numerical approximation but with certain variations. In spite of having different formulations, both methods are valid interpolation models used for prediction. These methods under study have been researched for accuracy and validity by various researchers in different themes. Performance of the IDW method has been compared with other methods like kriging, spline, etc. (Weber & Englund, 1992; Priyakant et al., 2000; Valley et al., 2005; Abdi & Nandipati, 2009) and also in a collection of methods including modified Shepard (Bakkali & Armani, 2008; Karabork et al., 2008; Yilmaz, 2009; Kirk, 2003). However, the performance of both methods has not been assessed in comparison with one another in varying sample densities and patterns. Another issue of significance here is sample density and pattern. It is a well-established fact that sample density and sample pattern influences the performance of interpolation methods. Various scholars have researched the impact of these, using varying sample densities and on a lesser scale of sample patterns (Chaplot et al., 2006; Gonçalves, 2006; Binh & Thuy, 2008). However, though it may be inconsequential in some cases, the impact of sampling on interpolation has never been negated. So, in the present paper, the performance of the two interpolation models is compared not only in relation to each other but also in terms of differing sampling schemes.

The main objectives of the present paper are to evaluate the effects of differing sample densities and sample patterns on the performance of IDW and modified Shepard's interpolation method. The area selected is complex/rugged terrain of Southern Rajasthan. Point data has been accumulated in clustered, regularly and randomly spaced data. In terms of sample density, three different datasets with sampling sizes of 100, 75 and 50 per cent coverage are taken. Evaluation of these models is done on the basis of the mean absolute percentage error method.

14.2 STUDY AREA

The study area selected for the present research is the complex/rugged terrain of Southern Rajasthan. As an example of discontinuous land surface, a small part of Southern Rajasthan, measuring 100 km^2, is taken as the case of the present experiment. The area is selected because of its discontinuity in elevation due to the presence of the Aravalli mountains running from northeast to southwest. Under extreme arid conditions of great heat and dry climate, the granite rocks in this part of the Aravalli range have developed typical morphological features of 'jagged outline with the granites, are exfoliated and honey combed into domes and tors, with perched blocks and boulders like cannonballs and sacks' (Heron, 1938).

The area selected shares a complex topographical character, as observed in the 3D view of the selected area in Figure 14.1. Abrupt change and discontinuity characterize the surface. The elevation in the selected area ranges from 624 to 1,101 metres above mean sea level (AMSL).

14.3 METHODOLOGY

In order to generate relevant information for the purposes of the present research the experiment is executed in the following manner. The initial values were measured

FIGURE 14.1 3D view of the Rajasthan region selected for study, depicting the discontinuity of the surface

on the Google Earth™ image of selected discontinuous surface. Nowhere has the surface been observed to be higher than 1,100 and lower than 600 metres. For the purpose, point datasets to use in interpolation models from Google Earth™ are generated using three distinct sampling designs and sample sizes.

To compare the performance of both methods on the basis of changing sampling designs and sample density, three sample patterns (regular, cluster and random) and three levels of sample density (100 per cent, 75 per cent and 50 per cent) have been considered for control points. The control points are first marked in the GIS environment and subsequently transferred onto Google Earth™ to record elevation values for these control points. The control points are interpolated using the methods – i.e., IDW and modified Shepard's, using Surfer 9.0.

The main parameter of the IDW method is the power assigned to distance variables among sample points. The power of 2 is assigned to it, as this is suggested to give more satisfactory results (Declercq, 1996). In modified Shepard's method, smoothing factor, quadratic neighbour and weighting neighbour, default values are adopted for these parameters. The smoothing parameter is given as 0, while the quadratic and weighting neighbour values are those recommended by Renka (1988) – i.e., 13 and 19, respectively.

Apart from control points, some reference points amounting to almost 15 per cent of the control points respectively have been randomly placed, but avoiding proximity to the control points. Subsequently, their value is also extracted from Google Earth™. The elevation values of these reference points are recorded for evaluation of accuracy of the interpolation models (Englund et al., 1992). The control point and reference point value is compared using mean absolute percentage error and correlation for evaluating performance of the methods.

$$MAPE = \frac{100}{n}\sum_{1}^{n}\frac{|A_i - P_i|}{A_i}$$

where, A_i is the actual value and P_i is the predicted value.

Mean absolute percentage error (MAPE) is a widely used statistical technique in model testing, due to its properties in that it uses all observations and has the smallest variability from sample to sample (Levy & Lemeshow, 1991). The correlation method has been used to find out the degree of relation between the actual and generated control point values.

14.4 RESULTS AND DISCUSSION

The results of the generated surface were compared using the visual similarity of the generated images and also using statistical methods like MAPE and correlation between the actual and generated surfaces. The statistical results of the comparison are shown in Table 14.1.

TABLE 14.1
Statistical Results of Comparison between IDW and Modified Shepard's

Sampling Strategy	MAPE		Correlation	
	IDW	MS	IDW	MS
Cluster 50	8.62	8.69	0.44	0.54
Cluster 75	4.22	3.82	0.89	0.91
Cluster 100	3.74	3.94	0.88	0.86
Regular 50	8.18	8.19	0.51	0.54
Regular 75	3.07	2.69	0.95	0.96
Regular 100	3.24	2.97	0.89	0.91
Random 50	8.98	8.46	0.48	0.66
Random 75	3.22	4.74	0.95	0.86
Random 100	3.85	3.99	0.87	0.87

14.5 COMPARISON OF RESULTS IN RELATION TO DIFFERENT SAMPLING STRATEGY

Cluster sampling: The images generated using the cluster sampling technique using IDW and modified Shepard's are shown in Figure 14.2. In the case of the IDW method, it is observed that the cluster sampling plans give rectangular-shaped patches which broaden with the increase in space between the sample points or decrease in the sample size and give block-like contours. The modified Shepard method, however, has generated surfaces which are smooth in appearance, but the general topography of the land is preserved. A surface generated with 75 per cent cluster sampling technique using modified Shepard method is accurate in this sampling technique, which is visible in the generated image, and the MAPE and correlation values also support this.

Regular sampling: The surfaces generated using a regular sampling plan is shown in Figure 14.2. The surfaces generated using the IDW method have a wave-like appearance. These waves are more pronounced with the increase in sample size and have a smooth appearance in the case of smaller sample size. However, the general topography of the region is maintained. In the case of the modified Shepard's method, the generated surfaces in the case of the regular sampling strategy with 100 per cent, 75 per cent and 50 per cent sample size show a smooth generated surface. The detail in the generated surfaces is more pronounced as the sample size increases and thus the accuracy is maintained.

Random sampling: The surfaces generated using IDW and modified Shepard's is shown in Figure 14.3. In the case of the IDW method, the generated surfaces are smooth in appearance with some degree of similarity with the original surface. However, in the case of the modified Shepard's method, the surface generated using 50 per cent sample size shows exaggeration of topographic detail whereas the 75 per cent sample size shows over-smoothened surface (Figure 14.4).

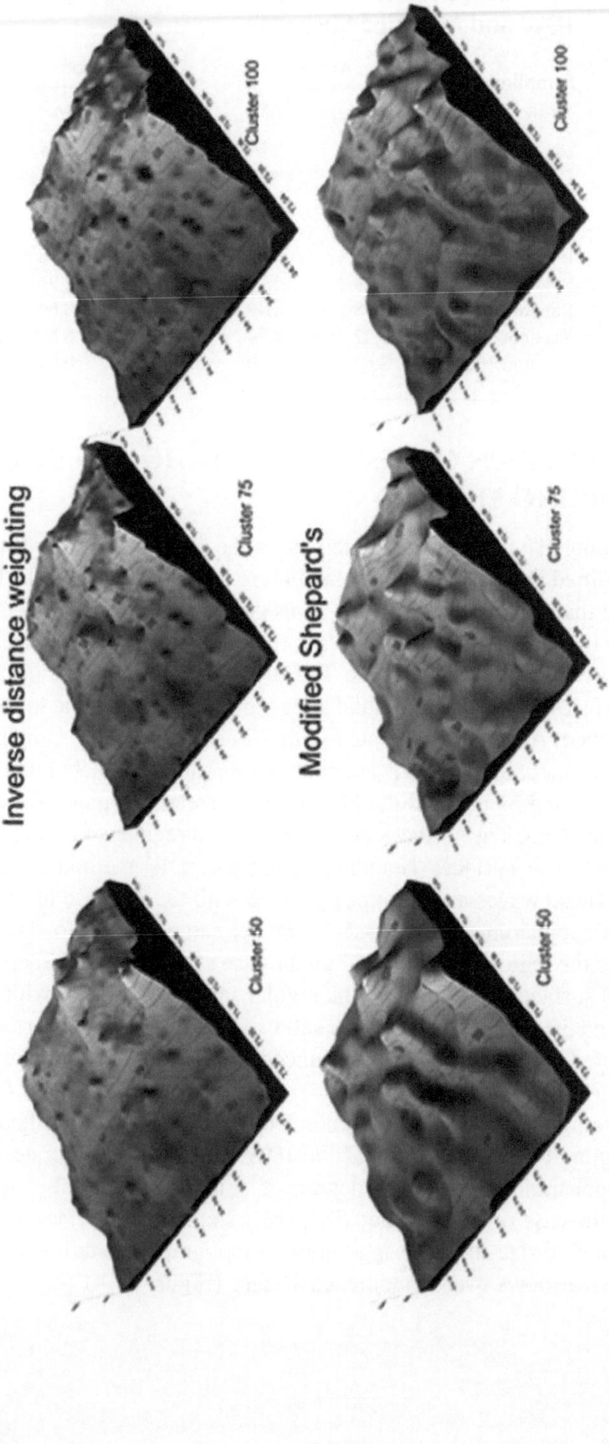

FIGURE 14.2 Surfaces generated using cluster sampling technique, using IDW and modified Shepard's

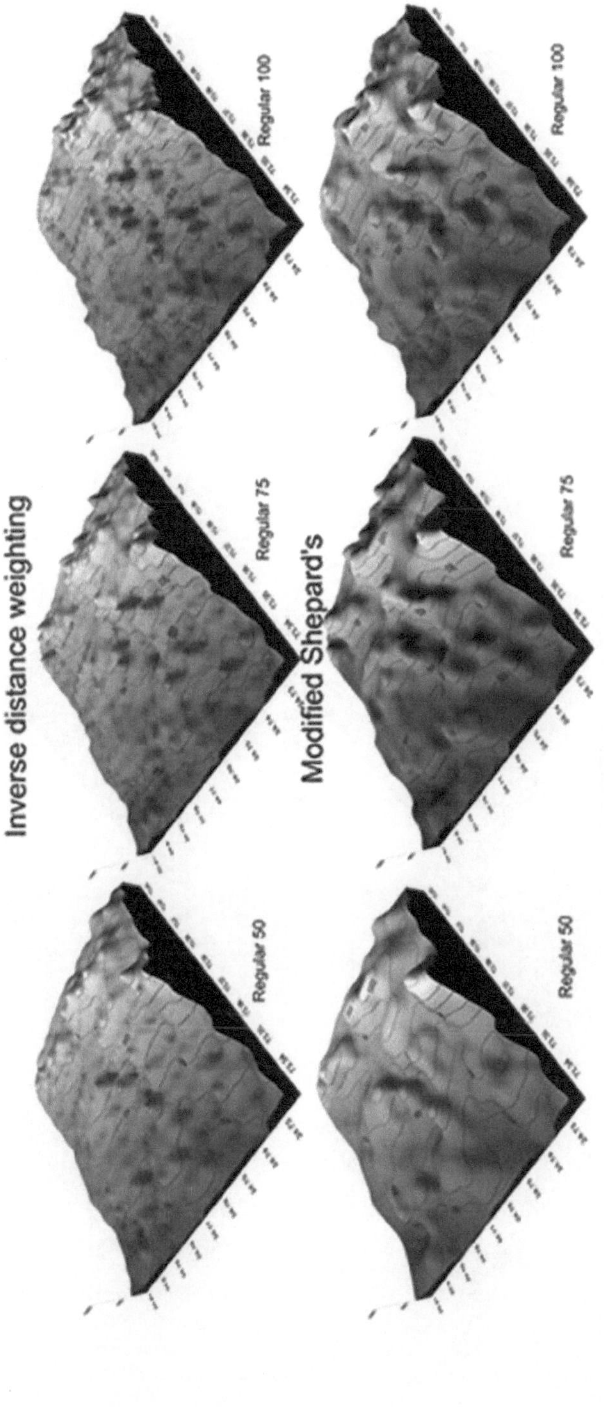

FIGURE 14.3 Surfaces generated using regular sampling technique, using IDW and modified Shepard's

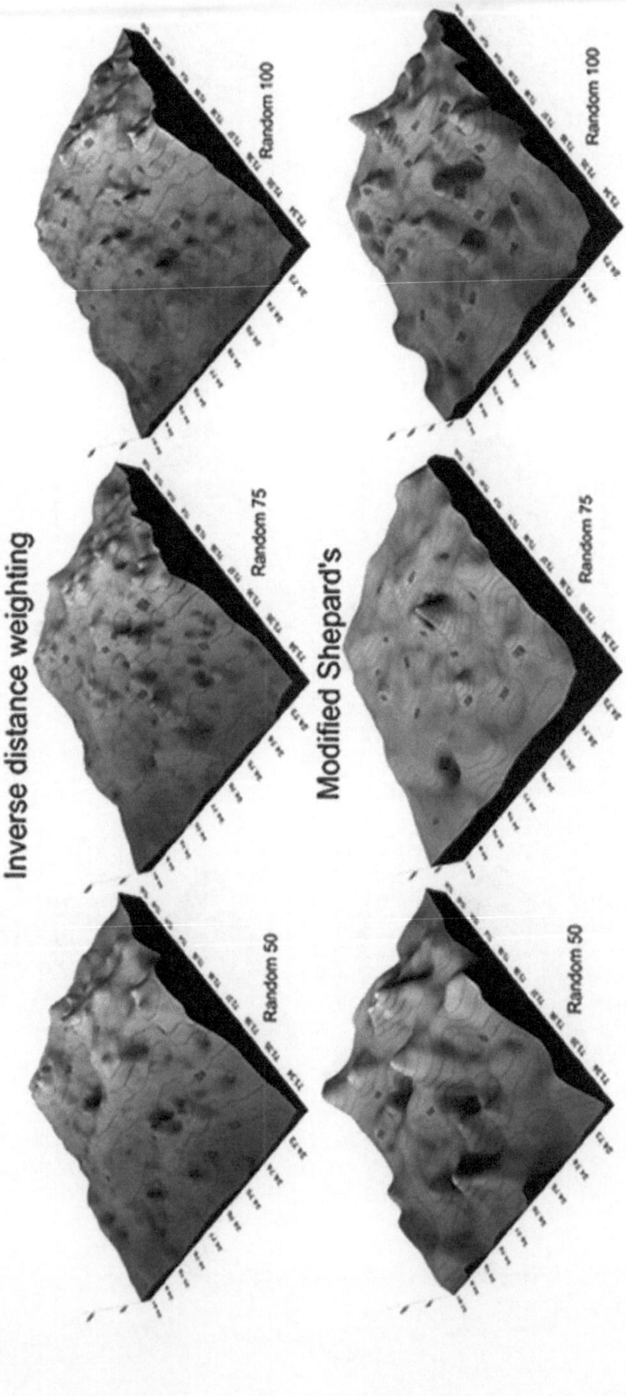

FIGURE 14.4 Surfaces generated using random sampling technique, using IDW and modified Shepard's

14.6 CONCLUSION

The study reveals certain insights into the working of both interpolation methods in relation to different sampling strategies and sizes. In some cases, it is observed that the visual surfaces generated using interpolation and the error values obtained from MAPE and correlation values signify the same level of accuracy. The overall assessment of both models reveals that the IDW model is more or less consistent in all three types of sample strategies and sample sizes. Though the surfaces generated using IDW are not very accurate in terms of visual appearance and also in terms of MAPE and correlation values, the method does not display any random error. It is a dependable method which is not much influenced by the sample strategy. However, increased sample size may give improved results. Shepard's method is found to give more accurate results and maintains the general topography of the land. A similar observation is also made in the case of the model that increases in sample size will give more refined results. However, the model gives random errors by generating dips and exaggerated elevations in the case of random sampling plan. Both models are quite accurate with large sample sizes; however, with a smaller number of samples a generalized surface is generated.

REFERENCES

Abdi, A., and Nandipati, A. 2009 Bird diversity modeling using geostatistics and GIS. *12th AGILE international conference on geographic information science* 2009 Leibniz Universität Hannover, Germany.

Aguilar, F. J., Agüera F, Aguilar M. A., and Carvajal F 2005 Effects of terrain morphology, sampling density and interpolation methods on grid DEM accuracy. *Photogrammetric Engineering & Remote Sensing*, 71 (7), 805–816.

Ali, T. A. (2004) On the selection of an interpolation method for creating a terrain model (TM) from LIDAR data In *Proceedings of the American Congress on Surveying and Mapping (ASCM)*. Conference. 2004, Nashville. TN.

Baigorria, G. A. and Bowen, W. T. (2001). A process-based model for spatial interpolation of extreme temperatures and solar radiation. In *Methodologies for Interdisciplinary Multiple Scale Perspectives. Proceedings of the SAAD III Third International Symposium on Systems Approaches for Agricultural Development*, Lima, Peru, November 8–10, 1999.

Bakkali, S. and Armani, M. (2008) About the use of spatial interpolation methods to denoising moroccan resistivity data phosphate 'disturbances' map. *Acta Montanistica Slovaca* 13 216–222.

Binh, T. Q. and Thuy, N. T. (2008) Assessment of the influence of interpolation techniques on the accuracy of digital elevation model. *VNU Journal of Science Earth Sciences*, 24, 176–183

Burrough, P. A. and McDonnel, R. A. (1998) *Principles of geographical information systems* Oxford University Press, London.

Chaplot, V., Darboux, F., Bourennane, H., Leguédois, S., Silvera, N., and Phachomphon, K. (2006) Accuracy of interpolation techniques for the derivation of digital elevation models in relation to landform types and data density. *Geomorphology*, 77, 126–141.

Declercq, F. A. N., (1996) Interpolation methods for scattered sample data: accuracy, spatial patterns, processing time. *Cartography and Geographic Information Systems*, 23 (3), 128–144

Englund, E., Weber, D., and Leviant, N. 1992. The effects of sampling design parameters on block selection. *Mathematical Geology*, 24(3), 329–343.

Franke, R., and Neilson, G. (1980) Smooth interpolation of large sets of scattered data. *International Journal of Numerical Methods in Engineering*, 15, 1691–1704.

Gonçalves, G. (2006) Analysis of interpolation errors in urban digital surface models created from LIDAR data. Proceedings of the *7th International Symposium on Spatial Accuracy Assessment in Resources and Environment Sciences.* M. Caetano and M. Painho (eds.), Sana Hotel, July 5th–7th, Lisbon, Portugal, pp. 160–168.

Heron, A. M. (1938) Proceedings of the twenty-fifth indian science congress calcutta Pt. II Presidential Addresses 122.

Karabork, H., Baykanb, O. K., Altuntasa, C. and Yildza, F. (2008) Estimation of unknown height with artificial neural network on digital terrain model. *The International Archives of the Photogrammetry Remote Sensing and Spatial Information Sciences.* 37Part B3b 180.

Kim, S., Kim, T., Park, W. and Lee, H. (1999) An optimal interpolation scheme for producing a DEM from the automated stereo-matching of full-scale SPOT images *Proceedings of ISPRS conference on "Sensors and Mapping from Space".* Hanover, Germany (CD-ROM Proceedings) Available at http://www.isprs.org/proceedings/xxxiii/congress/part3/705_XXXIII-part3.pdf

Kirk, K. (2003) Spatial sampling and interpolation methods-comparative experiments using simulated data Master's thesis Aalborg University.

Kurtzman, D. and Kadmon, R. (1999) Mapping of temperature variable in Israel: A comparison of different interpolation methods. *Climatic Research*, 7 (13), 33–43.

Levy, P. and Lemeshow, S. (1991) *Sampling of populations: Methods and applications.* New York: John Wiley.

Li, Z. (1992) Variation of the accuracy of digital terrain models with sampling interval. *Photogrammetric Record*, 14, 113–128.

Li, X., Chend, G. and Lu, L. (2005) Spatial analysis of air temperature in the Qinghai Tibet plateau. *Arctic, Antarctic and Alpine Research*, 37 (2), 246–252.

Mitas, L., & Mitasova, H. (1999). Spatial interpolation. *Geographical Information Systems: Principles, Techniques, Management and Applications*, 1, 481–492.

Nikolova, N. and Vassilev, S. (2006) Mapping precipitation variability using different interpolation methods, *Proceedings of the Conference on Water Observation and Information System for Decision Support (BALWOIS).* 25–29 May 2006, Ohrid, Macedonia. Available on http://balwois.com/balwois/administration/full_paper/ffp-631.pdf.184

Priyakant, N., Verma, V., Rao, L. I. M. and Singh, A. N. (2000) *Surface approximation of point data using different interpolation techniques – a GIS approach.* Available at http://gisdevelopment.net/technology/survey/techgp0009pf.htm

Renka, R. J. (1988) Multivariate interpolation of large set of scattered data. *ACM Transactions Mathem Software*, 14, 139–148.

Rossi, R. E., Dungan, J.L., and Beck, L. R. (1994) Kriging in the shadows: Geostatistical interpolation for remote sensing, *Remote Sensing of Environment*, 49, 1, 32–40, ISSN 0034-4257, doi: 10.1016/0034-4257(94)90057-4.

Shi, W., Tian, Y. and Liu, K. (2007) An integrated method for satellite image interpolation, *International Journal of Remote Sensing*, 28:6, 1355–1371, doi: 10.1080/01431160600851876

Singh, R. (ed.) (2021) *Re-envisioning remote sensing applications: Perspectives from developing countries.* First Edition. CRC Press. doi: 10.1201/9781003049210

Teoh, K. K., Ibrahim, H. and Bejo, S. K., (2008). Investigation on several basic interpolation methods for the use in remote sensing application, *2008 IEEE Conference on Innovative Technologies in Intelligent Systems and Industrial Applications*, Cyberjaya, pp. 60–65, doi: 10.1109/CITISIA.2008.4607336.

Thacker, W. I., Zhang, J., Watson, L. T., Birch, J. B., Iyer, M. A. and Berry, M. W. (2010) Algorithm 905: modified shepard algorithm for interpolation of scattered multivariate data. *ACM Transactions on Mathematical Software – TOMS*, 37(3), 1–20.

Valley, R. D., Drake, M. T. and Anderson, C. S. (2005) Evaluation of alternative interpolation techniques for the mapping of remotely-sensed submersed vegetation abundance. *Aquatic Botany*, 81, 13–25.

Vicente-Serrano, S. M., Saz-Sánchez, M. A., and Cuadrat, J. M. (2003) Comparative analysis of interpolation methods in the middle Ebro valley (Spain): application to annual precipitation and temperature. *Climatic Research* 24 (2), 161–180.

Weber, D. and Englund, E. (1992) Evaluation and comparison of spatial interpolators. *Mathematical Geology*, 24, 381–391.

Yilmaz, I. (2009) A research on the accuracy of landform volumes determined using different interpolation methods. *Scientific Research and Essay*, 4(11), 1248–1259.

Zhang, H, Lu, L, Liu, Y, and Liu, W. (2015) Spatial sampling strategies for the effect of interpolation accuracy. *ISPRS International Journal of Geo-Information*, 4, 2742–2768.

15 Applications of Remote Sensing in the Groundwater Potential Analysis of Developing Countries

Sajad Ahmad Mir, Durdanah Mattoo, M Sultan Bhat and G M Rather
University of Kashmir, Srinagar, India

CONTENTS

15.1 INTRODUCTION

Water has quintessential significance in being the elixir of life for humans and life systems supporting the ecosystems on this planet. Groundwater has uneven distribution throughout the world, playing an imperative role in economic development and food security and being central to the sustainable development of any region. There is an increasing pattern of human population and related diversification of human footprint when it comes to the activities being carried out (Ayazi et al., 2010; Manap et al., 2013), including rampant disposal tendencies of wastes into the water resources, complicating appropriate and judicious water management (Neshat, 2013; Pradhan, 2013). Groundwater has been described as the presence of water in impermeable layers of hydrogeological formations (Fitts, 2002) with the evident porosity among crevices or in the rocks being fissured in grain-size geometries of rock strata. Precipitation and snow melt acts are the dominant processes for the formation

DOI: 10.1201/9781003224624-19

of groundwater which percolate down and seep through the substratum geological environments (Banks et al., 2002). Groundwater resources have been over-exploited on account of rapid urbanization and increasing population, causing a widespread reduction in the levels of groundwater signalling the exhaustion of groundwater resources (Singh et al., 2013). Fetter (2001) while presenting a comprehensive overview of groundwater scenario of the world estimated that around 97.5 per cent of the world's global water is unfit for drinking purposes while 70 per cent of the world's fresh water (2.59 per cent of global water) is based in glaciers/permafrost, of which 30.1 per cent and 0.4 per cent represent groundwater and surface/atmospheric water respectively. Groundwater covers about 50 per cent of water used for drinking purposes (Zektser & Everett, 2004), and most of the irrigation is drawn from groundwater in the USA (Kenny et al., 2009). Groundwater has gained much importance as a dependable source of water in all the hydro-meteorological regions of the world (Todd & Mays, 2005). Groundwater has turned out to be an essential resource for industrial, domestic and irrigation services, augmenting the demand for groundwater as the population increases and the need for water consumption is getting huge (Jha et al., 2010). Groundwater has turned out to be the mainstay in providing the source of supplying water around the world (Hoque et al., 2007), prompting us to recognize the spatiotemporal dynamics of groundwater and its fluctuating patterns for managing groundwater resources (Sreekanth et al., 2009). This has led to enormous pressure on aquifers contributing (Döll et al., 2012) to faster degradation of groundwater storativity reserves when the recharge and extraction equilibrium gets destabilized (Famiglietti, 2014). The Gravity Recovery and Climate Experiment (GRACE) experimental study carried out by NASA revealed that a vast amount of groundwater source regions are showing the signs of transforming into a global threat in terms of water storage and adaptation in the world (Richey et al., 2015). Groundwater in India represents the kernel in utilizing the water supply services, including for drinking purposes and agricultural activities; the World Bank (2009) estimated that 60 per cent irrigation and 85 per cent of the drinking water is being drawn from groundwater, cautioning us that as many as 60 per cent of aquifers will be in a 'critical' abstraction stage around the year 2030. Climate change has led to massive changes in the availability of water resources, frequency of storm events, droughts, floods, an intensification of warming behaviours and the occurrence of extreme weather events, exerting tremendous effects on the feedback mechanism of sustainable water-environmental interactions (Hagemann et al., 2012; Schewe et al., 2014). There is global concern for hydrogeologists amidst the changing character of environmental entropies posing a great threat to the world's biophysical systems. Anthropogenic stresses on the global climate have the tendency to intensify the magnitude of hydrological extremes (Trenberth, 2011; Donat et al., 2016), directly affecting human life and their consumption patterns of water (Van Loon, 2015). Hence monitoring of groundwater resources and its variability in storage distribution is vital for the mapping of stress regions which are facing groundwater storage loss connection with changing climatic patterns around the world. Although modern groundwater signifies a marginal proportion of the total on Earth, its volume is equal to a water body with a depth of about 3 m spread over the continents. Groundwater forms the major component of the hydrologic cycle, dwarfing all other resources as volume of groundwater in upper continental formations is about 22.6 million km^3, where 0.1–5.0 million km^3 is less than 45 years old (Gleeson et al., 2016). It has also helped in empowering the people by raising their minimum social standards and has common access to poor sections of society, being cheap and of more immediate utilization than the medium of canal supply (IWMI, 2001). Groundwater acts as a climate buffer and has low pollution levels, having attracted a large human population globally with its wide distribution (Arkoprovo et al., 2012). In drought-prone areas during previous decades groundwater has been seen as one of the most important natural resources (Cui & Shao, 2005).

Groundwater resources are experiencing depletion and there is a decreasing trend due to its excessive abstraction (Vaux, 2011; Page et al., 2012; Mukherjee et al., 2012). Barlow and Clarke (2002) carried out a study with the findings suggesting that there would be a severe crisis in freshwater reserves within 10 to 15 years in many countries of the world. There is a huge gap between the harnessing and abstracting rates of groundwater which makes the water table wither away in its recovery stage. Developing countries are facing tremendous pressure on groundwater everywhere and in India there is an alarming rate of groundwater demand (Black & Talbot, 2005; Holden, 2014) as more than 50 per cent of industrial and urban demand and about 80–90 per cent of water supply in rural areas depend on groundwater, besides catering to the water needs of more than 50 per cent of agricultural use (Central Ground Water Board, 2014). Groundwater extraction in South Asia and in the higher plains of the USA for irrigation purposes has led to the depletion and degradation of aquifer systems (Wada et al., 2010). Bredehoeft (2002) while analysing the sustainability criterion for groundwater considered it the 'no net reduction in groundwater storage over a year'. Groundwater resources have a vital role in realizing the sustainable water needs of Asia, being widely used for irrigation, industrial production and domestic consumption in both rural and urban communities, comprising about 25 per cent of total water use in Asia (FAO, 2016). The withdrawal of groundwater accounts for the most (72%) of global share, initiated by an intensive mode of agriculture practices and rapid growth of population in China, India, Bangladesh, Iran and Pakistan (Shah, 2005; Gleeson et al., 2012; Singh, 2021). Groundwater also has the primary role of feeding the base flow reserves of South Asian surface water, serving as an indispensable resource for sustaining many ecosystems that depend on it (Danielopol et al., 2003; Wösten et al., 2008; Kløve et al., 2014). In addition to providing environmental services, groundwater has significant socioeconomic associations as it was assessed that groundwater irrigation in Asia accounts for about US$10 to US$30 billion per annum of its economy (Shah et al., 2003; WWAP, 2016). Groundwater quality too has been an emerging threat to safe groundwater as there are many high-risk arsenic zones in Asia with high incidences in Bangladesh and the bordering Indian region of Western Bengal, transforming into a massive public health concern (Brikowski et al., 2014). Studies on arsenic poisoning have suggested that about 35 million people are exposed to this deadly threat through ingestion patterns (Harvey, 2008; Hasan et al., 2009; Shukla et al., 2010; Radloff et al., 2011). Thus, for having access to safe and sustainable groundwater resources there is a need for comprehensive and scientific management of groundwater reserves vis-à-vis the changing climatic and environmental conditions across the globe. Groundwater potential analysis carried out with the help of remote sensing has been very effective in keeping a check on both the variability and distribution of this precious natural resource.

15.2 MATERIALS AND METHODS

Bibliometric analysis was done for taking the purview of the relevance of remote sensing in groundwater potential analysis (Singh, 2021). Bibliometrics became mainstream in the twentieth century, developing into a robust method of finding a trend of significance of any phenomenon in the knowledge applications around the world (Chen, 2004; Merigó & Yang, 2017; Wang & Xu, 2016; Weiler et al., 2012; Benito et al., 2013; Yang et al., 2018). Recently, this analysis has been extensively used in evaluating the growth and expansion of a particular specialized domain, largely because of its distinct character, to validate the scholarly progress, developmental stages and highlighting research clusters derived from the temporal additions to the existing state of knowledge. The main thrust of this study is to systematically and analytically offer a bibliometric assessment in the applicability of remote sensing technologies in the identification, management and prospecting of groundwater resources in developing countries. CiteSpace analytics were used for more precise connections, in-depth assessment and appraisal of remote sensing vis-à-vis groundwater. Datasets required for understanding and visualizing the growing significance of remotely sensed imageries were applied in the study for the holistic and comprehensive audit in quantifying the potential groundwater resources. The overall research standing and drifts were examined in addition to related disciplines using a case study approach (Bengtsson & Larsson, 2012; Weber, 2012).The case study method based on bibliometric analysis was incorporated into the study in order to have an objective and comprehensive review of the remote sensing applications on groundwater potential. Further, the different hydrogeological parameters for groundwater potential are thoroughly analysed, given the significance of remote sensing technologies, using the integrative effect of these layers in defining and completing the overall prospecting of groundwater resources in the regions. For this evaluation, different datasets were tallied from various sources, and then approximations made in accordance with the reliability of the material available, such as IGRAC, FAOSTAT and AQUASTAT – the data-cataloguing sites of the Food and Agriculture Organization (FAO) of the United Nations, etc. The methodological flowchart describing the schematic structure of this study is represented in Figure 15.1.

15.3 REMOTE SENSING AND GROUNDWATER POTENTIAL ANALYSIS

The peculiarity of groundwater occurrence in complex sub-surface hydrogeological environments, coupled with its fluctuating character, has compounded the intricacies associated with assessing the spatiotemporal changes in this resource. Therefore, quantitatively managing, resourcefully evaluating and exploring groundwater resources has been challenging (Lee et al., 2012; Tahmassebipoor et al., 2016). Lately, geographic information systems (GIS) are witnessing rapid advances in the field of geospatial data processing and its relevance in the manipulation and management of earth-environmental data integration (Israil et al., 2004; Solomon & Quiel, 2006; Junge et al., 2010; Ozdemir, 2011; Ayele et al., 2014; Rahmati et al., 2015) Geospatial operations, especially remote sensing, are widely used in the assessment

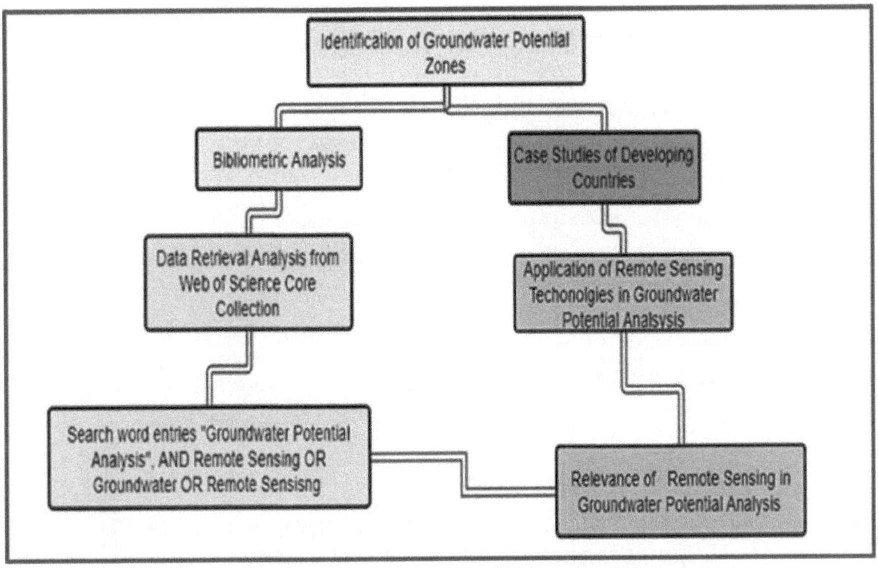

FIGURE 15.1 Methodological flowchart for the application of remote sensing in groundwater potential zonation

and sustainable management of groundwater resources (Dar et al., 2010; Magesh et al., 2011). This technology has been an effective tool in identifying and delineating groundwater potential areas (Chi & Lee, 1994; Sreedevi et al., 2005; Nobre et al., 2007; Manap et al., 2013). Currently, the all-encompassing delivery and application of satellite data, coupled with conventional mapping and enhanced image-processing techniques, have made it easily accessible and adapted to the groundwater potential baseline data repository (Tiwari & Rai, 1996; Harinarayana et al., 2000; Chowdhury et al., 2010). The remote sensing approach, in addition to supporting in the widespread area coverage across space-time observations, is also a cost-effective and robust technique for saving large amounts of time, in contrast to conventional prospecting studies (Leblanc et al., 2003; Tweed et al., 2007). Besides, this revolutionizing application of the geospatial and GIS-driven approach has been extensively applied for the perusal of characterizing the earth's geomorphological markers (lineaments, lithology, drainage, etc.) and to survey the groundwater rich areas (Sener et al., 2005).

Various studies have reported hydrogeomorphic and geological indicators as determining groundwater potential, such as land use/land cover, drainage density, soil cover, lineament pattern, precipitation, lithology, slope pattern and elevation, in a given spatiotemporal setting of the region (Sreedhar et al., 2009; Avtar et al., 2010, 2012; Fennta et al., 2015). These findings further stress the significance of remote sensing and satellite datasets as the source of preparing the baseline scheme of quantifying and precise positioning of groundwater source regions with the indicators related to the hydrological and hydrogeological configuration of the area. In this context, these geospatial tools have emerged as a rigorous medium for analysing and

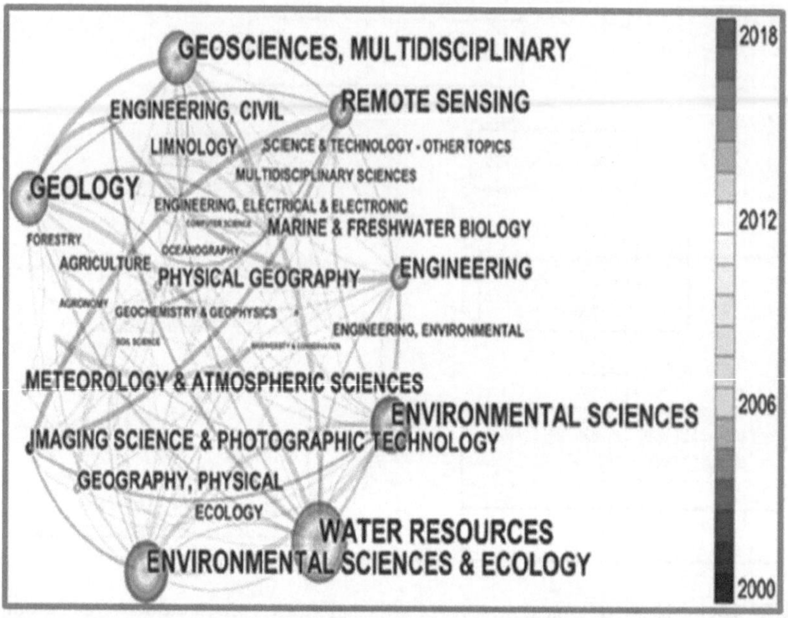

FIGURE 15.2 Intersection of geosciences and allied fields from 2000 to 2008. Adapted from Cui et al. (2019). (Accessed 07 May 2020)

handling voluminous data applied in building scenarios, providing accurate decision support strategies (Stafford et al., 2008). In recent times, there has been an increasing trend toward the integration of GIS techniques and remote sensing methods which has helped in mapping and prospecting the potential areas for carrying out comprehensive geophysical and hydrogeological surveys in the groundwater and geo-environmental settings, as seen in Figures 15.2 and 15.3). With the advance of methodological and computational inclusion of machine learning and hybrid artificial intelligence, data mining techniques like the frequency ratio (FR), AHP, weights-of-evidence (WOE) and evidential belief function (EBF) have been used in groundwater potential analysis derived from the remote sensing and GIS data integration. Chen et al. (2019) in their groundbreaking study used the novel hybrid approach for groundwater potential analysis by examining and investigating the different geospatial data from hydrogeological indicators and had a 90 per cent success rate in validating and pinpointing the target areas of immense groundwater potential.

15.4 GROUNDWATER POTENTIAL ANALYSIS IN AFRICA USING REMOTE SENSING

Demand for water in Africa is expected to rise significantly in the next few decades, resulting from increasing population growth and the intensive use of irrigation for agriculture. Presently, there are around 350 million people (many of whom are poor and vulnerable) in the African continent without access to fresh drinking water

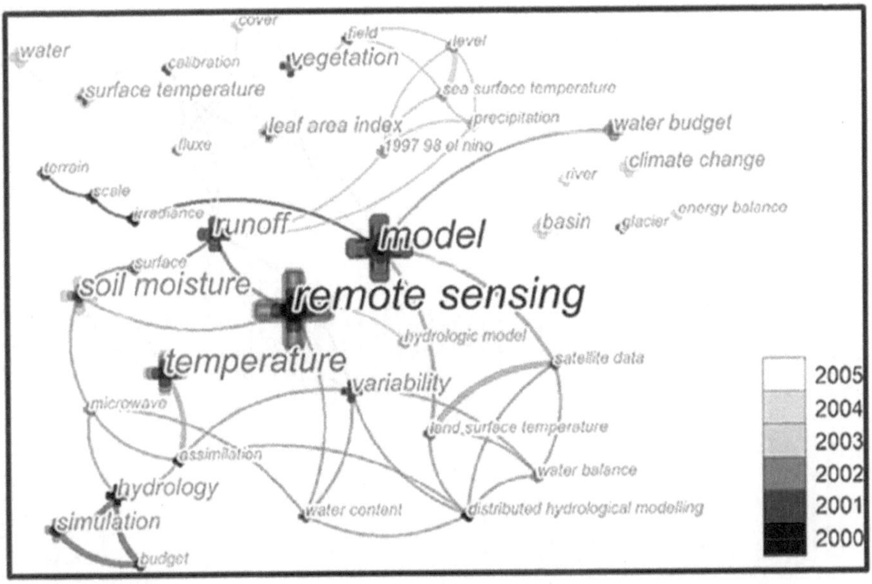

FIGURE 15.3 Network mapping of remote sensing and groundwater from 2000 to 2005. Adapted from Cui et al. (2019). (Accessed 10 May 2020)

(Hunter et al., 2010). Consequently, increased use of improved and safe drinking water is a global urgency. In Africa, studies have shown that domestic water use needs to be increased extensively for socioeconomic reasons and given the low development stage of most areas (Grey & Sadoff, 2007). There are growing concerns about food security issues and productivity too (Funk & Brown, 2009) compounded by the threat of impending climate change. Only 5 per cent of the arable land comes under irrigation (Siebert et al., 2010) and discussions about intensification in irrigation trends have increased immensely for meeting the growing demands in food productivity with a decreasing precipitation regime (UNEP, 2010; Pfister et al., 2011).

Increasing constant and judicious access to drinking water across Africa will depend on the availability of groundwater in the area (Giordano, 2009). Groundwater, however, is neither a general solution to water stress nor protected from depletion. A vibrant resource-driven approach is needed to examine the operative framework for managing and minimizing environmental impacts on groundwater degradation and widespread abstraction. Bates et al. (2008) assessed the IPCCs Technical Report on Climate Change and Water and discovered limited data availability and a lack of information in terms of groundwater. Figures 15.4 and 15.5 depict the quantitative assessment of groundwater in Africa.

15.5 RESULTS AND DISCUSSION

Most of the data on groundwater has been deficient in reliable and precise information; remote sensing has proven to be the immediate source and medium of assessing and evaluating groundwater sources in Africa. Groundwater potential analysis

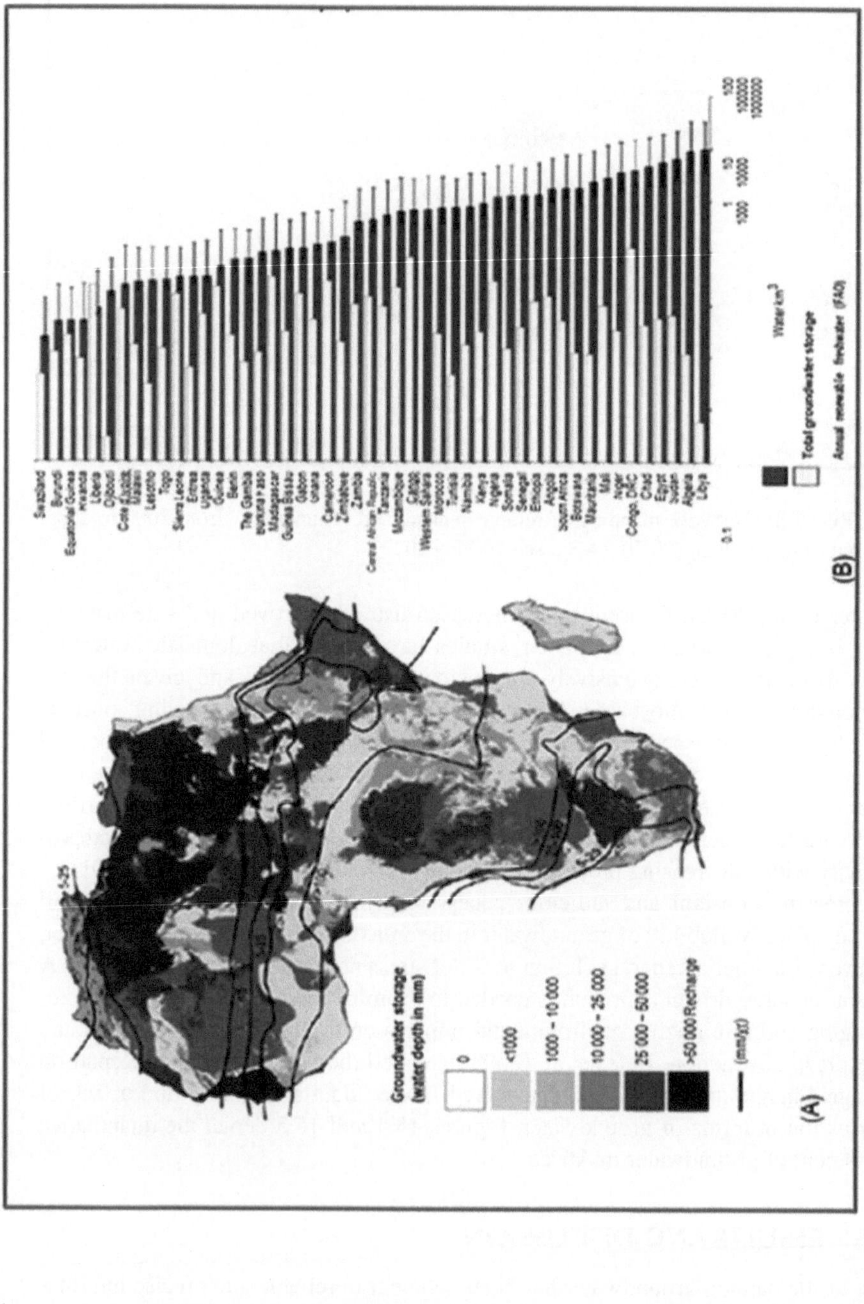

FIGURE 15.4 Groundwater in Africa (A) Groundwater storage (B) Country-wise groundwater status. (From UN-IGRAC, 2016) (Accessed 09 May 2020)

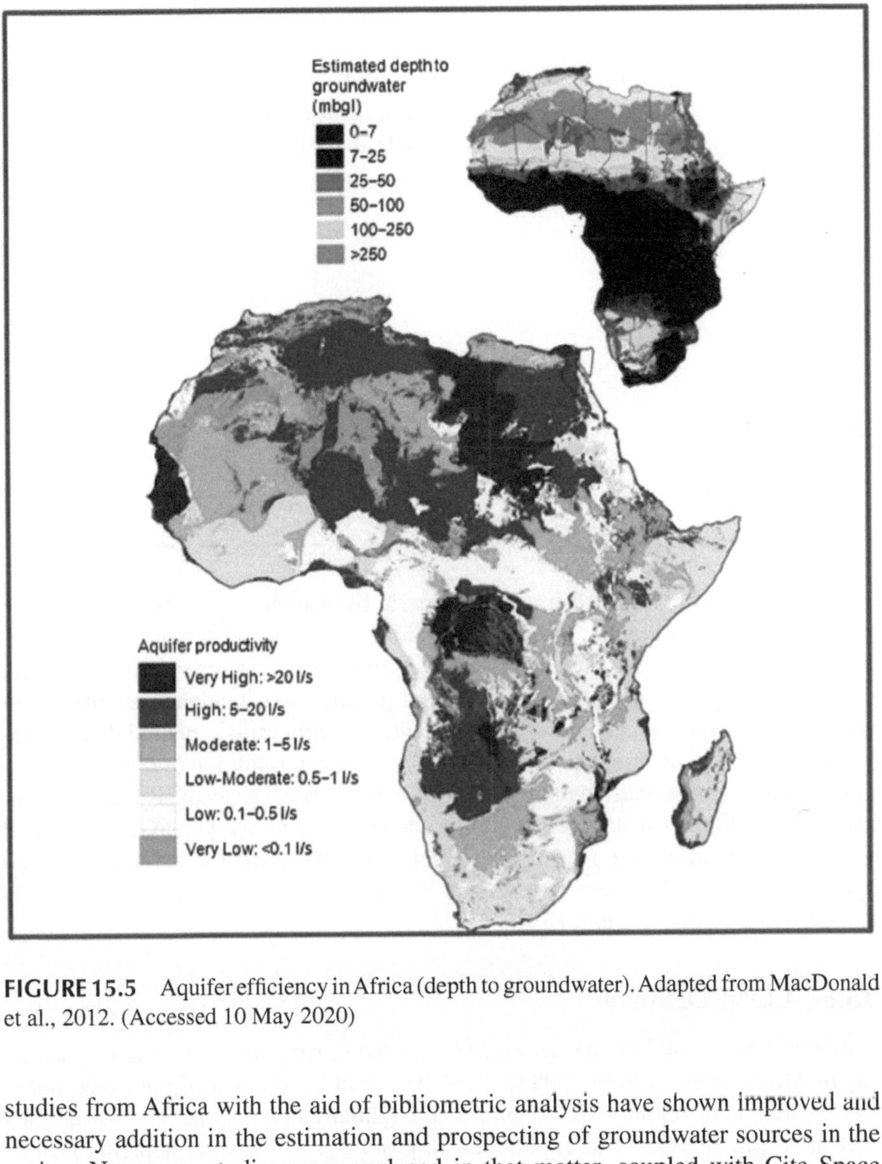

FIGURE 15.5 Aquifer efficiency in Africa (depth to groundwater). Adapted from MacDonald et al., 2012. (Accessed 10 May 2020)

studies from Africa with the aid of bibliometric analysis have shown improved and necessary addition in the estimation and prospecting of groundwater sources in the region. Numerous studies were analysed in that matter, coupled with Cite Space analytics on groundwater potential ranging from Ghana to Nigeria, which concur on the growing relevance of remote sensing in groundwater studies. Gumma and Pavelic (2013) in their study on groundwater potential in Ghana used the remote sensing-based approach which stressed the need for replicating the same in African regions. The study further demonstrated that the zonation derived from geospatial modelling should be more directed towards the high potential regions. Nigussie et al. (2019) using satellite imagery (Sentinel) and thematic layer data mapped the Ethiopian rift, showing the occurrence of high potential zones primarily in fracture extents of basalts and ignimbrites, followed by alluvial deposits and lacustrine sediments.

TABLE 15.1
Bibliometric Analysis of Groundwater Potential Analysis and Remote Sensing

Year	Keywords/Search Entries: Groundwater Potential Analysis OR Remote Sensing and Groundwater OR Groundwater OR Remote Sensing	Total Publications
2008		04
2009		07
2010		10
2011		12
2012		13
2013		15
2014		17
2015		21
2016		25
2017		28
2018		31
2019		39

The groundwater potential analysis of Nigeria by Anudu et al. (2011) proves the remotely sensed interpretation of hydrogeological layers and the accurate modelling of highly prospective regions in terms of groundwater occurrence and potential. The study mainly focused on lineament and groundwater incidence relation, with areas having high lineament density representing highly ranked potential zones for groundwater. Bibliometric analysis showed the increasing addition of remote sensing in Africa with studies carried out between 2008 and 2019. The analysis further vindicates the fact that, on account of being a resource deficient region and the paucity of information in Africa, there are reassuring observations of the remote sensing-based approach aiding in the effective and scientific appraisal of groundwater resources in the region, as shown in Table 15.1.

15.6 CONCLUSION

This study deliberated on the relevance of remote sensing to groundwater prospecting in African regions from 2008 to 2019. The total of interrelated published material showed an increase of 10% annually throughout this time, indicative of the fact that the depth and extent of exploration in groundwater potential applying remote sensing have enhanced significantly. Moreover, this area traverses a vast breadth of disciplines with the inclusive coverage of a wide range of interests from hydrology, geology, agriculture and ecology to engineering and geospatial sciences. However, the lack of trans-boundary collaborations between regions across the world limits the augmentation of remote sensing data access. Enhanced international collaboration will constitute an important asset. Quantitative and spatial cataloguing of information on groundwater in Africa is essential for utilizing it in many ways practically for the structure and formulation of strategies in adapting to the increasing demands for water connected with both the emerging trend of population pressure and the incidence of extreme weather events. Currently, the mapping of Africa on a continental scale provides both qualitative evidence and quantitative evaluation of groundwater storativity and potential groundwater productivity in Africa. Thus, remote sensing has become the new integration mechanism for having comprehensive and precise assessment of groundwater and water resource management in developing countries at both country and sub-continental/continental scales.

REFERENCES

Anudu, G.K., Essien, B.I., Onuba, L.N., Ikpokonte, A.E. (2011) Lineament analysis and interpretation for assessment of groundwater potential of Wamba and adjoining areas, Nasarawa state, northcentral Nigeria. *Journal of Applied Technology in Environmental Sanitation*, 1(2), pp. 185–198.

Arkoprovo, B., Adarsa, J., & Prakash, S. S. (2012). Delineation of groundwater potential zones using satellite remote sensing and geographic information system techniques: A case study from Ganjam district, Orissa, India. *Research Journal of Recent Sciences*, 2277, 2502.

Avtar R, Sing CK, Shashtri S, Sing A, Mukherjee S (2010) Identification and analysis of groundwater potential zones in Ken-Betwa river linking area using remote sensing and geographic information system. *Geocarto International* 25(5):379–396

Ayazi, M.H., Pirasteh, S., Arvin, A.K.P., Pradhan, B., Nikouravan, B., Mansor, S., 2010. Disasters and risk reduction in groundwater: Zagros Mountain Southwest Iran using geoinformatics techniques. *Disaster Adv* 3 (1), 51–57.

Ayele S, Raghuvanshi TK, Kala PM (2014) Application of remote sensing and GIS for Landslide Disaster Management: A Case from Abay Gorge, Gohatsion–Dejen Section, Ethopia, Landscape and Water Management Japan

Banks, D., Robins, N.S., Robins, N., 2002. An Introduction to Groundwater in Crystalline Bedrock. Norges Geologiske Undersokelse.

Barlow, M., & Clarke, T. (2002). *Blue gold: The fight to stop the corporate theft of the world's water*. New York: The New Press.

Bates, B C, Kundzewicz Z W, Wu S and Palutikof J P (ed) 2008 *Climate change and water Technical Paper of the Intergovernmental Panel on Climate Change* (Geneva: IPCC Secretariat)

Bengtsson, L. and Larsson, R. (2012) Researching mergers & acquisitions with the case study method: Idiographic understanding of longitudinal integration processes. Working paper, 2012/4, Centre for Strategic Innovation Research.

Benito, G., Meseguer-Lloret, S., & Torres-Cartas, S. (2013) Sensitive determination of Fenamiphos in water samples by flow injection photoinduced chemiluminescence *International Journal of Environmental Analytical Chemistry*, 93 (2), 152–165, DOI: 10.1080/03067319.2012.663755

Black, M. and Talbot, R. (2005) *Water – a matter of life and health: Water supply and sanitation in village India*. Oxford University Press, New Delhi.

Bredehoeft, J. D. (2002) The water budget myth revisited: Why hydrogeologists model *Ground Water*, 40 (4), pp. 340–345. doi: 10.1111/j.1745-6584.2002.tb02511.x

Brikowski, T.H., A. Neku, S.D. Shrestha, L.S. Smith (2014), Hydrologic control of temporal variability in groundwater arsenic on the Ganges floodplain of Nepal. *Journal of Hydrology*, 518 pp. 342–353. doi: 10.1016/j.jhydrol.2013.09.021.

Central Ground Water Board (2014). Dynamic ground water resources of India. Ministry of Water Resources, River Development and Ganga Rejuvenation, Government of India.

Chen, C.M. (2004). Searching for intellectual turning points: progressive knowledge domain visualization. *Proceedings of the National Academy of Sciences* 101, 5303–5310.

Chen, C. et al. (2019). China and India lead in greening of the world through land-use management. *Nature Sustainability*, 2, 122–129. https://doi.org/10.1038/s41893-019-0220-7

Chi, K.H., Lee, B.J. (1994) Extracting potential groundwater area using remotely sensed data and GIS techniques. In: *Proceedings of the regional seminar on integrated application of remote sensing and GIS for land and water resource management*. Bangkok, pp. 64–69

Chowdhury, A., Jha, M.K., Chowdary, V.M. (2010) Delineation of groundwater recharge zones and identification of artificial recharge sites in West Medinipur district, West Bengal, using RS, GIS and MCDM techniques *Environment and Earth Science*, 59 pp. 1209–1222.

Cui, X., Guo, X., Wang, Y., et al. (2019). Application of remote sensing to water environmental processes under a changing climate. *Journal of Hydrology*, 574, 892–902.

Cui, Y., & Shao, J. (2005). The role of ground water in arid/semiarid ecosystems. Northwest China. *Groundwater*, 43(4), 471–477.

Danielopol, D.L., C. Griebler, A. Gunatilaka, J. Notenboom (2003) Present state and future prospects for groundwater ecosystems. *Environmental Conservation*, 30 pp. 104–130. doi: 10.1017/S0376892903000109.

Dar, I.A., Sankar, K., Dar, M.A. (2010) Deciphering groundwater potential zones in hard rock terrain using geospatial technology. *Environmental Monitoring and Assessment*, 173 pp. 597–610.

Döll, P., Hoffmann-Dobrev, H., Portmann, F. T., Siebert, S., Eicker, A., Rodell, M., et al. (2012). Impact of water withdrawals from groundwater and surface water on continental water storage variations. *Journal of Geodynamics*, 59–60 (2012), 143–156. doi: 10.1016/j.jog.2011.05.001.

Donat, M.G., Lowry, A.L., Alexander, L.V., O'Gorman, P.A., Maher, N., 2016. More ex-treme precipitation in the world's dry and wet regions. *Nature Climate Change* 6 (5), 508–513. doi: 10.1038/nclimate2941

Famiglietti, J. S. (2014). The global groundwater crisis. *Nature Climate Change*, 4(11), 945–948. doi: 10.1038/nclimate2425.

FAO UASTAT Main Database, Food and Agriculture Organization of the United Nations (FAO). (2016) http://www.fao.org/nr/water/aquastat/water_use. Accessed 12 May 2020.

Fennta, AA, Kifle, A, Hailu, G, Gebreyohannes, T (2015) Spatial analysis of groundwater potential using remote sensing and GIS-based multi-criteria evaluation in Raya, Northern Ethiopia. *Hydrogeology Journal* 23:195–206

Fetter, C (2001) *Applied hydrogeology*. 4th edn. Prentice Hall, Upper Saddle River

Fitts, C.R., (2002). *Groundwater Science*. Academic Press.

Funk, C. and Brown, M.E. (2009). Declining global per capita agricultural production and warming oceans threaten food security. *Food Security* 1 271–289.

Giordano, M 2009 Global groundwater? Issues and solutions. *Annual review of Environment and Resources* 34 153–178

Gleeson, T., Befus, K., Jasechko, S. et al. (2016). The global volume and distribution of modern groundwater. *Nature Geoscience* 9, 161–167 doi: 10.1038/ngeo2590

Gleeson T., Wada, Y., Bierkens, M.F.P., van Beek, P.H. (2012) Water balance of global aquifers revealed by groundwater footprint. *Nature*, 488 pp. 197–200. doi: 10.1038/nature11295.

Grey D and Sadoff C W 2007 Sink or swim? Water security for growth and development *Water Policy* 9, 545–571

Gumma, M.K., Pavelic, P. (2013). Mapping of groundwater potential zones across Ghana using remote sensing, geographic information systems, and spatial modeling. *Environmental Monitoring and Assessment* 185, 3561–3579. doi: 10.1007/s10661-012-2810-y

Hagemann, S., Chen, C., Clark, D.B., Folwell, S., Gosling, S.N., Haddeland, I., Hanasaki, N., Heinke, J., Ludwig, F., Voss, F., Wiltshire, A.J., 2012. Climate change impact on available water resources obtained using multiple global climate and hydrology models. *Earth System Dynamics* 4 (1), 129–144. doi: 10.5194/esd-4-129-2013

Harinarayana P., G.S. Gopalakrishna, A. Balasubramanian (2000) Remote sensing data for groundwater development and management in Keralapura watersheds of Cauvery basin, Karnataka, India *The Indian Mineralogists*, 34 pp. 11–17.

Harvey, C.F. (2008) Environmental science: poisoned waters traced to source *Nature*, 454 pp. 415–416. doi: 10.1038/454415a.

Hasan, M.A., P. Bhattacharya, O. Sracek, K.M. Ahmed, M. von Brömssen, G. Jacks (2009) Geological controls on groundwater chemistry and arsenic mobilization: hydrogeochemical study along an E–W transect in the Meghna basin. *Bangladesh Journal of Hydrology*, 378 pp. 105–118. doi: 10.1016/j.jhydrol.2009.09.016.

Holden, J. (2014). Water fundamentals. In *Water resources – An integrated approach*. Routledge, New York.

Hoque, M.A., Hoque, M.M., & Ahmed, K.M. (2007). Declining groundwater level and aquifer dewatering in Dhaka metropolitan area, Bangladesh: causes and quantification. *Hydrogeology Journal*, 15(8), 1523–1534. doi: 10.1007/s10040-007-0226-5.

Hunter, P.R., MacDonald, A.M. and Carter, R.C. 2010 Water supply and health. *PLoS Medicine*. 7 e1000361.

Israil, M, Al-Hadithi M., Singhal D.C., Bhishm K., Rao M.S., Verma S.K. (2004) *Groundwater resources evaluation in the piedmont zone of Himalaya, India, using isotope and GIS techniques*. Roorkee, National Institute of Hydrology Roorkee

IWMI (2001) The strategic plan for IWMI 2000–2005. International Water Management Institute (IWMI), Colombo, pp 28–52

Jha, M.K., Chowdary, V.M., Chowdhury, A. (2010) Groundwater assessment in Salboni Block, West Bengal (India) using remote sensing, geographical information system, and multicriteria decision analysis techniques *Hydrogeology Journal*, 18 (7), pp. 1713–1728.

Jha, M.K., Chowdhury, A., Chowdary, V.M., Peiffer, S. (2007) Groundwater management and development by integrated remote sensing and geographic information systems: Prospects and constraints. *Water Resources Management*, 21 (2), pp. 427–467

Junge, B., Alabi, T, Sonder, K, Marcus, S, Abaidoo, R, Chikoye, Stahr K (2010) Use of remote sensing and GIS for improved natural resources management: case study from different agroecological zone of West Africa. *International Journal of Remote Sensing* 31:5116–6141

Kenny, J.F., Barber, N.L., Hutson, S.S., Linsey, K.S., Lovelace, J.K. Maupin, M.A. (2009) *Estimated use of water in the United States in 2005, USGS Circular US Geological Survey Reston*, VA [online] Available from: http://pubs.usgs.gov/circ/1344/.

Kløve B., P. Ala-Aho, G. Bertrand, J.J. Gurdak, H. Kupfersberger, J. Kværner, T. Muotka, H. Mykrä, E. Preda, P. Rossi, C.B. Uvo, E. Velasco, M. Pulido-Velazquez (2014) Climate change impacts on groundwater and dependent ecosystem. *Journal of Hydrology*, 518 (10), pp. 250–266

Leblanc M., Leduc, C., Razack, M., Lemoalle, J., Dagorne, D., Mofor, L. (2003) Application of remote sensing and GIS for groundwater modeling of large semiarid areas: example of the Lake Chad Basin, Africa Hydrology of Mediterranean and Semiarid Regions Conference, Montpieller, France. Red Books Series, 278, IAHS, Wallingford pp. 186–192.

Lee, S., Kim, Y.S., Oh, H. 2012. Application of a weights-of-evidence method and GIS to regional groundwater productivity potential mapping. *Journal of Environmental Management* 96:91–105. doi: 10.1016/j.jenvman.2011.09.016

MacDonald, A.M., et al., 2012. Quantitative maps of groundwater resources in Africa. *Environmental Research Letters* 7, 024009.

Magesh, N.S., Chandrasekar, N., Soundranayagam, J.P. (2011). Morphometric evaluation of Papanasam and Manimuthar watersheds, parts of Western Ghats, Tirunelveli district, Tamil Nadu India: A GIS approach. *Environment and Earth Science*, 64, pp. 373–381.

Manap M.A., Sulaiman WNA, Ramli M.F., Pradhan B, Surip N. 2013. A knowledge-driven GIS modeling technique for groundwater potential mapping at the Upper Langat Basin Malaysia, *Arabian Journal of Geosciences* 6, 1621–1637.

Merigó, J.M., Yang, J.B., 2017. A bibliometric analysis of operations research and management science. *Omega* 73, 37–48.

Mukherjee P, Singh CK, Mukherjee S (2012) Delineation of groundwater potential zones in arid region of India—a remote sensing and GIS approach. *Water Resources Management* 26:2643–2672

Nigussie, W., Hailu, B. T., & Azagegn, T. (2019). Mapping of groundwater potential zones using sentinel satellites (–1 SAR and –2A MSI) images and analytical hierarchy process in Ketar watershed, Main Ethiopian Rift. *Journal of African Earth Sciences*, 103632. doi: 10.1016/j.jafrearsci.2019.103632

Nobre, R.C.M., Filho, O.C.R., Mansur W.J., Nobre M.M.M., Cosenza C.A.N. (2007). Groundwater vulnerability and risk mapping using GIS, modeling and a fuzzy logic tool. *Journal of Contaminant Hydrology* 94:277–292. doi: 10.1016/j.jconhyd.2007.07.008

Ozdemir A. 2011. Using a binary logistic regression method and GIS for evaluating and mapping the groundwater spring potential in the Sultan Mountains (Aksehir, Turkey). *Journal of Hydrology* 405:123–136. doi: 10.1016/j.jhydrol.2011.05.015

Page ML, Berjamy B, Fakir Y, Bourgin F et al (2012) An integrated DSS for groundwater management based on remote sensing. The case of a semi-arid aquifer in Morocco. *Water Resources Management*, 26:3209–3230

Pfister S, Bayer P, Koehler A and Hellweg S 2011 Projected water consumption in future global agriculture: Scenarios and related impacts *Science of the Total Environment* 409 4206–4216.

Pradhan, B., 2013. A comparative study on the predictive ability of the decision tree, support vector machine and neuro-fuzzy models in landslide susceptibility mapping using GIS. *Computational Geosciences* 51, 350–365. doi: 10.1016/j.cageo.2012.08.023

Radloff, Y. Zheng, H.A. Michael, M. Stute, B.C. Bostick, I. Mihajlov, M. Bounds, M.R. Huq, I. Choudhury, M.W. Rahman, P. Schlosser, K.M. Ahmed, A. (2011) Arsenic migration to deep groundwater in Bangladesh influenced by adsorption and water demand *Nature Geoscience*, (4), pp. 793–798. doi: 10.1038/ngeo1283.

Rahmati O, Nazari Samani A, Mahdavi M, Pourghasemi HR, Zeinivand H. 2015. Groundwater potential mapping at Kurdistan region of Iran using analytic hierarchy process and GIS. *Arabian Journal of Geosciences* 8:7059–7071.

Richey, A.S., Thomas, B.F., Lo, M.H., Reager, J.T., Famiglietti, J.S., Voss, K., Swenson, S. and Rodell, M., 2015. Quantifying renewable groundwater stress with GRACE. *Water Resources Research*, 51(7), 5217–5238. doi: 10.1002/2015WR017349.

Schewe, J., Heinke, J., Gerten, D., Haddeland, I., Arnell, N.W., Clark, D.B., Dankers, R., Eisner, S., Fekete, B.M., Colón-González, F.J., Gosling, S.N., Kim, H., Liu, X.C., Masaki, Y., Portmann, F.T., Satoh, Y., Stacke, T., Tang, Q.H., Wada, Y., Wisser, D., Albrecht, T., Frieler, K., Piontek, F., Warszawski, L., Kabat, P., 2014. Multimodel assessment of water scarcity under climate change. *Proceedings of the National Academy of Sciences USA* 111 (9), 3245–3250. doi: 10.1073/pnas.1222460110

Sener E., A. Davraz, M. Ozcelik (2005) An integration of GIS and remote sensing in groundwater investigations: a case study in Burdur, *Turkey Hydrogeology Journal*, 13 pp. 826–834.

Shah, T. (2005) Groundwater and human development: challenges and opportunities in livelihoods and environment. *Water Science and Technology*, 51 (8) pp. 27–37

Shah, T, Roy, A.D., Qureshi, A.S., Wang, J. (2003) Sustaining Asia's groundwater boom: an overview of issues and evidence. *Natural Resources Forum*, 27 (2) pp. 130–141

Shukla, D.P., Dubey, C.S., Singh, N.P., Tajbakhsh, M. and Chaudhry, M., (2010) Sources and controls of arsenic contamination in groundwater of Rajnandgaon and Kanker district, Chattisgarh Central India. *Journal of Hydrology*, 395 pp. 49–66. doi: 10.1016/j.jhydrol.2010.10.01

Siebert, S., Burke, J., Faures, J.M., Frenken, K., Hoogeveen, J., Döll, P. and Portmann, F.T. 2010 Groundwater use for irrigation—a global inventory. *Hydrology and Earth System Sciences* 14 1863–1880.

Singh, R. (ed.) (2021) *Re-envisioning Remote Sensing Applications: Perspectives from Developing Countries*. First Edition. CRC Press. doi: 10.1201/9781003049210

Singh, A. K., Raj, B., Tiwari, A. K., & Mahato, M. K. (2013). Evaluation of hydrogeochemical processes and groundwater quality in the Jhansi district of Bundelkhand region, India. *Environment and Earth Science*, 70(3), 1225–1247. doi: 10.1007/s12665-012-2209-7.

Solomon, S., Quiel, F. (2006) Groundwater study using remote sensing and geographic information systems (GIS) in the central highlands of Eritrea. *Hydrogeology Journal* 14:729–741

Sreedevi, P.D., Subrahmanyam, K. and Ahmed, S. 2005. Integrated approach for delineating potential zones to explore for groundwater in the Pageru River basin, Cuddapah District, Andhra Pradesh, India. *Hydrogeology Journal* 13:534–543. doi: 10.1007/s10040-004-0375-8

Sreedhar, G., Kumar, G.T., Krishna, I.V., Ercan, K., Cuneyd, D.M. (2009) Mapping of ground water potential zones in the Musi basin using remote sensing data and GIS. *Advances in Engineering Software* 40:506–518

Sreekanth, P. D., Geethanjali, N., Sreedevi, P. D., Ahmed, S., Ravi Kumar, N., Kamala Jayanthi, P. D. (2009). Forecasting groundwater level using artificial neural networks. *Current Science* 96(7):933–939. https://www.jstor.org/stable/24104683.

Stafford, K.W., Rosales-Lagarde, L., Boston, P.J. (2008) Castile evaporate karst potential map of the Gypsum Plain, Eddy County, New Mexico and Culberson County, Texas: a GIS methodological comparison. *J Cave Karst Stud* 70(1):35–46.

Tahmassebipoor, N., Rahmati, O., Noormohamadi, F., Lee, S. 2016. Spatial analysis of groundwater potential using weights-of evidence and evidential belief function models and remote sensing. *Arabian Journal of Geosciences* 9:1–18

Tiwari, A., Rai, B. 1996 Hydromorphological mapping for groundwater prospecting using landsat – MSS images—a case study of Part of Dhanbad District, Bihar *Journal of the Indian Society of Remote Sensing*, 24, pp. 281–285.

Todd, D. and Mays, L. (2005) *Groundwater Hydrology*. 3rd Edition, John Wiley and Sons, Inc., Hoboken.

Trenberth, K.E., 2011. Changes in precipitation with climate change. *Climate Research* 47 (1–2), 123–138. doi: 10.3354/cr00953

Tweed, S.O., Leblanc, M., Webb, J.A. and Lubczynski, M.W. (2007) Remote sensing and GIS for mapping groundwater recharge and discharge areas in salinity prone catchments, southeastern Australia. *Hydrogeology Journal*, 15 pp. 75–96.

UNEP. 2010 Africa Water Atlas (Nairobi: Division of Early Warning and Assessment (DEWA), United Nations Environment Programme)

UN-IGRAC. 2016. *Africa Groundwater Atlas*. Retrieved from: https://www2.bgs.ac.uk/africagroundwateratlas/index.cfm

Van Loon, A.F., 2015. Hydrological drouaght explained. *Wiley Interdisciplinary Reviews* 2 (4), 359–392.

Vaux H (2011) Groundwater under stress: the importance of management. *Environment and Earth Science* 62:19–23.

Wada, Y., Van Beek, L.P., Van Kempen, C.M., Reckman, J.W., Vasak, S. and Bierkens, M.F. (2010), Global depletion of groundwater resources. *Geophysical Research Letters*, 37 (20), 114–122.

Wang, H., Xu, Z.S., 2016. Admissible orders of typical hesitant fuzzy elements and their application in ordered information fusion in multi-criteria decision making. *Information Fusion* 29, 98–104.

Weber, Y. (Ed.), (2012) *Handbook for Mergers and Acquisitions Research*, Edward Elgar, Cheltenham, UK, pp. 172–202.

Weiler, B., Moyle, B., & McLennan, C.L. (2012). Disciplines that influence tourism doctoral research: The United States, Canada, Australia and New Zealand. *Annals of Tourism Research*, 39(3), 1425–1445

World Bank Report. (2009). India groundwater: a valuable but diminishing resource. Available at: http://www.worldbank.org/en/news/feature/2012/03/06/india-groundwater-critical-diminishing.

Wösten, J.H.M., Clymans, E., Page, S.E., Rieley, J.O. and Limin, S.H., 2008. Peat–water interrelationships in a tropical peatland ecosystem in Southeast Asia. *Catena*, *73*(2), pp.212–224. doi: 10.1016/j.catena.2007.07.010

WWAP (United Nations World Water Assessment Programme) 2016. The United Nations World Water Development Report: Water and Jobs UNESCO, Paris.

Yang, S., Sui, J., Liu, T., Wu, W.J., Xu, S.Y., Yin, L.H., Pu, Y.P., Zhang, X.M., Zhang, Y., Shen, B., 2018. Trends on PM2.5 research, 1997–2016: A bibliometric study. *Environemental Science and Pollution Research* 25 (13), 12284–12298.

Zektser, I.S., Everett, L.G. (2004) *Groundwater resources of the world and their use; 2004, IHP-VI series on groundwater UNESCO*, paris [online] Available from: http://www.unesco.org/ulis/cgi-bin/ulis.pl?catno=134433&set=51BF4847_1_322&gp=1&lin=1&ll=1 (Accessed 17 March 2020).

16 Remote Sensing Applications in Identification of Villages on the Basis of Intensity of Problems in Physical Resources in the Kandi Region of Punjab

Rajesh Jolly
Lovely Professional University, Punjab, India

Ripudaman Singh
Amity University, Noida, Uttar Pradesh, India

CONTENTS

DOI: 10.1201/9781003224624-20

16.1 INTRODUCTION

The Kandi region is a sub-mountainous undulating region, elevated from its surrounding and so termed a 'kandi'. This region lies below the Shiwalik hills and is located in the northeast portion of the state of Indian Punjab. The word Kandi originated from the local Punjabi dialect 'kanda' which means an edge, so being located on the edge of the Shiwalik hills, this region came to be called the 'Kandi region'. It is mainly confined to five districts, namely Pathankot, Hoshiarpur, Rupnagar, SBS Nagar and SAS Nagar, covering an area of 419,900 hectares, i.e., 8 per cent of the state's total area. The region stretches from 30° 21′ 48″ to 32° 30′ 30″ north latitudes and from 75° 32′ 12″ to 75° 56′ 00″ east longitudes. It consists of 21 blocks and 1,533 villages (Govt of India, 2011).

Physiographically speaking, the Kandi region has structurally originated dissected hills, piedmont alluvial plains and alluvial fans (Singh & Singh, 2019a, 2019b, 2020). The Shiwalik hills of the Kandi region are marked by a number of problems due to physical obstacles, such as the severe rate of soil erosion and frequent flooding along the river beds (Jain, 1998; Chowdary et al., 2003; Jolly, 2003).

16.2 METHODOLOGICAL PLAN

For accomplishing the task of identifying the intensity of problems in the physical resource base of the Kandi region, the following steps have been taken.

16.2.1 Step 1. Preparation of Spatial Database (Drainage, Slope, Soil Erosion Layer of Kandi Region of Punjab)

The spatial database of the study region has been prepared from the free online source of satellite data, i.e., the Bhuvan official site. The satellite data of IRS Resourcesat-1 LISS-3 has been utilized to prepare a drainage map of the study region (Figure 16.1). The spatial layers with different categories consisting of seasonal stream, canal, pond and reservoir database have been generated in the ArcGIS 10.1 environment.

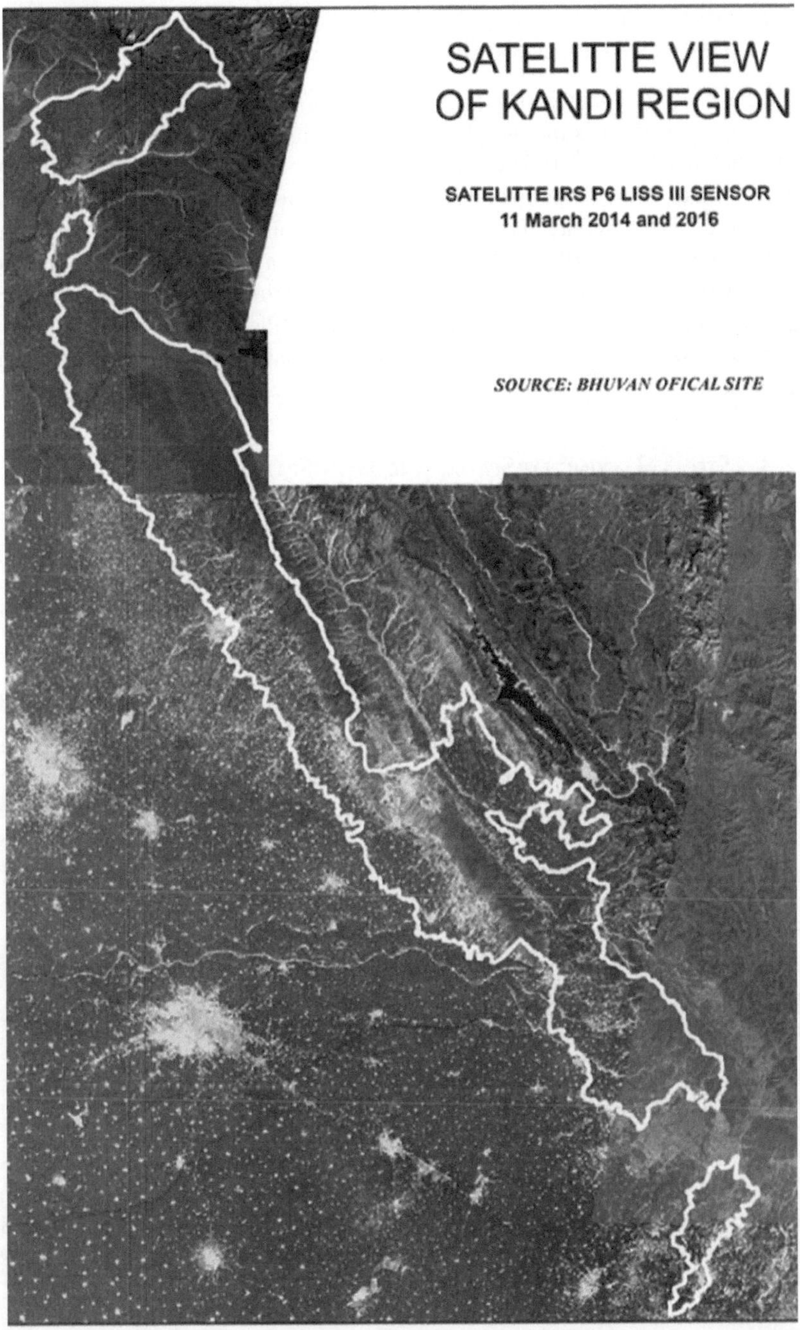

FIGURE 16.1 Satellite image of Kandi region of Punjab

16.2.2 Step 2. Preparation of Spatial Database (Village Maps of Kandi Region of Punjab)

After the successful creation of the digital layers of drainage of the study region, the second important layer is the base layer – that is, the village boundary of the Kandi region. The village boundary in digital format has been procured from the Survey of India, Dehradun. The village boundary was further edited and topologically rectified in the ArcGIS 10.1 software. The village boundary collaborated at block and district level. The creation of attribute data has been done in a Microsoft Access and Excel database which could be easily exported to any environment for further analysis – e.g., the identifications of Kandi and non-Kandi villages in each of the districts were made based on the list of Kandi villages available with the Punjab planning department.

16.2.3 Step 3. Linking of Spatial and Non-Spatial Database in the GIS Environment

The spatial and non-spatial databases were created in different environments, then interlinked in the GIS environment for further spatial analysis of the database and mapping of the different layers of the region.

16.2.4 Step 4. Identification of Severity of the Problem in the Region

First, on the basis of the drainage network, the villages that were suffering due to active floods along the river beds were identified. These villages were given a score card based on the proximity of the river bed. Secondly, the erosion factors considered for the villages with a severe rate of soil erosion were identified and a score card was assigned to them. Lastly, the villages with cumulative score cards having the highest number were assigned with a high intensity of problems in the physical resource base of the region (Singh, 2021).

For the analysis of the physical resource base, the physical aspects such as the geomorphology, drainage, slope, soil erosion and vegetation cover have been derived from the LISS-3 data available at the Bhuvan official site.

16.3 DRAINAGE MAP OF KANDI REGION

The Kandi region has high drainage frequency and drainage density. There are several seasonal streams flowing from east to west of the region. There are three main perennial rivers flowing from the region. The Ravi river makes a natural boundary of the region in the northern part of Kandi, while the Dangri stream demarcates its extent in the southern parts.

16.3.1 Ravi River

The Ravi river is a part of the Indus river system. The Ravi river originates from the Kullu district in the Rohtang Pass and follows a northwest course and then enters the

Punjab plains near Madhopur. It is a perennial river originating from a glacier at an elevation of 14,000 feet.

The Ravi river enters the Kandi region from the Dharkalan block of the Pathankot district and flows in a southwest direction, making the northern boundary of the Sujanpur block. It covers an area of 2.02 km^2 (0.7 per cent of the total drainage area).

16.3.2 Beas River

The Beas river originates from the Rohtang Pass in the Kullu district at a height of 4,062 m above mean sea level. After crossing the Middle Himalayas, it enters Punjab near Mirthal, where it joins the northward-coming Chaki stream. The Beas river covers an area of 10.2 km^2 (3.7 per cent of the total drained area). The Beas river enters the Kandi region from the northern portion of the Mukerian and Talwara blocks of the Hoshiarpur district (Gosal, 2020) (Figure 16.2).

16.3.3 Sutlej River

The Sutlej river is the main river of the Kandi region. It originates from the eastern part of the Mansarovar lake of Tibet at an elevation of 4663 m above mean sea level. The Sutlej river is an antecedent stream of the Himalayas, so during its journey it makes deep gorges. It enters the Punjab from the Shiwalik hills of Nangal city of the Anandpur Sahib block. From Nangal it turns in a northeast and southwest direction. It enters Rupnagar district from the Jaswan Dun valley of the Shiwalik hills. From a westward direction, it enters Ropar through Nurpur Bedi and then flows towards the plains of the Rupnagar block. The Bhakra dam on the Sutlej river not only provides hydroelectricity in the state but also saves people from floods in the region. The Nangal barrage canal located at Nangal diverts two canals, the Sirhind and Bhakra canals. The Sutlej river is noted for changing its course from ancient times. The river bed is 500 m to 1.5 km wide in some parts of the region. From the Nurpurbedi block it flows southwest and enters the Ludhiana district. The Sutlej river covers an area of 23.2 km^2 (8.4 per cent of the total drained area) (Table 16.1).

16.3.4 Choes (Seasonal Stream/Dry Nadi)

In addition to main rivers, some seasonal streams also flow through the region. These seasonal streams feed the area immediately beyond the Shiwalik hills. Most of these streams flow south of the Shiwalik hills. These seasonal streams dissect the way of the foothill zone into 10 to 20 kms and slightly disappear and merge into the ground or submerge into the nearby rivers. They are dry in nature and filled with water only during the monsoon. They make a wide bed filled with stone, gravels and sand. They are closely spaced at intervals of 3 to 5 kms on average. In the Hoshairpur district, these seasonal streams are closely spaced to each other – i.e., one choe per kilometre. These seasonal streams have a strong impact on the soil erosion activity. These seasonal streams are now being channelized and embankments have been constructed with reinforced boulders and stones. Small check dams/earthern dams have also been constructed to lessen the severity of floods in the region (Gosal, 2020).

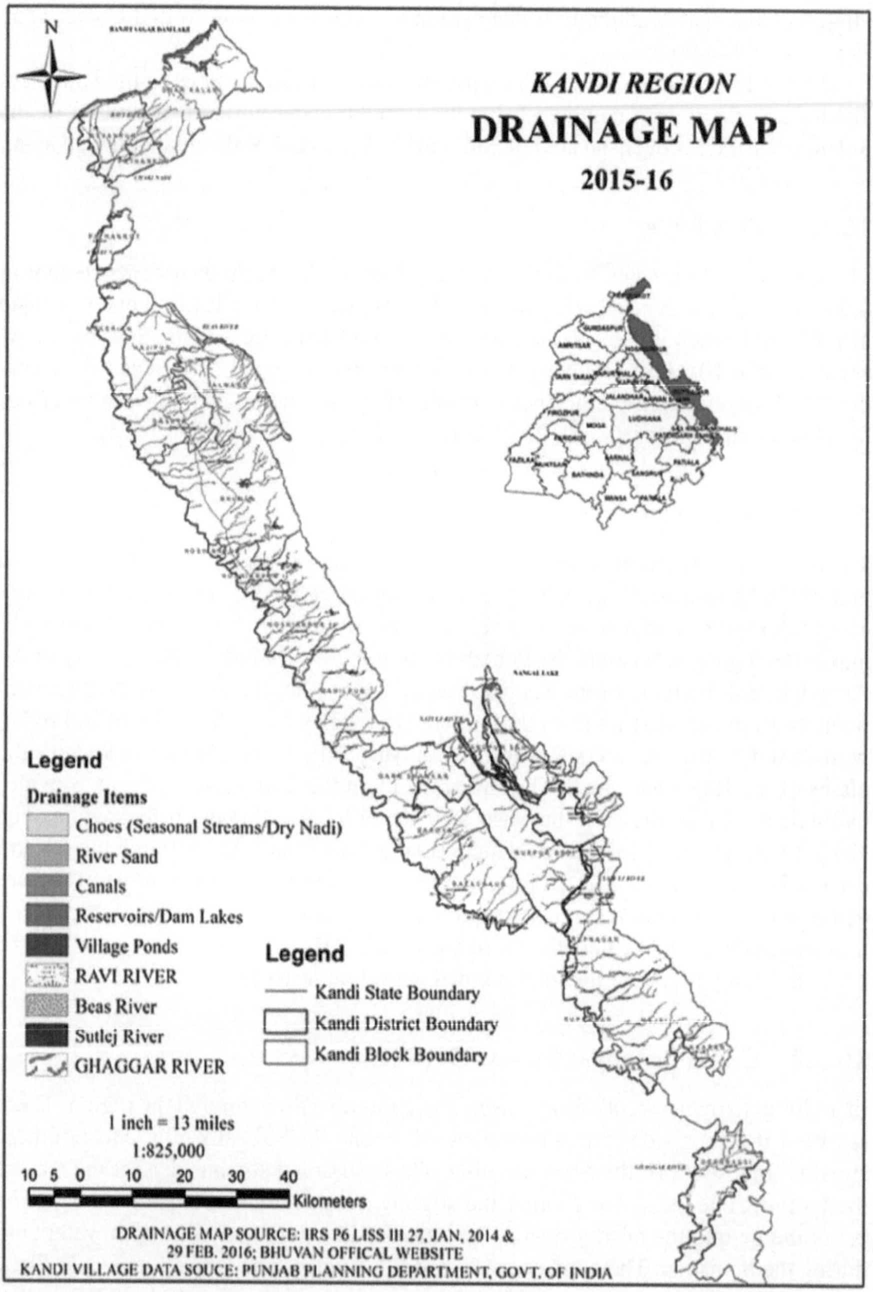

FIGURE 16.2 Drainage map of Kandi region

TABLE 16.1
Kandi Region: Drainage Network, 2014–16

Sr. No.	Drainage Item	Area in km^2	Percentage of Total
1	Choe (Seasonal stream/dry Nadi)	160	57.8
2	River sand	25.4	9.2
3	Canals	20.3	7.3
4	Reservoir/dams lake	23.8	8.6
5	Village pond	0.45	0.2
6	Ravi river	2.02	0.7
7	Beas river	10.2	3.7
8	Sutlej river	23.2	8.4
9	Ghaggar river	11.3	4.1
	Total	**276.7**	**100.0**

Source: IRS P6, LISS III, Jan. to Feb., 2014 & 2016, Bhuvan official website

16.4 SOIL EROSION OF KANDI REGION

Soil erosion is defined as the physical removal of the topsoil or particles by the physical dissemination of rocks or soil. The major factors responsible for soil erosion are deforestation, overgrazing, wrong cropping practices, high cropping intensity and large development projects, such as dams and mining sites (Sharma et al., 2004, 2007; Jolly & Singh, 2019a, 2019b).

16.4.1 Types of Soil Erosion

16.4.1.1 Slight Rate of Soil Erosion

Slight rate of soil erosion has a soil erosion rate of 0 to 5 Mg/hect./year. In the Kandi region slight erosion covers an area of 1,121 km^2 which is 27 per cent of the total geographical area of the region. This ratio is quite low as compared to the state average, but slightly high as compared to the national average. The main blocks coming under this category are Mukerian, North Hajipur, Sujanpur, Hoshiarpur I, Nurpur Bedi and Anandpur Sahib. The soils in these areas are coarse and fine loamy calcareous soils.

16.4.1.2 Moderate Rate of Soil Erosion

Moderate rate of soil erosion ranges from 5 to 10 Mg/hect./year. In the Kandi region moderate rate of soil erosion covers an area of 1,755 km^2 which is 42 per cent of the total geographical area of the region. This ratio is quite high as compared to the state average, but slightly low as compared to the national average. The main blocks coming under this category are West Dharkalan, Pathankot, Dasuya, East Bhunga, Mahilpur, Saroya, Balachaur, South Rupnagar and Majiri. The soils in these areas are coarse loamy, calcareous soils, coarse loamy with sandy soils, coarse loamy or sandy soils with moderate and slight flooding, sandy soils, sandy coarse loamy with calcareous soils (Figure 16.3, Table 16.2).

16.4.1.3 Severe Rate of Soil Erosion

The areas close to the river beds have a severe rate of soil erosion – i.e., ravine areas especially along the Beas river, Ravi river, Sutlej river and Ghaggar river (Department of Soil and Water Conservation, Punjab, 2002). A severe rate of soil erosion ranges from 40 to 80 Mg/hect./year loss of soil particles. In the Kandi region a severe rate of soil erosion covers an area of 1,222 km^2 which is 29 per cent of the total geographical area of the region. This ratio is quite high as compared to the state as well as the national average. It mainly covers the hilly track of the Shiwalik hills. The main covering blocks are east Majiri, Rupnagar and Anandpur Sahib, west Nurpur Bedi, east Garhshankar, east Mahilpur, east Hoshiarpur II, east Bhunga and Talwara.

16.4.1.4 Very Severe Rate of Soil Erosion

A very severe rate of soil erosion is more than 80 Mg/hect./year loss of soil particles. In the Kandi region, a very severe rate of soil erosion covers an area of 101 km^2 which is 2.4 per cent of the total geographical area of the region. This ratio is quite high as compared to the state and national average. It mainly covers the highly and moderately dissected hills and valley of the Kandi region.

16.5 CLASSIFICATION OF VILLAGES ON THE BASIS OF INTENSITY OF PROBLEMS IN PHYSICAL RESOURCES

From the physical resource base, three problems – that is, floods, soil erosion and deep drinking-water level – were identified at the village level (Figure 16.4). In the Kandi region 431 villages face an acute problem of active floods, particularly in the monsoon period, and these are named as flood-affected villages. Secondly, there are the villages facing an acute problem of soil erosion. Thirdly, in the Kandi region some of the villages are facing a problem of acute shortage of drinking groundwater level, which is below 20 metres (Table 16.3).

These identified villages were scored in ascending order according to the severity of the problems – e.g., in the case of flood-affected areas the villages were selected on the closest vicinity of the river. The villages which come under the 1 km vicinity of the river catchment were scored as 4 (severe). The villages coming under the 2 km vicinity were scored as 3 (high), the villages coming under 3 km were scored as 2 (moderate) and last the villages come under the 4 km vicinity of the river catchment were scored as 1 (low) (Table 16.3).

The second criterion was soil erosion, which is quite prominent in the villages of the Kandi region. The villages have been identified and scored on the basis of the severity of the soil erosion. For example, a village having a very severe rate of soil

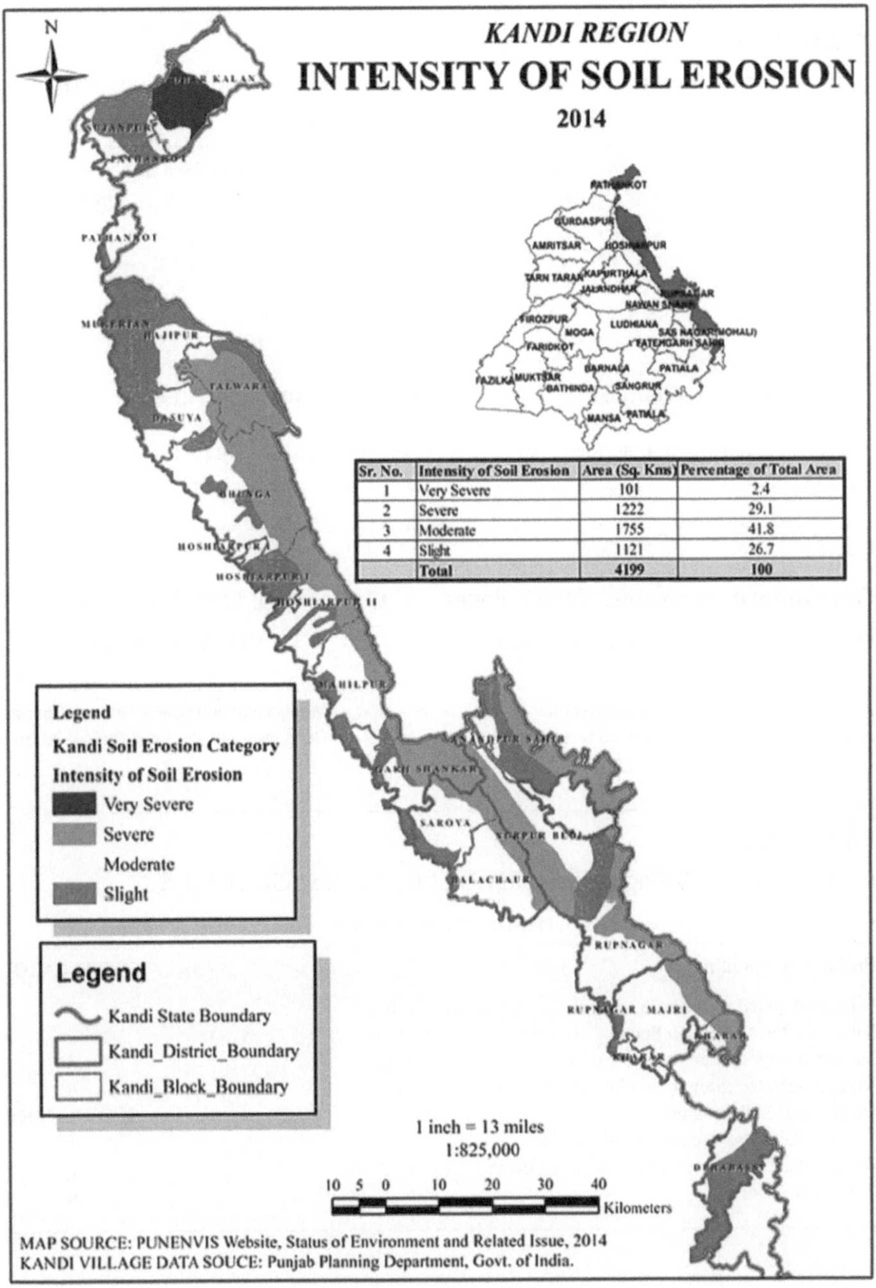

Sr. No.	Intensity of Soil Erosion	Area (Sq. Kms)	Percentage of Total Area
1	Very Severe	101	2.4
2	Severe	1222	29.1
3	Moderate	1755	41.8
4	Slight	1121	26.7
	Total	4199	100

FIGURE 16.3 Soil erosion map of Kandi region

TABLE 16.2
Soil Erosion: Comparison between India, Punjab and Kandi Region, 2014

		India		Punjab		Kandi Region	
Sr. No.	Intensity of Soil Erosion	Area in km^2	Percentage of Total Area	Area in km^2	Percentage of Total Area	Area in km^2	Percentage of Total Area
1	Slight	801,350	24.4	44,072	87.5	1,121	26.7
2	Moderate	1,405,640	42.8	3,178	6.3	1,755	41.8
3	High	805,030	24.5	1,087	2.2	Nil	Nil
4	Very High	160,050	4.9	443	0.9	Nil	Nil
5	Severe	83,300	2.5	866	1.7	1,222	29.1
6	Very Severe	31,895	1.0	716	1.4	101	2.4
	Total	**3,287,265**	**100**	**50,362**	**100**	**4,199**	**100**

Source: PUNENVIS Website, Status of Environment and Related Issues, 2014

TABLE 16.3
Development Problems of the Villages and Operational Definitions Used

Sr. No.	Name of Problem	Definition of Problem
1.	Floods	Floods arise along the river beds
2.	Soil Erosion	Very severe and severe soil erosion rate
3.	Groundwater Level	Deep groundwater level (below 20 m)

TABLE 16.4
Classification of Villages on the basis of Cumulative Score Card

Physical Resource Base	
Problem of Flooding	SCORE CARD
Villages affected through flooding along the 4 km of riverbed	1
Villages affected through flooding along the 3 km of riverbed	2
Villages affected through flooding along the 2 km of riverbed	3
Villages affected through flooding along the 1 km of riverbed	4
Problem of Soil Erosion	**SCORE CARD**
Villages having severe rate of soil erosion	1
Villages having very severe rate of soil erosion	2
Total Score Card	**6**

erosion was scored as 2, while a village with a severe rate of soil erosion was scored as 1 (Table 16.4).

16.6 RESULT AND DISCUSSION

The total score card for the physical resources was 6. These villages were further categorized in ascending order to obtain a cumulative score card of intensity of problems with physical resource. The higher the score, the more severe the problems. In physical resource base there were nine villages which have a severe rate of problems in physical resources. These nine villages are Thara Jhikala, Kot and Dhureti (Dhar Kalan block) of Pathankot district; Chak Dhera, Miani, Nehon, Inder pura, Saina Majra and Malkpur of Rupnagar block, Rupnagar district. These villages face an acute problem of soil erosion as well as active floods during the monsoon season. Sixty-nine villages were identified as having a high intensity of problems. These villages were found in Talwara, Anandpur Sahib, Nurpur Bedi and Rupngar blocks of the Kandi region. A total of 277 villages were identified with moderate intensity of problems. These villages are mainly located in the Dhar Kalan, Sujapnpur, Pathankot, Mukerian, Hajipur and Talwara Block, Garh Shankar, Anandpur Sahib, Rupnagar, Nurpur Bedi and Dera Bassi blocks. A total of 840 villages are free from the problems of soil erosion and active flooding (Figure 16.4).

16.7 CONCLUSION

The Kandi region has a sub-mountainous badland topography, which is numerously affected by the severity of soil erosion and active flooding. The region has a wide potential for vegetable growing, horticulture and dairy farming. It has a wide scope in the earth-filled dams and rainwater harvesting structures which will reduce the problem of water scarcity in the region. Similar conclusions have also been drawn by Krishan et al. (1998), Singh et al., (2011) and Singh and Khanduri (2011). The ravine lands which are inaccessible for agriculture can be put under forestry and orchard plantations. Soil erosion can be prevented by the practice of mixed cropping, which reduces land denudation by providing a maximum canopy and avoiding the beating action of raindrops. Dhillon and Dhillon (2003) have similarly referred to preventing soil erosion in such conditions. To conclude, it may be suggested that micro-level planning through constructing check dams and introducing mixed cropping would be helpful in controlling soil erosion and managing the water resources of the region.

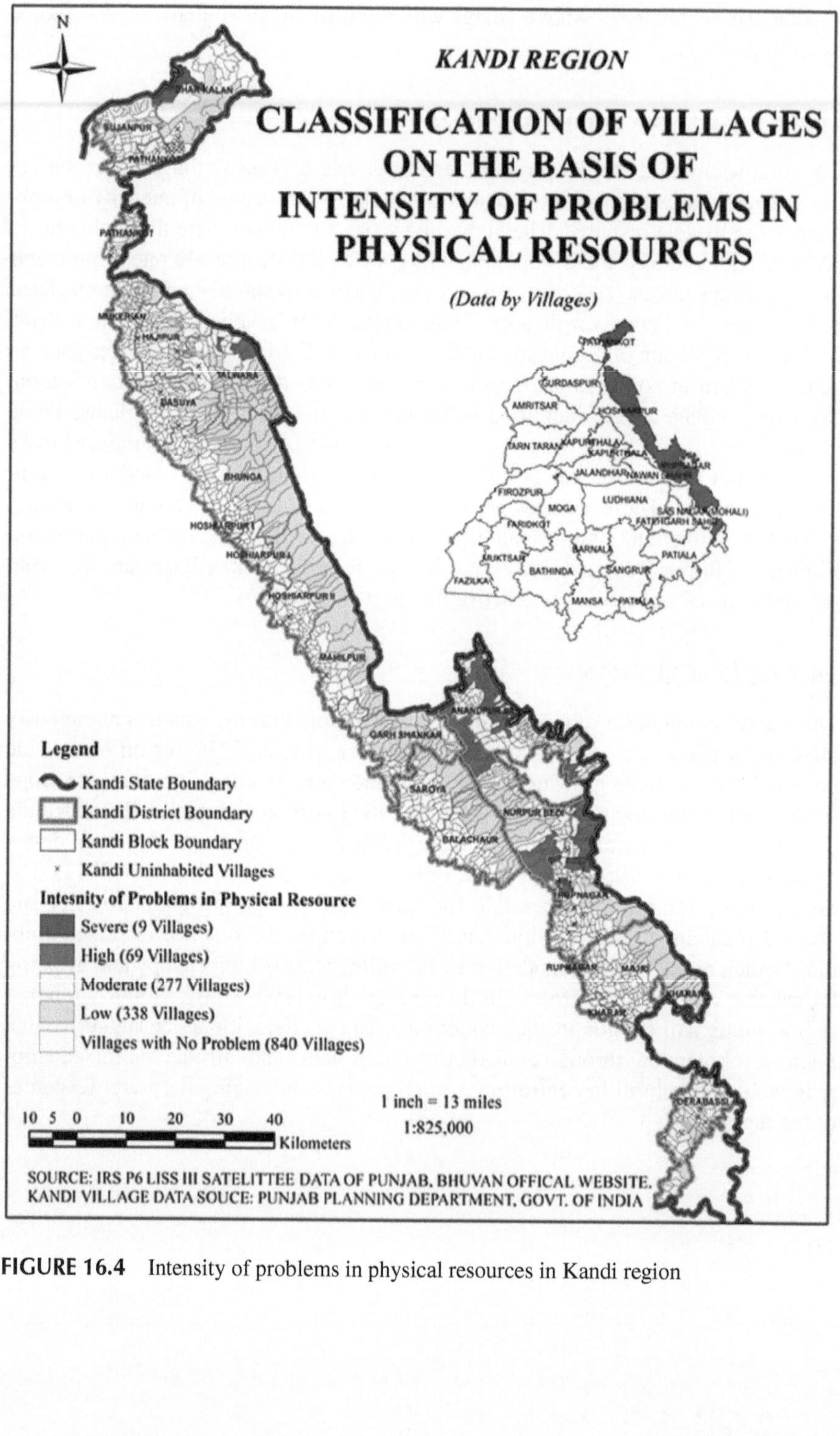

FIGURE 16.4 Intensity of problems in physical resources in Kandi region

REFERENCES

Chowdary, V. M., Rao, N. H. and Sarma, P. B. S. (2003). GIS-based decision support system for groundwater assessment in large irrigation project areas. *Agricultural Water Management*, 62: 229–252.

Dhillon, K.S. and Dhillon, S.K. (2003). Distribution and management of seleniferous soils. *Journal of Advances in Agronomy*, 79: 119–184.

Gosal, G. S. (2020). Physical Geography of Punjab. *Journal of Punjab Studies*, 11(1): 19–37.

Government of India, (2011). *Primary Census Handbook, Pathankot, 2011.* Government of India, New Delhi.

Jain. P.K. 1998. Remote sensing techniques to locate groundwater potential zones in upper Urmil River basin, district Chatarpur-central India. *Journal of Indian Society of Remote Sensing*, 26. pp. 135–147.

Jolly, R. (2003). *Application of Remote sensing and GIS technology for micro level planning in the dissected foothill zone: A case study of Raipur Rani Block, District Panchkula, Haryana.* Postgraduate Diploma in Remote Sensing and GIS dissertation, Department of Geography, Panjab University, Chandigarh

Jolly, R. and Singh, R. (2019a). RS & GIS application for Choe infested Kandi area in Dhar Kalan and Sujanpur block of Pathankot. *International Journal of Engineering Development and Research*, 7(2): 273–280.

Jolly, R. and Singh, R. (2019b). Application of Geographical Information System to study the socio-economic infrastructural facilities in the Kandi region of Punjab. *National Conference on Big Geospatial Data: Analytics, Modelling & Applications*, 25–26 September, 2019.

Krishan, G., Julka, A.C. and Singh, M. (1998). *Integrated Development Plan for Anandpur Sahib Block, Rupnagar District, Punjab. Block Development Plan*, submitted to Government of Punjab, Chandigarh.

Sharma, G. N., Bath, S. K., Jolly, R. and Dhiman, S. (2007). Mapping of S&T needs Inventorization and documentation of location-specific problems requiring scientific/technical interventions in Punjab. Project submitted to Punjab State Council for Science and Technology, Chandigarh.

Sharma, P.K. et al. (2004). Landuse/Landcover cum drainage mapping in the selected sub-watershed in the Kandi area of Punjab. Scientific Note PUNSEN-TR 2004. Punjab Remote Sensing Centre, P.A.U., Ludhiana.

Singh, C. K. et al. (2011). Application of GWQI to assess effect of land use change on groundwater quality in lower Shiwaliks of Punjab: Remote sensing and GIS based approach. *Water Resource Management*, 25: 1881–1898.

Singh, P. and Khanduri, K. (2011). Land use and land cover change detection through remote sensing & GIS technology: A case study of Pathankot and Dhar Kalan Tehsils, Punjab. *International Journal of Geomatics and Geosciences*, 1(4). 839–846.

Singh, S. and Singh, R. (2020). Geo-spatial topology based morphometric analysis for soil and water conservation in Dholbaha watershed of Kandi region. *Journal of Physics: Conference Series*, 1531 (2020), pp. 1–14. doi: 10.1088/1742-6596/1531/1/012087

Singh, R. and Singh, S., (2019a). Morphometric analysis based conservation measures for Dholbaha watershed in Shivalik foothills of Punjab using geo-spatial technology. *Our Heritage*, 67(6) pp. 1–9.

Singh, R. and Singh, S. (2019b). Role of geo-informatics for investigating causes and impacts of gully erosion in Kandi region of Punjab. *Our Heritage*, 67(6), pp. 10–17.

Singh, R. (ed.) (2021) *Re-envisioning Remote Sensing Applications: Perspectives from Developing Countries.* First Edition. CRC Press. doi:10.1201/9781003049210

REFERENCES

[illegible]

Conclusion
Outlook to Future Research Agenda

Ripudaman Singh
Amity University, Noida, Uttar Pradesh, India

CONTENT

Moving towards concluding this volume, it is not only expected that it will inspire researchers about new approaches to remote sensing applications, but it also explores some ideas about new frontiers of remote sensing. It is anticipated that the sixteen chapters presented in three broad areas in this book are just the continuation of the bigger domain of remote sensing and its varied applications. Thus, the new knowledge and information being created by remote sensing and geospatial technologies will certainly be helpful in solving socioeconomic, environmental and geographical problems and challenges. The present book has embraced papers from three major fields of remote sensing applications: urbanization and its impacts; geospatial technology for disaster management; and remote sensing applications in models and planning. The Introduction to this book envisaged the evolution and growth of remote sensing and its culmination with Indian remote sensing accomplishments.

From the taking of the first picture by Gaspard-Félix Tournachon (1858) from a balloon above Paris, the foundations of aerial photography and remote sensing were laid. That early period of remote sensing went at a very slow pace, when early black and white photos were being taken from balloons, kites and pigeons. During the First and Second World Wars and up to the 1950s, remote sensing moved to a moderately advanced level, when aircrafts started taking panchromatic aerial photographs dealing within the visible light spectrum only. With the commencement of satellites (1960s onwards), remote sensing reached a higher level of advances with multispectral satellite images being taken in infrared (IR), near infrared (NIR) and thermal infrared (TIR) regions of the electromagnetic spectrum. The 1990s witnessed further progression in remote sensing through radars and lidars advancing in hyperspectral and microwave remote sensing. Evolving higher, the twenty-first century has ushered in vast technological improvements, which have fostered gigantic advances in remote sensing applications as well. Developments in satellites, drones and unmanned aerial

DOI: 10.1201/9781003224624-21

vehicles (UAVs), and simultaneous advances in superior spatial resolutions have been helpful in advancing remote sensing and its applications further. Additionally, the developments in information technologies and data sciences are paving the way for advances in 4D and fusion remote sensing. Advancing further, the future frontiers of remote sensing merge with big data and artificial intelligence and may be moving further with the post remote sensing (PRS) phase (Figure 17.1).

The future of remote sensing lies in its interface with advances in information and communication technologies (ICT) and emerging data sciences. Similarly, developments in data sciences and related fields will propel advances in geospatial technologies and remote sensing (Khorram et al., 2016). Over the time, there have also been advances in hardware and software. The sizes of satellites have decreased from larger and bigger ones to mini/nano satellites, to pico and femto satellites. Simultaneously, there have been improvements in the sensors, harnessing various regions of the electromagnetic spectrum (Pagano & Kampe, 2001), as well as spectrum and spatial resolutions (Hartley, 2003). These advances have further led to major cost cutting and wider reach of these satellites, geospatial technologies and availability of geospatial data (Datta, 2017). Additionally, these advances and the continued decrease in the size of sensors will certainly drive towards increased processing and storage capacities (Chao et al., 2002; Hartley, 2003). Such advances in remote sensing and related geospatial technologies are, by now, providing intensive resource inventories, quicker monitoring, more efficient emergency responses, precision-based map products, improved navigation and better geospatial information products for the mass market and a vast variety of professional fields (Peri et al., 2001).

Improved internet access has already flooded with information the general public and professionals alike. Remotely sensed data provisions through the internet and its

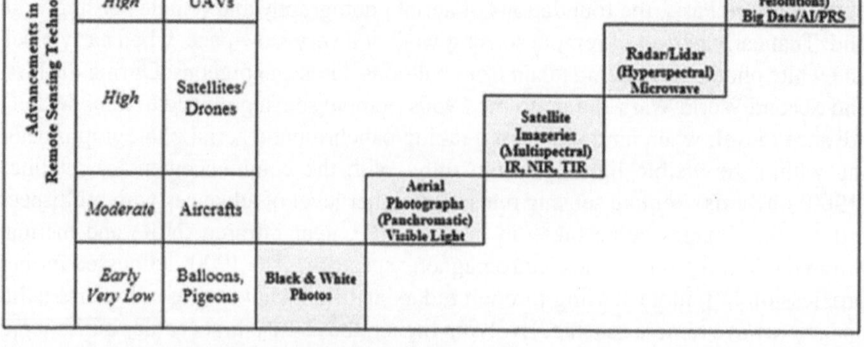

FIGURE 17.1 Advances in Remote Sensing

Source: Singh, 2021

availability to potential customers, news agencies, websites and other visual media will continue to benefit from the advances in remote sensing and the provision of relevant and near real-time information regarding events around the world (Lurie, 1999). Everywhere Google Earth presents such a case of providing near real-time remotely sensed data usage in daily life. Consequently, freely available remote sensing data has further increased its usage and resultant applicability.

The last decade has witnessed a rapid growth in 'big data', referring to the growth in data creation which lies beyond traditional databases and software in relation to data acquisition, storage, processing and its usage (Manyika et al., 2011). Thus, big data has become a state-of-the-art technical term in academia, industry and business, as well as politics. However, it is presenting significant challenges as well (Chen et al., 2013). Nevertheless, big data has established new opportunities (Stantic & Pokorny, 2014). Remote sensing is among those subjects which are harnessing remote sensing (RS) big data, and has emerged as a systematic platform for collecting, storing, organizing, analysing and applying RS big data, based on cloud storage and computing (Ma et al., 2015; Mulyono & Fanany, 2015; Sun et al., 2015; Wang et al., 2015; Li & Huang, 2017; Pekturk & Unal, 2017; Wang et al., 2018; Xiaochuang & Guoqing, 2018).

Re-envisioned advances in remote sensing, depicting the future research agenda really lies in its interface with data sciences and ever increasing spatial and non-spatial data interactions. As discussed previously, the future research agenda of remote sensing applications can be demonstrated through Figure 17.2. The future frontiers of remote sensing envision intensive interface with artificial intelligence (AI), big data, cloud computing, internet of things (IoT), open source, Python, R; terrestrial/extra-terrestrial and post-pandemic remote sensing; and creating combinations of fusion remote sensing, smart remote sensing and post remote sensing (Singh, 2021) (Table 17.1).

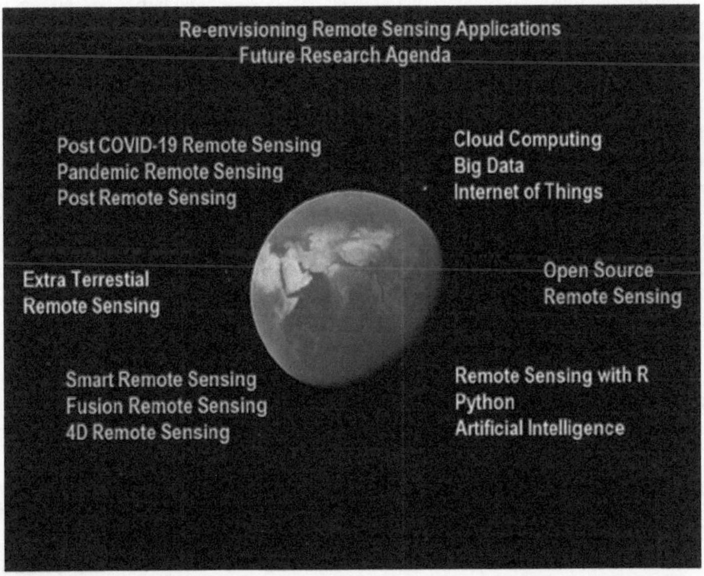

FIGURE 17.2 Re-envisioning Advances in Remote Sensing Applications

Source: Singh, 2021

TABLE 17.1
Re-envisioning Advances in Remote Sensing: Future Research Agenda

Big Data Remote Sensing (*Ma et al., 2015; Mulyono and Fanany, 2015; Sun et al., 2015; Wang et al., 2015; Li and Huang, 2017; Pekturk & Unal, 2017; Dey et al., 2018; Sedona et al., 2019)*
Remote sensing applications with big data, data mining and machine learning are emerging as a novel and forthcoming field.
Cloud Computing and Remote Sensing (*Wang et al., 2015; Wang et al., 2018; Wang et al., 2020)*
Cloud computing, cloud data and cloud storage in relation to remote sensing data is imminent in enhanced applications of remote sensing in the fields of disaster management, hazard monitoring, climate change, environmental and urban & regional planning and other related fields.
Internet of Things and Remote Sensing (*Mathew et al., 2019; Liu & Dhakal, 2020)*
Remote sensing's interface with IoT will be very useful in monitoring and controlling various activities with reference to smart learning, ecological modelling, mining and related domains.
Open-Source Remote Sensing (*Wegmann, 2016; Grizonnet et al., 2017; Cresson, 2020)*
Open-sourced and openly accessed geospatial data availability is helping in various applications through deep learning and machine learning.
Remote Sensing with R ((*Ghosh & Hijmans, 2019; Kamusoko, 2019; Rani et al., 2020; Rani et al., 2021)*
Programming languages such as R and others are also being used for processing and classifying remotely sensed data and satellite images through machine learning algorithms.
Remote Sensing with Python (*Morton, 2014; Lakshmi et al., 2019)*
Varied applications for image analysis classification using Python and other programming languages are envisaged.
Remote Sensing with Artificial Intelligence (*Estes et al., 1986; Martine, et al., 2019)*
Artificial intelligence and its use with remote sensing data is helpful in generating actionable intelligence for business, industries and organizations.
Smart Remote Sensing (*Cavallaro et al., 2014; Mihail et al., 2020)*
Smart data analytics examining large quantities of spatial data to analyse hidden patterns and unknown correlations.
Similarly, smart remote sensing is intended with broader applications and intricacies through the usage of smart technologies and devices.
Fusion Remote Sensing (*Taxt & Solberg, 1997; Alparone et al. 2015)*
Fusion of data and its integration with remote sensing is envisioned as a promising means for getting high quality geospatial and temporal datasets.
4D Remote Sensing (*Waters et al., 1999; Yang et al., 2016)*
Moving further, remote sensing will be recreated through 4D intricacies and technologies.
Extra-Terrestrial Remote Sensing (*Athanassas et al., 2018; Liu et al., 2020 and Wu et al., 2020)*
Extra-terrestrial remote sensing is anticipated to search and monitor other heavenly bodies for the examination of their mineral wealth, availability of air, water and beyond.
Pandemic Remote Sensing (*Zaragoza, 2020; Bhattacharjee & Bhattacharjee, 2021; Dukiya, 2021.*
The need for remote sensing and remotely sensed data and their applications has greatly increased during the current pandemic.
Post COVID-19 Remote Sensing (*Reeves et al., 2020; Wolters, 2020; Bustamante-Calabria et al., 2021.*
Changes taking place with reference to improved environments, less polluted waters and air during global lockdowns and again increasing pollution and similar analysis are expected to be the themes of post-COVID remote sensing applications.
Post Remote Sensing (*Singh, 2021)*
The future of remote sensing will be ushering in the post-remote sensing phase, where the traditional methodologies of remote sensing will be transforming into innovative arenas of very big data usage, extra-terrestrial remote sensing on the one extreme and ultra-smart phone-based nano processors creating remotely sensed data.

Lastly, the COVID-19 pandemic could be considered as a watershed in the advances of remote sensing and geospatial technologies (Singh, 2021). Improved environmental conditions and reduced levels of pollution were observed all over the world during the global lockdowns, and various studies with remote sensing applications have already analysed the pre-lockdown, lockdown and post-lockdown scenarios across the world. Wolters (2020) analysed such changes for European regions and global trends. For India, various such studies are also coming up. Based on INSAT-3D data, reduced aerosol levels across Indian states during pre-lockdown (mid-March 2020) and lockdown (early April 2020) have been analysed. The results confirm a reduction in air aerosol levels ranging from proportions of less than one third up to two fifths (IIRS, 2020). Earth observation (EO) data may provide new perspectives on the socioeconomic impacts of the COVID-19 pandemic; however, restraints on the sharing of large amounts of EO and remote sensing data, particularly in poor and middle-income developing countries, need to be examined. This pandemic may also be an eye-opener in improving such restrictions in the sharing of remote sensing data in countries like India, which is emerging as a hub of cheap satellite launchers and affordable remote sensing data (Ashok & Basu, 2020). Certainly, this pandemic has, on the other hand, altogether augmented the need for monitoring and surveillance all over the world (GEO, 2020).

One thing which has been prominently observed during the last year is that the levels of surveillance and monitoring have increased many times over, along with the resultant data creation. During the year 2020, 64 zettabytes of digital data have been produced, and it is expected that the data created in the next five years will be more than double the size of all the data produced since the beginning of computers (IDC, 2021). This vast creation of spatial and non-spatial digital data has similarly enlarged the applications of remote sensing in the current time (Zaragoza, 2020). And these EO and remote sensing data play a major role in achieving various sustainable development goals (Figure 17.3) (GEO, 2017). Ferreira et al. (2020) have reviewed the monitoring of sustainable development goals by means of EO and machine learning. It has also been found advantageous to support official statistics with EO data and satellite images, which are very cost effective, useful and complement the traditional socioeconomic and environmental data sources (United Nations, 2017). In addition, Gail (2006) remained optimistic over the rapid growth of geospatial services and easy access of EO data, with businesses, consumers and governments all benefitting in the coming decade and beyond. However, technological risks and market uncertainties cannot be ruled out. He further perceives a dramatic confrontation between the envisioned remote sensing accomplishments, availability of funds and commitments, on the parts of organizations and governments. Long-term implications for the remote sensing community as well as society as a whole are also anticipated. Alderton (2015) thus perceived the darkening of the skies with more and more satellites revolving around the Earth. And this being the bright future for remote sensing and Earth observation, as these satellites are now clicking countless images every minute for every inch of the Earth. So, as with Alderton (2015), it may be concluded that the skies may be darkening, but the next generation of commercial remote sensing promises to shed greater light on Earth and its environs than ever before.

FIGURE 17.3 Sustainable Development Goals and Remote Sensing Applications

Source: GEO, 2017. https://www.earthobservations.org/documents/publications/201703_geo_eo_for_2030_agenda.pdf

REFERENCES

Alderton, M. (2015). Dark Skies, Bright Future: The next generation of commercial remote sensing has arrived, and GEONIT will never be the same. Trajectory, 20 August, 2015. Retrieved from: https://trajectorymagazine.com/dark-skies-bright-future/

Alparone, L. et al. (2015). *Remote Sensing Image Fusion (Signal and Image Processing of Earth Observations)*. Florida: CRC Press, Taylor & Francis Group.

Ashok, G. V. and Basu, P. (2020). COVID-19 Pandemic – Eye opener for better Remote Sensing Policies in India? *Geospatial World*. Retrieved from: https://www.geospatialworld.net/blogs/covid-19-pandemic-eye-opener-for-better-remote-sensing-policies-in-india/

Athanassas, C.D. et al. (2018). Remote Sensing of Mars: Detection of Impact Craters on the Mars Global Surveyor DTM by Integrating Edge-and Region-Based Algorithms. *Earth Moon Planets*, Vol. 121: 59–72. doi:10.1007/s11038-018-9515-3

Bhattacharjee, T., and Bhattacharjee, I. (2021). A Review – How Space Technology can help in COVID-19 Pandemic with reference to Remote Sensing and GIS. *Journal of Remote Sensing and GIS*, Vol. 10 (3): 286. Retrieved from: https://www.longdom.org/open-access/a-review-how-space-technology-can-help-in-covid19-pandemic-with-reference-to-remote-sensing-and-gis.pdf

Bustamante-Calabria, M., Sanchez de Miguel, A., Martin-Ruiz, S., Ortiz, J.-L., Volchez, J. M., Pelegrina, A., Garcia, A., Zamorano, J., Bennie, J., Gaston, K. J. (2021). Effects of the COVID-19 Lockdown on Urban Light Emissions: Ground and Satellite Comparison. *Remote Sensing*, Vol. 13 (2): 258. doi:10.3390/rs13020258

Cavallaro, G., Riedel, M., Benediktsson, J. A., Goetz, M., Runarsson, T., Jonasson, K., Lippert, T. (2014). Smart Data Analytics Methods for Remote Sensing Applications. *Conference Proceedings of the IEEE International Geoscience and Remote Sensing Symposium* (IGARSS, 2014), July 13–18, Quebec City, Canada.

Chao, T.-H., Zhou, H., Reyes, G., Dragoi, D. and Hanan, J. (2002). High Density High Speed Holographic Memory. In: *Proceedings of the Earth Science Technology Conference*, June 11–13, 2002, Pasadena, CA.

Chen, J., Chen, Y., Du, X., Li, C., Lu, J., Zhao, S., and Zhao, X. (2013). Big Data Challenge: A Data Management Perspective. *Frontiers of Computer Science*, Vol. 7 (2): 157–164. doi:10.1007/s11704-013-3903-7

Cresson, B. (2020). *Deep Learning for Remote Sensing Images with Open Source Software*. Boca Raton: CRC Press.

Datta, A. (2017). The NewSpace Revolution: The emerging commercial space industry and new technologies. Geospatial World: Advancing Knowledge for Sustainability. Retrieved from: https://www.geospatialworld.net/article/emerging-commercial-space-industry-new-technologies/

Dey, N., Bhatt, C.and Ashour A.S. (2018). *Big Data for Remote Sensing: Visualization, Analysis and Interpretation: Digital Earth and Smart Earth*. Cham, Switzerland: Springer.

Dukiya, J. J. (2021). The Role of Remote Sensing in Epidemiological Studies and the Global Pandemic Surveillance. *Journal of Atmospheric & Earth Sciences*, Vol. 5:024. Retrieved from: https://www.heraldopenaccess.us/openaccess/the-role-of-remote-sensing-in-epidemiological-studies-and-the-global-pandemic-surveillance

Estes, J. E., Sailer, C., and Tinney, L. R. (1986). Applications of Artificial Intelligence Techniques in Remote Sensing. *The Professional Geographer*, Vol. 38 (2): 133–141. doi:10.1111/j.0033-0124.1986.00133.x

Ferreira, B., Iten, M., and Silva, R.G. (2020). Monitoring Sustainable Development by Means of Earth Observation Data and Machine Learning: A review. *Environmental Sciences Europe*, 32: 120. doi:10.1186/s12302-020-00397-4

Gail, W. B. (2006). Remote sensing in the coming decade: the vision and the reality. Proceeding SPIE 6298, Remote Sensing and Modelling of Ecosystems for Sustainability III, 629801. doi:10.1117/12.694379

GEO (2017). *Earth Observations in support of the 2030 Agenda for Sustainable Development*. Retrieved from: https://www.earthobservations.org/documents/publications/201703_geo_eo_for_2030_agenda.pdf

GEO (2020). *GEO Community Response to COVID-19*. Group on Earth Observations. Retrieved from: https://earthobservations.org/covid19.php

Ghosh, A. and Hijmans, R. J. (2019). *Remote Sensing Image Analysis with R*. Rspatial.org. Retrieved from: https://rspatial.org/rs/rs.pdf

Grizonnet, M. et al. (2017). Orfeo ToolBox: Open-Source Processing Of Remote Sensing Images. *Open Geospatial Data, Software and Standards*, Vol. 2 (15): 1–8. doi: 10.1186/s40965-017-0031-6

Hartley, J. (2003). Earth Remote Sensing Technologies in the Twenty-First Century. In: *Proceedings of the International Geoscience and Remote Sensing Symposium*, July 21–25, 2003, Toulouse, France, Vol. 1, pp. 627–629.

International Data Corporation (2021). Data Creation and Replication Will Grow at a Faster Rate than Installed Storage Capacity, IDC Global Data Sphere and Storage Sphere Forecasts. International Data Corporation, United States Retrieved from: https://www.idc.com/getdoc.jsp?containerId=prUS47560321&utm_medium=rss_feed&utm_source=alert&utm_campaign=rss_syndication

IIRS (2020). *Space Based Observation on Changes in Air Quality During COVID-19 Lockdown Period.* Dehradun: IIRS Update, Indian Institute of Remote Sensing. Retrieved from: https://www.iirs.gov.in/iirs_slide_page?s=67

Kamusoko, C. (2019). *Remote Sensing Image Classification in R.* Singapore: Springer Nature.

Khorram, S., Cynthia, F. W., Frank, H. K., Stacy, A. C. N., and Matthew, D. P. (2016). *Principles of Applied Remote Sensing*. New York: Springer International Publishing.

Lakshmi, J. V. N., Hemanth, K., Bharath, J. (2019). Optimizing Quality and Outputs by Improving Variable Rate Prescriptions in Agriculture using UAVs. *International Conference on Computational Intelligence and Data Science (ICCIDS 2019), Procedia: Computer Science.*

Li, G., and Huang, Z. (2017). Data Infrastructure for Remote Sensing Big Data: Integration, Management and on-Demand Service. *Jisuanji Yanjiu yu Fazhan/ Computer Research and Development*, Vol. 54 (2): 267–283. doi:10.7544/issn1000-1239.2017.20160837

Liu, Y., and Dhakal, S. (2020). Internet of Things Technology in Mineral Remote Sensing Monitoring. *International Journal of Circuit Theory and Applications*, Vol 48 (12): 2065–2077. doi:10.1002/cta.2890

Liu, D. et al. (2020). An Empirical Abundance of Nanophase Metallic Iron (npFe0) in Lunar Soils. *Remote Sensing*, Vol. 12 (6), 1047. doi:10.3390/rs12061047

Lurie, I. (1999). The Commercial Future: Making Remote Sensing a Media Event. In: Fujisada, H., and Lurie, J. B. eds. *SPIE Proceedings, Vol. 3870, Sensors, Systems and Next Generation Satellites III*, pp. 601–610. doi:10.1117/12.373224

Ma, Y., Wu, H., Wang, L., Huang, B., Ranjan, R., Zomaya, A., and Jie, W. (2015). Remote Sensing Big Data Computing: Challenges and Opportunities. *Future Generation Computer Systems*, Vol. 51: 47–60. doi:10.1016/j.future.2014.10.029

Manyika, J., Chui, M., Brown, B., Bughin, J., Dobbs, R., Roxburgh, C., and Byers, A. H. (2011). Big data: The next frontier for innovation, competition, and productivity. McKinsey Global Institute, Washington DC. Retrieved from: https://www.mckinsey.com/~/media/mckinsey/business%20functions/mckinsey%20digital/our%20insights/big%20data%20the%20next%20frontier%20for%20innovation/mgi_big_data_full_report.pdf

Martinez del Horno, M. et al. (2019). Calibration of Wi-Fi-based Indoor Tracking Systems for Android Based Smartphones. *Remote Sensing*, Vol. 11 (9): 1072 doi:10.3390/rs11091072

Mathew, C., Mathukutty, R., and Madhanan, P. (2019). Controlling of Greenhouse Parameters based on IoT and Remote Sensing. *International Journal of Engineering Research & Technology (IJERT)*, Vol. 7 (5): 1–3. Retrieved from: https://www.ijert.org/research/controlling-of-greenhouse-parameters-based-on-iot-and-remote-sensing-IJERTCONV7IS05011.pdf

Mihail, S., Elena, S., and Al-Qatrany, A.C.D. (2020). *The Use of Smart Remote Sensing Technologies in the Development of Master Plans of Cities on the Example of the City of Basra. IOP Conf. Series: Materials Science and Engineering*, Vol. 869 (2020) 022013. doi:10.1088/1757-899X/869/2/022013

Morton, J. C. (2014). *Image Analysis, Classification and Change Detection in Remote Sensing with Algorithms for ENVI/IDL and Python*. CRC Press: Boca Raton.

Mulyono, S., and Fanany, M. I. (2015). Remote Sensing Big Data Utilization for Paddy Growth Stages Detection. Paper presented at *the IEEE International Conference on Aerospace Electronics and Remote Sensing Technology*, Bali, Indonesia, Dec. 3-5.

Pagano, T. and Kampe, T. (2001). The Spaceborne Infrared Atmospheric Sounder (SIRAS) Instrument Incubator Program Demonstration. In: *Proceedings of the Earth Science Technology Conference*, August 28–30, 2001, College Park, MD.

Pekturk, M. K., and Unal, M. (2017). A Review on Real Time Big Data Analysis in Remote Sensing Applications. Paper presented at the *25th Signal Processing and Communication Applications Conference (SIU)*, Antalya, Turkey, May 15–18.

Peri, Jr., F., Hartley, J. B., and Duda, J. L. (2001). The Future of Instrument Technology for Space Based Remote Sensing for NASA's Earth Science Enterprise. In: *Proceedings of the International Geoscience and Remote Senisng Symposium*, July 9-13, Sydney, Australia, Vol. 1, pp. 432–435.

Rani, A., Kumar, N., Singh, S.K., Sinha, N.K., Jena, R.K., Patra, H. (2020). *Remote Sensing Data Analysis in R*. New Delhi: New India Publishing Agency.

Rani, A., Kumar, N., Singh, S.K., Sinha, N.K., Jena, R.K., Patra, H. (2021). *Remote Sensing Data Analysis in R*. CRC Press.

Reeves, M., Carlsson-Szlezak, P., Whitaker, K., and Abraham, M. (2020). Sensing and Shaping the Post-COVID Era. BCG Henderson Institute, Boston. Retrieved from: https://image-src.bcg.com/Images/BCG-Sensing-and-Shaping-the-Post-COVID-Era-Apr-2020-rev_tcm9-244426.pdf

Sedona, R. et al. (2019). Remote Sensing Big Data Classification with High Performance Distributed Deep Learning. *Remote Sensing*, Vol. 11 (24) 3056. doi:10.3390/rs11243056

Singh, R. (Ed.). (2021). *Re-envisioning Remote Sensing Applications: Perspectives from Developing Countries* (1st ed.). CRC Press. doi:10.1201/9781003049210

Stantic, B., and Pokorny, J. (2014). Opportunities in Big Data Management and Processing. *Frontiers in Artificial Intelligence & Applications*, 270: 15–26. doi:10.3233/978-1-61499-458-9-15

Sun, Z., Chen, F., Chi, M. and Zhu, Y. (2015). A Spark-Based Big Data Platform for Massive Remote Sensing Data Processing. Paper presented at the *2nd International Conference on Data Science (ICDS)*, University of Technology, Sydney, Australia, Aug. 8-9.

Taxt, T., Solberg, A. H. S. (1997). Information Fusion in Remote Sensing. *Vistas in Astronomy*, Vol. 41 (3): 337–342. doi:10.1016/S0083-6656(97)00036-6.

United Nations (2017) *Earth Observations for Official Statistics: Satellite Imagery and Geospatial Data Task Team Report*. Retrieved from: https://unstats.un.org/bigdata/task-teams/satellite/UNGWG_Satellite_Task_Team_Report_Whitecover.pdf

Wang, L., Geng, H., Liu, P., Lu, K., Kolodziej, J., Ranjan, R., and Zomaya, A. Y. (2015). Particle Swarm Optimization Based Dictionary Learning for Remote Sensing Big Data. *Knowledge-Based Systems*, Vol. 79: 43–50. doi:10.1016/j.knosys.2014.10.004

Wang, L. Ma, Y., Yan, J., Chang, V., and Zomaya, A. Y. (2018). pipsCloud: High Performance Cloud Computing for Remote Sensing Big Data Management and Processing, *Future Generation Computer Systems* Vol. 78: 353–368. doi:10.1016/j.future.2016.06.009

Wang, L., Yan, J. and Ma, Y. (2020). *Cloud Computing in Remote Sensing*. Florida: CRC Press, Taylor & Francis Group.

Waters, C. R., Sommese, T., Essel, M. and Mark, S. (1999). Signal and Subspace Processing of 4D Remote Sensing Data. *IEEE Aerospace Conference Proceedings, 1999*, Vol. 4: 281–288. doi:10.1109/AERO.1999.792096

Wegmann, M., Leutner, B. and Dech, S. (2016). *Remote Sensing and GIS for Ecologists using Open Source Software*. Exeter UK: Pelagic Publishing.

Wolters, E. (2020). *Air Quality before, during and after COVID-19 Lockdown*. Vito Remote Sensing. Retrieved from: https://blog.vito.be/remotesensing/air-quality-monitoring-before-during-and-after-covid-19-lockdown

Wu, K. et al. (2020). Simulation Study of Moon-Base InSAR Observation for Solid Earth Tides. *Remote Sensing*, Vol. 12 (1), 123. doi:10.3390/rs12010123

Xiaochuang, Yao and Guoqing, Li (2018). Big Spatial Vector Data Management: A Review. *Big Earth Data*, Vol. 2 (1): 108–129. doi:10.1080/20964471.2018.1432115

Yang, C-H., Kenduiywo, B. K., and Soergel, U. (2016). 4D Change Detection Based on Persistent Scatterer Interferometry. Pattern Recognition in Remote Sensing (PRRS), 2016 9th IAPR *Workshop on Pattern Recognition in Remote Sensing*, Cancun, Mexico. doi:10.1109/PRRS.2016.7867016

Zaragoza, A. (2020). Remote Sensing in times of COVID-19: just what the doctor ordered. *Worldsensing*. Retrieved from: https://blog.worldsensing.com/industrial-iot/remote-sensing-covid19/

Index

T

U

V

W

Z